SEISMIC DESIGN OF BUILDINGS

SEISMIC DESIGN OF BUILDINGS

JAMES AMBROSE
Professor of Architecture
University of Southern California

DIMITRY VERGUN
Consulting Structural Engineer
Lecturer in Architecture
University of Southern California

A Wiley-Interscience Publication
JOHN WILEY & SONS
New York Chichester Brisbane Toronto Singapore

Library of Congress Cataloging in Publication Data:

Ambrose, James E.
Seismic design of buildings.

"A Wiley-Interscience publication."
Bibliography: p.
Includes index.
1. Earthquake resistant design. I. Vergun, Dimitry, 1933– II. Title.
TA658.44.A44 1985 693.8′52 84–26990
ISBN 0-471–88979-2

Printed in the United States of America

10 9 8 7 6 5 4 3 2 1

Preface

The purpose of this book is to assist persons who lack experience in the design of buildings for resistance to earthquakes. The reader may be generally experienced in building design but lack specific experience with earthquakes, or the reader may be without design experience of any kind. The development of the presentations in this book has been done with both types of reader in mind. The study of the materials in this book will provide the reader with a general understanding of the problems of designing for resistance to the effects of earthquakes and a familiarity with the common techniques for producing structures with resistance to lateral loads.

Much of the material in this book should be understandable to the average reader. However, the discussions of building construction and structural behaviors are developed with the assumption that the reader has some background experience in these areas. The undergraduate course work in a collegiate program in architecture should be adequate in this regard.

The design examples are based on current practices, using the requirements of the *Uniform Building Code* (*UBC*)—the most widely used reference for seismic design in the United States. Most of the portions of the 1982 edition of the *UBC* that deal with seismic design are reprinted in this book. We are grateful to the International Conference of Building Officials, publishers of the *UBC*, for their permission to reprint this material. We are also grateful to Inryco, Inc., the Simpson Company, and the Concrete Masonry Association of California and Nevada for permission to reprint materials from their publications.

While we claim authorship, we must acknowledge that the quality of this book owes much to the work of the editors and production staff at John Wiley & Sons in New York. Finally, we must acknowledge that the need to explain many things, and the effectiveness of various means for doing so, has been accumulated from many contacts with our students and professional clients.

James Ambrose
Dimitry Vergun

Los Angeles, California
January 1985

Contents

Nomenclature

The following is a list of symbols used in this book; it is in agreement with the notation used in the *Uniform Building Code*.

a	Acceleration
A	Amplitude, or cross-sectional area
C	Base shear coefficient
d	Depth of a beam
e	Eccentricity
f	Frequency
F	Force
g	Acceleration of gravity
G	Shear modulus
h	Height
I	Moment of inertia, or occupancy importance factor
J	Polar moment of inertia
K	Construction type coefficient, lateral load
L	Span length
m	Mass
M	Bending moment
r	Radius of gyration
R	Relative rigidity
S	Site–structure resonance factor
t	Time, or thickness
T	Period, or torsional moment
v	Velocity, or unit shear
V	Shear
w	Unit of uniformly distributed load
W	Weight
Z	Region coefficient (seismic risk)

Introduction

Over the years, earthquakes have been the cause of great disasters in the form of destruction of property and injury and loss of life to the population. It is quite proper, therefore, for the users of buildings to ask the designers to give some consideration to the possibility of damage due to earthquakes. Easier said than done. Sophisticated techniques for seismic design of buildings have a relatively short history. Just a few decades ago, design was done with very simplistic methods, using only a simple set of data. Extensive study and research and a great international effort of cooperation have vastly improved design procedures and the data and theory on which they are based. This effort is ongoing, and our designs of the future will surely be as much more intelligently done as ours of today are in comparison with those of past decades.

Unfortunately, the most valuable information comes from direct observations of the results of real, disastrous earthquakes, especially when they occur in highly populated areas. Ghoulish as it may seem, engineers and scientists have an obligation to pick over the rubble of such disasters in order to advance our appreciation of what types of construction are effective and what are not.

The essential aim of this book is to provide the serious designer of buildings with a general overview of the problem of seismic design and with what is available for use in terms of information, design criteria, design guidelines and procedures, design aids, and construction technology. The basic focus is design, rather than scientific inquiry or analytical investigation. By design we mean the collective mass of decisions, selections, and inventions that must be made in order to define what is to be built.

Design decisions that affect the seismic behavior of buildings range from broad to highly specific ones. Many of these decisions are in most cases out of the hands of the structural designer. Restrictions on the use of procedures for analysis of structural behavior and the selection of materials and systems for construction are established by building codes. Decisions about siting, building form and scale, and selection of materials and details of construction are usually made by building owners and architectural designers. Thus, quite often, before the structural designer ever gets involved, much of the design is already accomplished fact.

As limited as the scope of decision making may be, the task of the structural designer is nevertheless a serious one: to keep the building from falling down. Actually, although that is the bottom line, the job is bigger than that. While ultimate collapse of the structure is very serious, the building's performance during an earthquake must be considered in many other ways as well. If the structure remains intact, but the nonstructural parts of the building sustain critical damage, much injury to occupants can occur, and the building may be unusable after the quake.

The integrity of the structure is the principal concern of this book. We will, however, address the various concerns for nonstructural damage since they may in some cases affect decisions about the structure. Since these issues affect the work of architectural designers as much as that of the structural designer, it behooves all the members of the design team to be well aware of them. All of the decisions that affect the building's seismic response should be carefully made—regardless of who makes them.

In a book of this kind it is necessary to start from some point and to end at some point. We start with some assumptions as to the background of the reader in terms of knowledge about basic structural behavior, the materials and methods of modern building construction, and the general nature of earthquakes. It is our intent to present the work in a way that makes it accessible to both architects and engineers. This calls for some compromise since architects are usually more highly trained in the general problems of building planning and construction, while engineers are better prepared to deal with mathematical investigation of structural behaviors.

If we must visualize an ideal reader, it is either a graduate of an architectural program who has a slightly better than average preparation in engineering or a graduate of a civil engineering program who has some training or experience in building planning and construction. In either case we expect the reader to know something about earthquakes and about the dynamic behavior of building structures. For readers who may be less prepared in any of the areas mentioned, it may be advisable to strengthen their backgrounds with some extra study in order to get the most from the work in this book. A list of references is given at the back of this book, which includes many titles that can be useful for this purpose.

PART ONE

BUILDING DESIGN ISSUES

CHAPTER ONE

General Effects of Earthquakes

This chapter deals with the general relations of buildings and the building design process to the hazards created by ground movements during earthquakes.

1.1 EARTHQUAKE MOTIONS

An earthquake is the vibratory movement of the earth's surface that follows a sudden release of energy in the crust. This energy release may be the result of a sudden dislocation of some part of the crust, of a volcanic eruption, or of a man-made explosion of great magnitude. Vibrations, called seismic waves, emanate from the location of the energy release and travel throughout the earth's mass. On the surface of the earth these waves cause a vibratory motion having the general form of typical harmonic motion. They may thus be plotted in the form of a displacement–time or acceleration–time graph from which observations can be made of the frequency and amplitude of the vibrations.

Large earthquakes have the potential for causing a number of actions that may be hazardous. First is the simple motion of the ground surface—up and down, and back and forth in all directions. However, some of the most disastrous actions are those caused by giant tidal waves and avalanches and by various secondary effects on the ground mass or surface, such as cracks, subsidence, lateral slips, soil liquefaction, and massive earth slides. There is little that can be done to protect buildings from these latter actions, other than to avoid locating buildings in predictably vulnerable areas. The major attention in structural design for seismic effects is on the response to the vibratory motions of the ground surface.

It is possible to move buildings in some cases without causing significant damage to them. Many buildings, including multistory ones, have been picked up from their foundations and moved to new locations. What is harmful about the motion caused by earthquakes is the speed and the cyclic nature of the motion. The stress-producing forces that are exerted on a building during such motions are affected by the relative stiffness and mass of the building itself. Thus evaluation of the potential for damage must include considerations of the properties of the building, as well as the specific nature of the ground movements.

With respect to damaging effects on buildings, there are several aspects of ground motion that must be considered. These include the following:

The *direction* of the motion. Movement in different directions has different effects on the building, and all of these effects must be considered.

The *displacement* or actual dimension of the movement.

The *acceleration* of the motion.

The time pattern of the motion in terms of *duration* and of *frequency* or *period* of the cyclic action.

Although buildings actually move in all directions during an earthquake, it is useful to consider the effect of two distinct movements: up and down (vertical) and back and forth (horizontal). Horizontal motion is most significant in that it tends to disturb the stability of the building, causing it to topple or collapse sideways. Since buildings are basically constructed to resist gravity (vertical) effects, many ordinary systems of construction are not inherently resistive to horizontal forces. Thus seismic design is mostly considered as the problem of *bracing* buildings against sideways movement. However, vertical movements may also cause some problems. When the building moves rapidly downward and then suddenly reverses direction upward, the momentum of the building mass produces a downward jolting force that adds to the effects of gravity. The result is a sudden, dynamic increase in the gravity-induced stresses. Since the gravity-resistive structure is designed with some margin of safety in the form of redundant strength, this is usually not a critical concern for well-designed structures.

When the building moves up and then reverses direction downward, the upward momentum causes a reversal of the normal gravity effects. This may result in some stress reversals or in the lifting of the building from its supports. A type of failure that has occurred in some structures utilizing large elements of precast concrete is the bouncing of the structure off of its supports (called dancing).

Dimensional displacements during vibratory seismic motions are usually not of themselves critical. What is usually more significant is the displacement–time relationship, essentially as measured by the acceleration–time plot. This is evident from the simple equation for dynamic force: $F = m \times a$, or force equals mass times acceleration. Thus the force effect potential in the building's momentum is the product of its mass and the acceleration of the ground motion.

In most cases it takes a number of coincident factors to produce a damaging earthquake. The first of these is the magnitude of the energy released by the fault, or simply the size of the shock. This magnitude is currently usually measured by use of the *Richter scale*, a log-based scale that evaluates the total energy released along the rupture surface of the subterranean fault that triggers the quake. A rating of 4 or more on this scale indicates a quake of some potential significance. Because of the log-base, the significance goes up rapidly; a rating of 6 indicates 100 times as much energy as a rating of 4.

The point on the ground surface immediately above the initial rupture point is called the *epicenter* of the quake. When the epicenter is near a populated area and the fault is near the surface, a quake of significant magnitude can be expected to produce some damage to structures and some injury or loss of life. Depending on the location of the epicenter, it is possible to have a major quake in a remote area with little effect on buildings or people or to have a relatively moderate quake in a highly populated area with significant losses.

Besides the size of a quake and its location, the other major factor determining the potential for damage is the nature of the structures affected by the quake. If these are of types of construction that are highly vulnerable to earthquake damage, losses will be substantially greater. Examples of such structures are those with low-grade, unreinforced masonry and frameworks with light connections, poor anchorage, or minimal lateral bracing.

For the purpose of evaluating the *effects* of an earthquake, as opposed to the actual physical nature of the quake itself, the *modified Mercalli scale* is of more significance than the Richter scale. This scale measures degrees of damage and response and can be used to plot a contour map around the epicenter of a quake, defining zones of damage.

In recent times a major factor that has been included in the development of design criteria for a specific building is the seismic response character of the site on which the building is located. This may be significant with regard to potential for failure of the site itself in the form of subsidence, landslides, soil liquefaction, and so on. Also of concern, however, is the interaction of the building and the site during a quake. Thus, if a site is hazardous on its own, more must be done in the way of design to produce a safe building. On the other hand it may be possible to manipulate the building–site interaction by changes in the building structure in order to minimize the effects of the interaction.

1.2 BUILDING RESPONSE TO EARTHQUAKES

As discussed in the preceding section, the response of a building to an earthquake is related to various properties of the building itself. The most notable of these are its mass, its fundamental period of free vibration, and the presence or absence of various factors that may cause damping or resonance of its vibration. The magnitude of actual forces sustained by a building, as well as its ability to absorb forces without sustaining critical damage, will be affected by these properties. In addition, the structure of the building—more specifically the elements of the structure directly involved in seismic resistance—must function within some acceptable limits (margin of safety) in terms of stress and strain behavior.

The strength of a structure, both in terms of the strength of individual parts and of the collective strength of the whole structure, is reasonably quantifiable. Given the specific nature of an earthquake and the measurable dynamic properties of a building, it is possible to evaluate the building's response to the quake, both in nature and magnitude. If accurately performed, this is usually a complex investigation due to the complex nature of seismic motions and the relative complexity of most building forms. However, with the present capabilities for performing analytical investigations, the existing knowledge of structural behaviors, and the large library of recordings of recent major quakes, such investigations can be quite reasonably performed.

Frankly, our confidence in building engineering design is based mostly on feedback from experience, rather than on theoretical projections. Observations of buildings that have endured various effects over time without significant failure give us the most assurance about the proper way to build. (And conversely, observed failures teach us most convincingly how *not* to build.) This applies to the effects of gravity, climate, fire, wind, and use of buildings, as well as to earthquakes. Indeed, most buildings are built from designs based heavily on the confidence of experience, and with relatively little in the way of complex engineering investigation.

It is likely that no building ever built

could sustain the hardest jolt of a major earthquake without some kind of damage. It thus becomes necessary to define various kinds of damage and the tolerable levels of damage for them. The worst form of damage, of course, is the total collapse of a building, or even of some significant part of it. Short of this is the kind of damage that renders a portion of the structure unusable, but does not result in collapse of the building or injury to the occupants. Finally is the relatively minor, though still undesirable, damage, such as cracked plaster, jammed doors, or broken window glass.

Actually much damage from earthquakes tends to consist of damage to nonstructural elements of buildings. Thus it is possible that the building structure sustains virtually no real damage, while the rest of the building is turned to junk. In the complete sense, therefore, seismic design of buildings must consider more than the effects on the structure. If it is to be done thoroughly, this requires bringing many other members of the building design team into the seismic design effort, in addition to the structural designer.

Preventing all forms of damage has other than property value implications. Injury to building occupants is the real major concern, and the various ways in which injury may occur must be considered. Obviously the collapsed structure is a major concern, but so are falling glass, plaster, or light fixtures. It is possible that the various parts of the building remain intact, but that the building sways so much that the motions may produce some form of injury. Ideally we want buildings to respond to earthquakes with a minimal risk of unrepairable damage and with the lowest possible risk of injury or loss of life for the occupants.

1.3 DESIGN RESPONSIBILITIES

To the extent possible, we are obliged to use our knowledge and experience to reduce the potential hazard for injury to building occupants and for property loss in regions where seismic risk is high. Only part of that effort is within the power of control by structural designers. They usually cannot control where buildings are built or what people decide to do with them after they are built. Still, the improvement of the seismic resistance of *any* building—new or existing—through better design is a step forward.

The *task* of seismic design of structures is usually performed by professional structural designers, who produce the computations, specifications, and construction details for proposed buildings. The *responsibility* for design, however, extends beyond the structural designer, who usually works as an employee of an engineering or architectural firm. Building owners and investors, architects, code-enforcing agencies, state registration boards, publishers of building codes, building financing organizations, and insurance companies must share the responsibility.

Assuming that what constitutes good design can be reasonably defined, the problem arises as to how to make it happen. What kinds of legal requirements or restrictions can be established and how can they be enforced? Building codes represent a major type of control for design work, but they have many limitations. Code requirements must usually be on the side of *minimal*, rather than *optimal*, resulting in a sort of legislated least acceptable, rather than best possible, design. Furthermore, the best of codes is of limited use unless it is vigorously and thoroughly enforced.

To the extent that good design means more design effort and time, more careful review for granting permits, more inspection during construction, and some additions to the construction, there are inevitable increases in the cost of buildings. If these increases are not accepted by owners, investors, financers, insurers, and the public as a whole, the cause can hardly be won by the structural designers alone.

In some instances cost and design time are not at issue. Placing a wall here instead of there, moving a window over, making a column this shape instead of that—these matters can have significant effect on seismic response without an increase in design effort or in construction cost. Careful and intelligent design does not automatically mean more expensive design work or higher cost for buildings.

A major issue of concern is the level of awareness and the general knowledge and design skills that should be expected of building designers—both architects and structural designers. They cannot all be expected to be experts in the field of seismic design. In a sense building codes and the recommendations of various industry and professional groups constitute directives for use by the less knowledgeable, less *expert*, individuals to whom the task of actual design work mostly falls. To the extent that these directives are accepted, understood, and used by designers, they become major controls for design work, which places a major responsibility on those who produce them.

Two highly influential groups in the United States at present are the publishers of the *Uniform Building Code (UBC)*, the International Conference of Building Officials, and the professional association of structural engineers in California, presently constituted as SEAOC (Structural Engineers Association of California). Recommendations of the Seismology Committee of SEAOC usually appear with reasonable promptness in new editions of the *UBC*, which appear every three years. Most of the design work illustrated in this book is based on recent editions of the *UBC*.

There are, of course, many other organizations, including international ones, and many individuals and research institutes engaged in work aimed at the development of better seismic design of buildings. Ideas, propositions, and conclusions are continuously brought forth and shared in circulated journals and in conferences. The body of knowledge and experience grows; design practices are refined; construction technology is improved; and the whole field exists in a dynamic state. Seismic design is not the same today as it was a few years ago, and changes will certainly be made in the years to come.

In the end, laws, design directives, new design procedures, and computer-aided techniques cannot guarantee good design. Judgment, skill, and care on the part of designers is essential. The best that can be expected is that design work is done in accordance with the latest recommendations of respected and experienced groups, and that it is performed with competence, thoroughness, and some insight by responsible designers.

CHAPTER TWO

Effects of Building Form

The form of a building has a great deal to do with the determination of the effects of seismic activity on the building. This chapter discusses various aspects of building form and the types of problem commonly experienced.

2.1 RELATIONS OF BUILDING FORM TO SEISMIC RESPONSE

Most buildings are complex in form. They have plans defined by walls that are arranged in complex patterns. They have wings, porches, balconies, towers, and roof overhangs. They are divided vertically by multilevel floors. They have sloping roofs, arched roofs, and multiplane roofs. Walls are pierced by openings for doors and windows. Floors are pierced by stairways, elevators, ducts, and piping. Roofs are pierced by skylights, vent shafts, and chimmneys. The dispersion of the building mass and the overall response of the building to seismic effects can thus be complicated, difficult to visualize, let alone to quantitatively evaluate.

Despite this typical complexity, investigation for seismic response may often be simplified by the fact that we deal mostly with those elements of the building that are directly involved in the resistance of lateral forces; what we refer to as the *lateral resistive system*. Thus most of the building construction, including parts of the structure that function strictly for resistance of gravity loads, may have only minimal involvement in seismic response. These nonstructural elements contribute to the load (generated by the building mass) and may offer damping effects to the structure's motion, but may not significantly contribute to the development of resistance to lateral force.

A discussion of the issues relating to building form must include consideration of two separate situations: that of the form of the building as a whole and that of the form of the lateral resistive system. Figure 2.1 shows a simple one-story building, with the general exterior form illustrated in the upper figure. The lower figure shows the same building with the parapet, canopy, window wall, and other elements removed, leaving the essential parts of the lateral resistive system. This system consists primarily of the horizontal roof surface and the portions of the vertical walls that function as shear walls. The whole building must be considered in determining the building mass for the lateral load, but the stripped down structure must

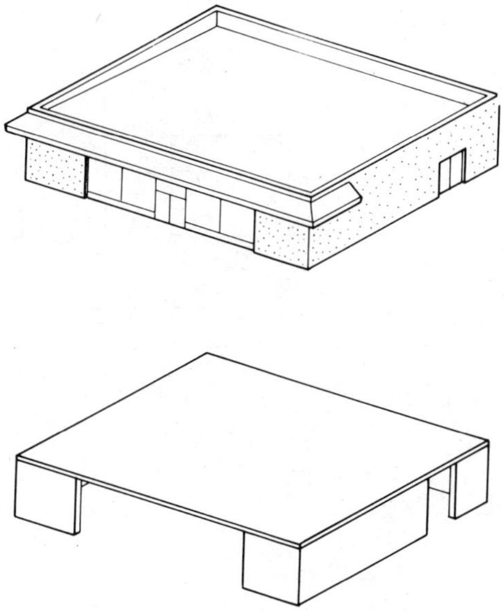

FIGURE 2.1

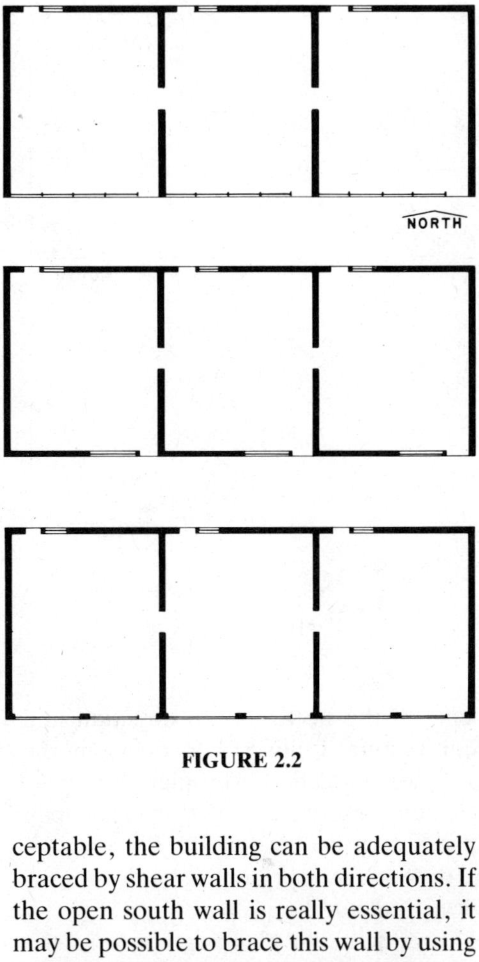

FIGURE 2.2

be visualized in order to investigate the effects of lateral forces.

In developing building plans and the building form in general, architectural designers must give consideration to many issues. Seismic response has to take its place in line with the needs for functional interior spaces, control of traffic, creation of acoustic privacy, separation for security, energy efficiency, and general economic and technical feasibility. In this book we dwell primarily on the problems of seismic response, but it must be kept in mind that the architect must deal with all of these other concerns.

Development of a reasonable lateral resistive structural system within a building may be easy or difficult, and for some proposed plan arrangements may be next to impossible. Figure 2.2 shows a building plan in the upper figure for which the potentiality for development of shear walls in the north–south direction is quite reasonable but in the east–west direction is not so good as there is no possibility for shear walls on the south side. If the modification shown in the middle figure is acceptable, the building can be adequately braced by shear walls in both directions. If the open south wall is really essential, it may be possible to brace this wall by using a column and beam structure that is braced by trussing or by rigid connections, as shown in the lower figure.

In the plan shown in Fig. 2.3*a* the column layout results in a limited number of possible bents that may be developed as moment-resistive frames. In the north–south direction the interior columns are either offset from the exterior columns or the bent is interrupted by the floor opening; thus the two end bents are the only ones usable. In the east–west direction the large opening interrupts two of the three interior bents, leaving only three usable bents that are not disposed symmetrically in the plan. The modification shown in Fig. 2.3*b* represents an improvement in seismic response, with six usable

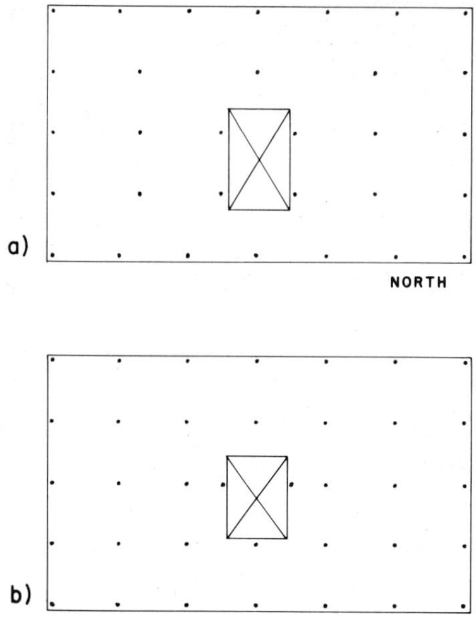

FIGURE 2.3

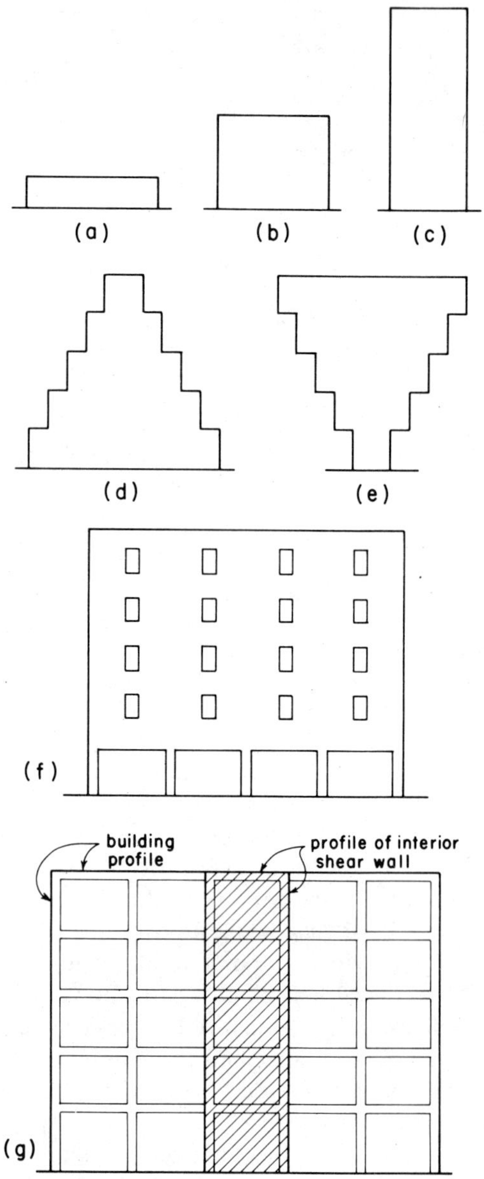

FIGURE 2.4

bents in the north–south direction and four symmetrically placed bents in the east–west direction. This plan, however, has more interior columns, smaller open spaces, and a reduced size for the opening, all of which may present some drawbacks for architectural reasons.

In addition to planning concerns the vertical massing of the building has various implications on its seismic response. The three building profiles shown in Fig. 2.4*a, b,* and *c* represent a range of potential response with regard to the fundamental period of the building and the concerns for lateral deflection. The short, stiff building shown in *a* tends to absorb a larger jolt from an earthquake because of its quick response (short period of natural vibration). The tall, slender building, on the other hand, responds slowly, dissipating some of the energy of the seismic action in its motion. However, the tall building may develop some multimodal response, a whiplash effect, or simply so much actual deflection that it may have problems of its own.

The overall inherent stability of a building may be implicit in its vertical massing or profile. The structure shown in Fig. 2.4*d* has considerable potential for stability with regard to lateral forces, while that shown in Fig. 2.4*e* is highly questionable. Of special concern is the situation in which abrupt change in stiffness occurs in the vertical massing. The structure shown in Fig. 2.4*f* has an open form at its base, resulting in the so-called soft story. While this type of system may be designed adequately by the general requirements of the equivalent static force method, a true dynamic analysis will indicate serious problems, as borne out by some recent serious failures.

As with the building plan, consideration of the vertical massing must include concerns for the form of the lateral resistive system, as well as the form of the whole building. The illustration in Fig. 2.4*g* shows a building whose overall profile is quite stout. However, if the building is braced by a set of interior shear walls, as shown in the section, it is the profile of the shear walls that must be considered. In this case the shear wall is quite slender in profile.

2.2 ADVANTAGES OF SYMMETRY

Investigation of the seismic response of a complex building is, in the best of circumstances, a difficult problem. Anything done to simplify the investigation will not only make the analysis easier to perform but will tend to make the reliability of the results more certain. Thus, from a seismic design point of view, there is an advantage in obtaining some degree of symmetry in the building massing and in the disposition of the elements of the lateral resistive structure.

When symmetry does not exist, a building tends to experience severe twisting, as well as the usual rocking back and forth. The twisting action often has its greatest effects on the joints between elements of the bracing system. Thorough investigation and careful detailing of these joints for construction is necessary for a successful design. The more complex the seismic response and the more complicated and unusual the details of the construction, the more difficult it becomes to assure a thorough and careful design.

Most buildings are not symmetrical; sometimes on one axis, often on any axis. However, real architectural symmetry is not necessarily the true issue in seismic response. Of critical concern is the alignment of the net effect of the building mass (or the centroid of the lateral force) with the center of stiffness of the lateral resistive system—most notably the center of stiffness of the vertical elements of the system. The more the eccentricity of the centroid of the lateral force from the center of stiffness of the lateral bracing system, the greater the twisting effect on the building.

Figure 2.5 shows an extreme example: the so-called *three-sided building*. In this situation the lack of resistive vertical elements on one side of the building requires that the opposite wall take all of the direct effect of the lateral force that is parallel to it. Assuming the centroid of the building mass to be approximately at the center of the plan, this results in a large eccentricity between the load and the resisting wall. The twisting action that results will be partly resisted by the two end walls that are at right angles to the load, but the general effect on the building is highly undesirable. This type of structure is presently highly restricted for use in regions of high seismic risk.

When a building is not architecturally symmetrical, the lateral bracing system

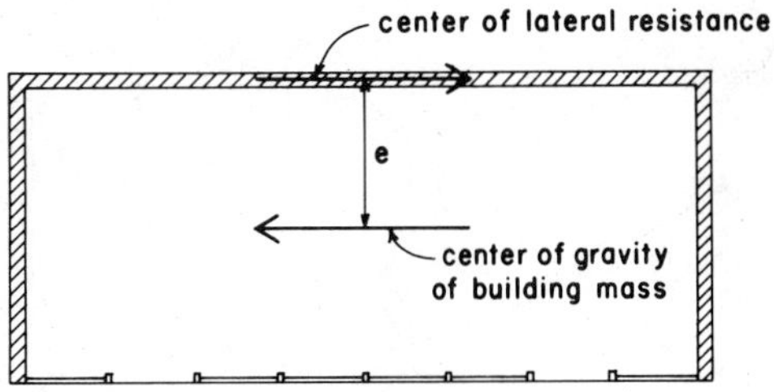

FIGURE 2.5

must either be adjusted so that its center of stiffness is close to the centroid of the mass or it must be designed for major twisting effects on the building. As the complexity of the building form increases, it may be necessary to consider the building to be multimassed, as discussed in the next section.

2.3 THE MULTI-MASSED BUILDING

Many buildings are multimassed, rather than consisting of a single geometric form. The building shown in Fig. 2.6 is multimassed, consisting of an L-shaped tower that is joined to an extended lower portion. Under lateral seismic movement the various parts of this building will have different responses. If the building structure is developed as a single system, the building movements will be very complex, with extreme twisting effects and considerable strain at the points of connection of the discrete parts of the mass.

If the elements of the tower of the building in Fig. 2.6 are actually separated, as shown in Fig. 2.7*a* or *c*, the independent movements of the separated elements will be different due to their difference in stiffness. It may be possible to permit these independent movements by providing structural connections that are detailed to tolerate the type and magnitude of the actual deformations. Thus the twisting effects on the building and the strain at the joints between the elements of the mass may be avoided.

There is also the potential for difference in response movements of the tower and the lower portion of the building, as shown in Fig. 2.7*e*. Actual separation may be created at this connection of the masses to eliminate the need for investigation of the dynamic interaction of the separate parts. However, it may not be feasible or architecturally desirable to make the provisions necessary to achieve either of the types of separation described. The only other option is therefore to design for the twisting effects and the dynamic interactions that result from having a continuous, single structural system for the entire building. The advisability or feasibility of one option over the other is often difficult to establish and may require considerable study of alternate designs.

Figure 2.8*a* shows an L-shaped building in which the architectural separation

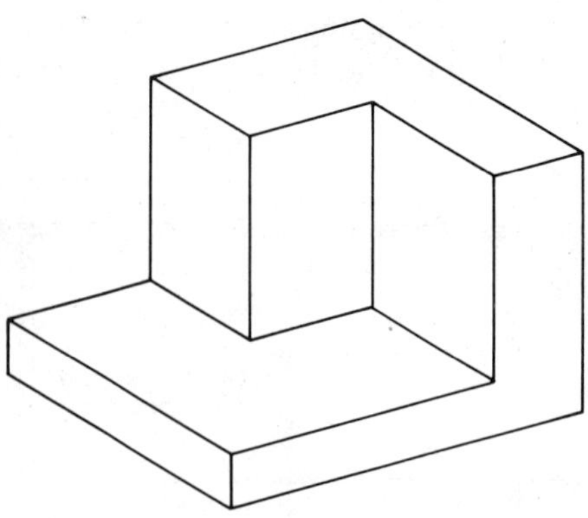

FIGURE 2.6

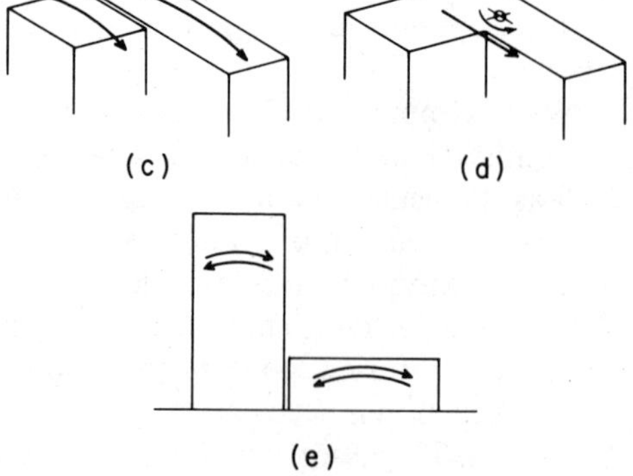

FIGURE 2.7

2.4 PROBLEMS OF JOINED MASSES

When individual parts of multimassed buildings are joined, there are many potential problems, some of which were described in the preceding section. Figure 2.9 shows three types of action that must often be considered for such structures. When moving at the same time, as shown in Fig. 2.9*a*, a problem for the separate masses becomes the actual dimension of the separation that must be provided to prevent them from bumping each other (called *battering* or *hammering*). If they are not actually separated, their independent de-

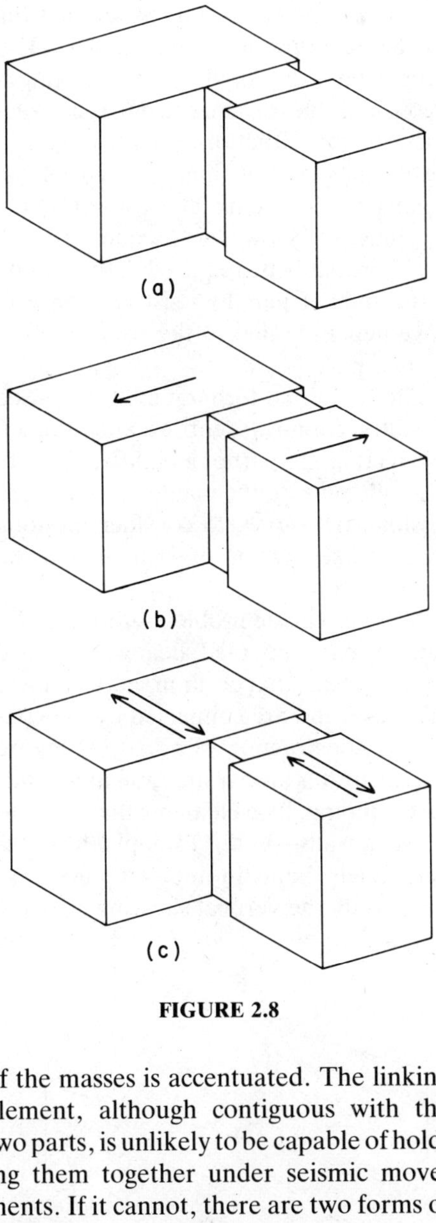

FIGURE 2.8

of the masses is accentuated. The linking element, although contiguous with the two parts, is unlikely to be capable of holding them together under seismic movements. If it cannot, there are two forms of differential movement that must be provided for, as shown in Figs. 2.8*b* and *c*. In addition to providing for these movements, it is also necessary to consider the bracing of the linkage element. If it is not capable of being independently braced, it must be attached to one or the other of the larger elements for support, making for a quite complex study of actions at the connection of the masses.

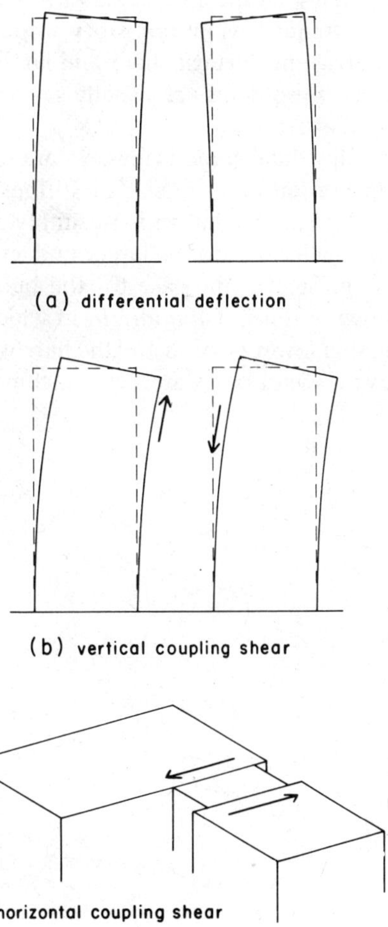

FIGURE 2.9

flections may be a basis for consideration of the forces that must be considered in preventing them from being torn apart.

Another potential action of separately moving parts is that shown in Fig. 2.9*b*. This involves a shearing action on the joint, similar to that which occurs in laminated elements subjected to bending. For the vertically cantilevered elements shown, both the shear and lateral deflection effects vary from zero at the base to a maximum at the top. The taller the structure, the greater the actual dimension of the critical movements near the top.

A third type of action is the horizontal shearing effect illustrated in Fig. 2.9*c*. This is probably the most common type of problem that must be dealt with, as it occurs frequently in one-story structures, whereas the vertical shear and lateral deflection problems are usually severe only in taller structures.

Individual joined masses are sometimes so different in size or stiffness that the indicated solution is to simply attach the smaller part to the larger and let it tag along. Such is the case for the buildings shown in Figs. 2.10*a* and *b*, in which the smaller lower portion and the narrow stair tower would be treated as attachments.

In some cases the tag along relationship may be a conditional one, as shown in Fig. 2.10*c*, where the smaller element extends a considerable distance from the larger mass. In this situation the movement of the smaller part to and away from the larger may be adequately resisted by the attachment. However, some bracing would probably be required at the far end of the smaller part to assist resistance to movements parallel to the connection of the two parts.

The tag along technique is often used for stairs, chimneys, entries, and other elements that are part of a building, but are generally outside the main mass. It is also possible, of course, to consider the total structural separation of such elements in some cases.

Another classic problem of joined elements is that of coupled shear walls. These are shear walls that occur in sets in a single wall plane and are connected by the continuous construction of the wall. Figure 2.11 illustrates such a situation in a multistory building. The elements that serve to link such walls—in this example the spandrel panels beneath the windows—are wracked by the vertical shearing effect illustrated in Fig. 2.9*b*. As the building

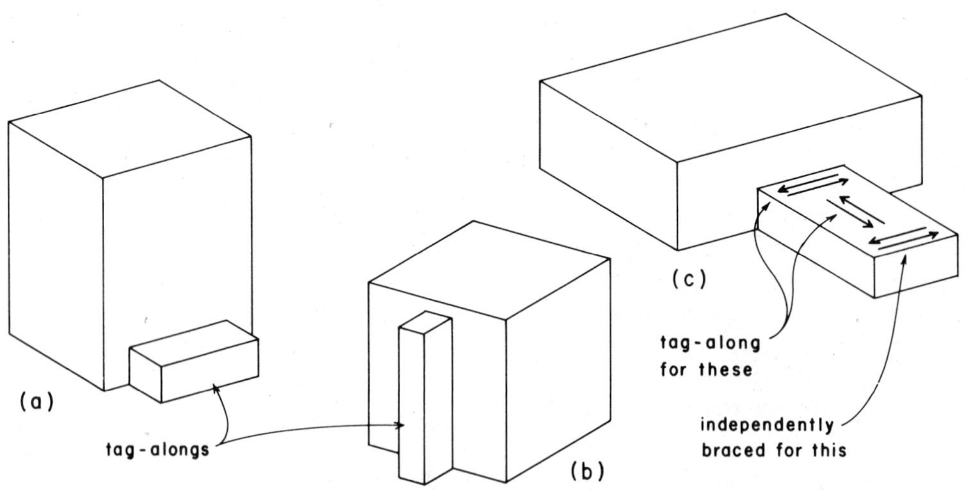

FIGURE 2.10

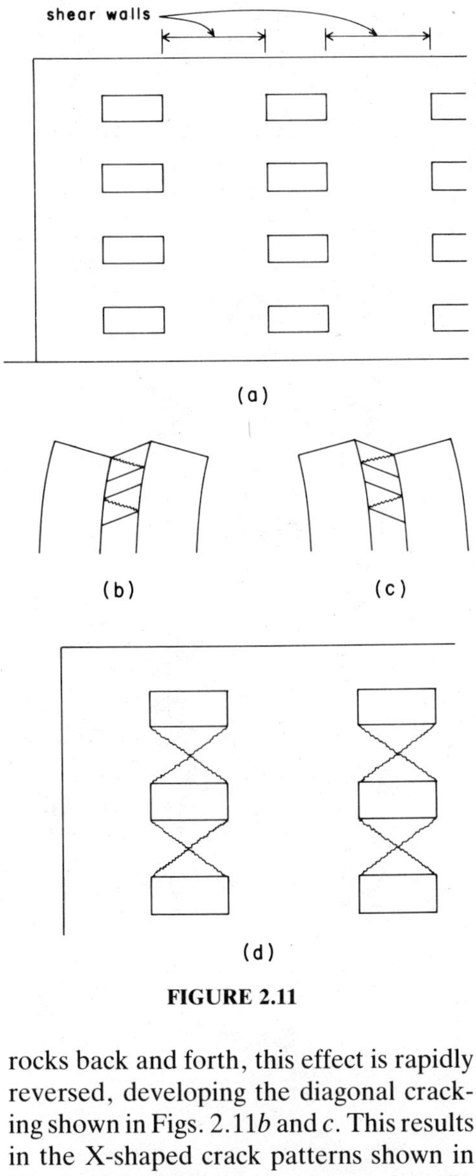

FIGURE 2.11

rocks back and forth, this effect is rapidly reversed, developing the diagonal cracking shown in Figs. 2.11*b* and *c*. This results in the X-shaped crack patterns shown in Fig. 2. 11*d*, which may be observed on the walls of many masonry, concrete, and stucco-surfaced buildings in regions of frequent seismic activity.

2.5 DISCONTINUITIES

Forces applied to buildings must flow with some direct continuity through the elements of the structure, be transferred effectively from element to element, and eventually be resolved into the ground. Where there are interruptions in the normal flow of the forces, problems will occur. For example, in a multistory building the resolution of gravity forces requires a smooth, vertical path; thus columns and bearing walls must be stacked on top of each other. If a column is removed in a lower story, a major problem is created, requiring the use of a heavy transfer girder or other device to deal with the discontinuity.

A common type of discontinuity is that of openings in horizontal and vertical diaphragms. These can be a problem as a result of their location, size, or even shape. Figure 2.12 shows a horizontal diaphragm with an opening. The diaphragm is braced by four shear walls, and if it is considered to be uninterrupted, it will distribute its load to the walls in the manner of a continuous beam. (See the discussion of flexibility of horizontal diaphragms in

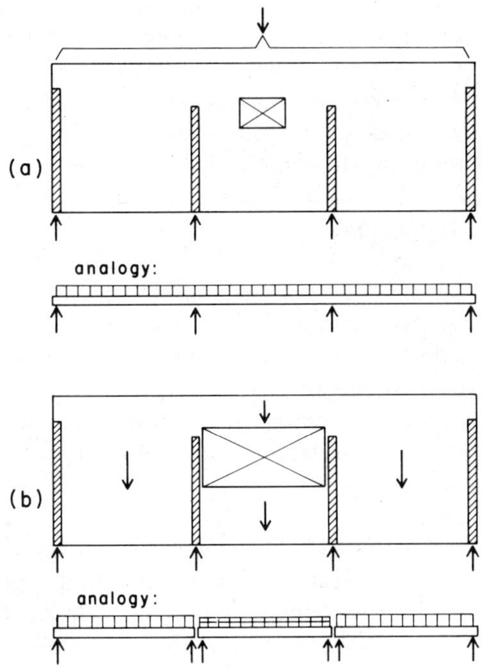

FIGURE 2.12

Section 15.4.) If the relative size of the opening is as shown in Fig. 2.12*a*, this assumption is a reasonable one. What must be done to assure the integrity of the continuous diaphragm is to reinforce the edges and corners of the opening and to be sure that the net diaphragm width at the opening is adequate for the shear force.

If the opening in a horizontal diaphragm is as large as that shown in Fig. 2.12*b*, it is generally not possible to maintain the continuity of the whole diaphragm. In the example the best solution would be to consider the diaphragm as consisting of four individual parts, each resisting some portion of the total lateral force. For openings of sizes between the ones shown in Fig. 2.12 judgment must be exercised as to the best course.

Another discontinuity that must sometimes be dealt with is that of the interrupted multistory shear wall. Figure 2.13 shows such a situation, with a wall that is not continuous down to its foundation. In this example it may be possible to utilize the horizontal structure at the second level to redistribute the horizontal shear force to other shear walls in the same plane. The overturn effect on the upper shear wall, however, cannot be so redirected, requiring that the columns at the ends of the shear wall continue down to the foundation.

It is sometimes possible to redistribute the shear force from an interrupted wall, as shown in Fig. 2.13, with walls that are sidestepped rather than in the same vertical plane of the upper wall. Again, however, the overturn on the upper wall must be accommodated by continuing the structure at the ends of the wall down to the foundation.

Figure 2.14 shows an X-braced frame structure with a situation similar to that of the shear wall in Fig. 2.13*a*. The individual panels of X-bracing are sufficiently similar in function to the individual panels of the shear wall to make the situation have the

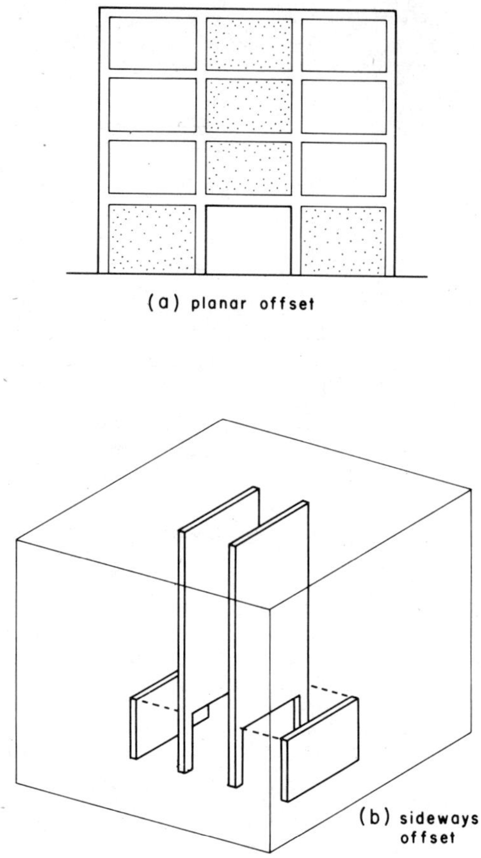

FIGURE 2.13

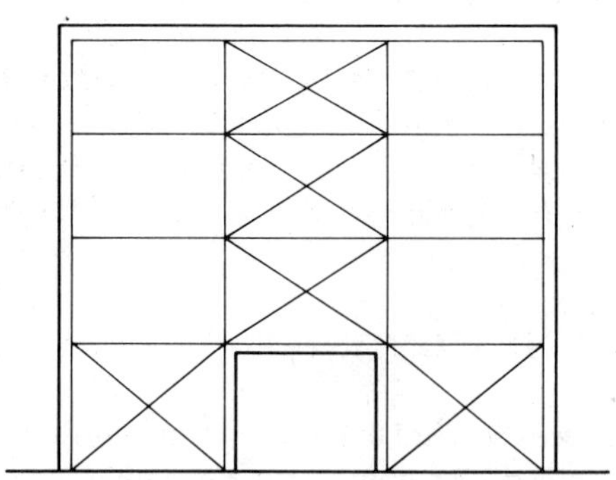

FIGURE 2.14

same general options and requirements for a solution.

Discontinuities are usually inevitable in multistory and multimassed buildings. They add to the usual problems of dissymmetry to create many difficult situations for analysis and design and require careful study for the proper assumptions of behavior and the special needs of the construction.

CHAPTER THREE

Materials and Methods of Construction

The fabric of a building is defined by the materials of its construction and by the common elements that are produced for use in building assemblages. This chapter presents some of the issues that must be dealt with as determined by the choice of materials and construction techniques.

3.1 SEISMIC BEHAVIOR OF MATERIALS

The nature of the response of building structures to various force actions has much to do with the materials of construction. This has partly to do with various properties of the materials themselves and partly to do with the products and assemblages that are made from them. In the case of seismic actions it has also to do with the density of the materials and the weight of the structures made from them. All of these factors are discussed in this chapter.

As with any force action, seismic action generates responses that relate to the fundamental stress–strain behavior of materials. Basic concerns are as follows:

Stress Resistance. Of various types (tension, compression, shear, combined stress, two- and three-dimensional stresses, etc.) and at various levels of magnitude.

Strain Resistance. Relative to other materials; related to stress magnitudes; affected by temperature, moisture, age, previous loadings, and so on.

Special Behaviors. Effects of stress direction, reversal, repetition, and qualitative responses such as ductility, brittle fracture, strain hardening, and so on.

For particular concerns in seismic analysis, the various dynamic response properties of materials are also important. These include the following:

Energy Absorption. Whereas resistance to static force is basically limited by stress magnitude, energy-absorbing capacity is essentially indicated by the area under the stress–strain graph (see Fig. 3.1). This results in a concern for the form of the graph, with special concern for various inelastic behaviors.

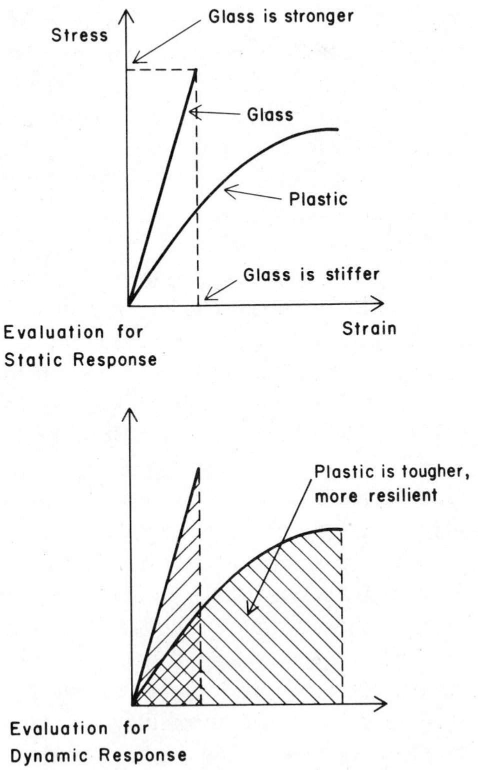

FIGURE 3.1. Static versus dynamic response.

Resilience. This is the limit of the energy absorption capacity which will not result in any significant permanent deformation or nonrepairable damage. It is important in determining the potential for reuse of the structure after the loading.

Toughness. This is essentially the ultimate limit of energy absorption, which usually infers some permanent damage in the form of ductile yielding, brittle fracture, splitting, shreading, delamination, and so on.

Response to Cyclic Loading. The repetition and reversal effects of seismic actions may result in progressive failure, fatigue, grinding, and so on.

Effect of Load Duration. The rapid-peaking impact or shock effect may have different effects on various materials. Materials with a yield character have more tolerance; those with a brittle nature are vulnerable.

Let us consider the various common materials of building construction and the products and assemblages ordinarily produced from them.

Wood

For structural purposes wood is used primarily in the form of boards, structural lumber, and glued-laminated products such as plywood, laminated timbers, and built-up wood plus plywood elements. For solid wood elements grain orientation and flaws such as knots, pitch pockets, and splits are of considerable importance. Lamination and building up of multiple-piece elements may reduce these effects.

Wood structures ordinarily consist of assemblages of many individual elements that must be fastened together to produce the complete structure. The response of wood structures to force actions, therefore, often has as much to do with the methods of fastening as with basic properties of the materials. Connections that rely on nails or bolts in loose holes may experience excessive deformation or loosening during seismic actions. It is generally desirable that connections remain *tight*, as well as simply resist the force magnitudes required of them.

A critical stress condition that should be avoided for seismic response is that of cross-grain bending, as shown in Fig. 3.2. The details of the construction of structures that utilize plywood (or other surfacing) on light wood framing to develop horizontal and vertical diaphragms must be carefully studied to avoid this situation.

The low modulus of elasticity of wood generally means that structures of wood experience considerable deformation. This is partly responsible for the high energy-absorbing character of wood struc-

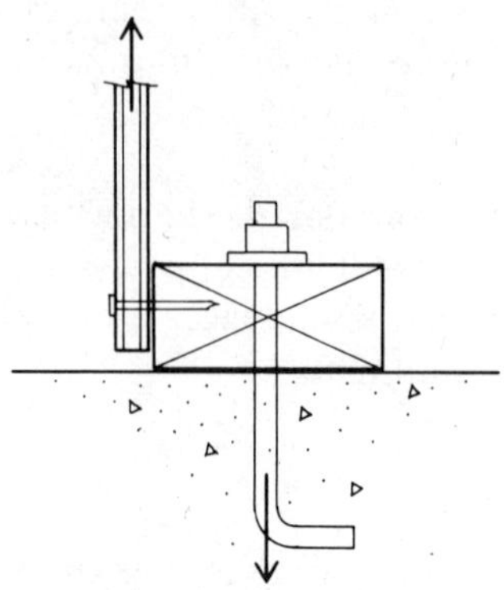

FIGURE 3.2. Cross-grain bending.

tures. However, the actual amount of defections, or the relative stiffness of the structure as compared to other parts of the construction, may present some problems. Stiff surfacing materials, such as tile or plaster, are likely to be damaged if the wood structure is permitted to deflect fully. In buildings with mixes of wood and masonry or concrete construction, the stiffer elements are likely to absorb dynamic loadings, or to fail if they must rely on bracing from the wood structure. In these situations careful analysis of deflections and conservative design is important.

The low density of wood (one fifth that of concrete, one fifteenth that of steel) is an asset in reducing the building mass for seismic load. Wood elements are usually bulky in form, however, so that wood structures are generally about the same in total weight as steel structures, steel elements typically being made of thin parts. The most drastic comparison may be made with structures of masonry and concrete, in which both density and bulkiness of parts is greater.

Steel

Steel is the most versatile structural material, being used in some form for almost every type of construction. It is essential for the assemblage of wood structures: for nails, screws, bolts, framing devices, and anchor bolts. It is essential in the form of wire or rods for reinforcement for structures of masonry and concrete. Steel structures themselves range from those of light, sheet metal elements to the heaviest of rolled and built-up elements.

An advantage of steel is its lack of weakness in terms of type, direction, reversal, or combinations of stresses. Its relative stiffness permits it to absorb a high percentage of total force when it is used in composite action with other materials. In concrete structures the steel reinforcement usually takes major parts of the internal forces, while ordinarily occupying only about 2% of the total volume.

A highly significant property is that of ductility, which is usually of considerable extent in most ordinary rolled sections and in concrete reinforcing rods. This is a major factor in development of energy-absorbing resistance, providing a means for partially overcoming the brittle nature of structures of masonry and concrete.

A principal concern in steel structures—due to the typical thinness of steel parts—is the problem of buckling. At upper limits of extreme slenderness, resistance is primarily a function of the stiffness of the material rather than its stress capabilities. This limits the potential for utilization of higher grades of steel, since the stiffness (modulus of elasticity) remains constant.

As with wood, steel structures are assembled from many separate elements. Thus the development of jointing methods and the detailing of the connections is a major part of the design work. Under seismic actions special problems are the loosening of bolts, brittle fracture in welds, tearing at locations of stress concentration, and twisting in eccentrically loaded joints. Many joint details that are sufficient for gravity or wind load situations are not acceptable for the rapid cyclic actions during earthquakes.

Although the modulus of elasticity of

the material is very high, the use of thin and slender parts means that steel structures are orindarily quite deformable. While adding to their energy absorption, these movements may be critical of themselves.

Concrete

The principal limitation of concrete as a structural material is its relatively low resistance to tensile stress, often as low as 10% of its resistance to compressive stress. For most structural tasks, steel reinforcement must be used to absorb internal tension forces or prestressing must be utilized to counteract load-developed tensile stresses.

While wood and steel can be used for small dimension elements, concrete must ordinarily be used in bulky form, resulting in structures with considerable dead weight. Due both to this increase in mass and to the typically high relative stiffness of the structures, concrete structures tend to absorb considerable force during seismic movement.

Concrete structures ordinarily experience a great deal of cracking. Some of this is due to the shrinkage of the material during the drying out process and to uneven expansion and contraction during changes in temperature. Shrinkage is more extensive when more water is used for mixing or when the wet mix is allowed to dry too rapidly after casting. Temperature stresses are more severe for exposed structures and where the climate range is greater.

Cracking is also inherent in the normal structural functioning of ordinary reinforced concrete. Tension induced by bending, torsion, and shear actions—although ultimately resisted by steel reinforcing—will often be at some level of strain beyond the fracture limit for the concrete. Thus the concrete must crack to permit the necessary level of stress to develop in the steel reinforcement.

When a concrete structure is subjected to the oscillatory actions of an earthquake, the rapid reversal of stresses will tend to operate on any existing cracks. This action may cause progression of the cracking and a grinding effect along the planes of the cracks. While this may be a principal means of energy absorption, it can also produce considerable unrepairable damage, especially when reinforcing is not adequate.

In most cases the useful strength limit of a reinforced concrete structure is that which is developed at the point of plastic yielding of the reinforcement. If this is the initial failure mode of the structure, it tends to give the structure a form of ductile behavior, in contrast to the normal brittle nature of the stiff, tension-weak material. Thus for both static and dynamic loading the design is usually carried out to develop this yielding mode as the initial failure condition.

In large concrete structures there are ordinarily a considerable number of so-called cold joints. These occur at points where the continuous pouring is halted during the construction process. Provision must be made for structural continuity of both the steel and concrete through these joints. Doweling of the steel bars usually assures that the joint will not pull apart. However, the concrete-to-concrete relationship is essentially the same as at a crack, and grinding during seismic action must be studied.

As has been mentioned, concrete structures are heavy and generally quite stiff, and seismic actions typically produce major forces in them. They must therefore be proportionately stronger than framed structures of wood or steel. However, ordinary techniques of construction will often result in major natural resistance to seismic effects so that little need be done in many cases to provide for seismic resistance.

Structures requiring especially careful study for seismic response are those employing elements of precast concrete, as opposed to concrete that is site poured into forms. These structures are similar to wood and steel structures in being vulnerable to the integrity of the jointing methods. In contrast to wood and steel, however, the structural elements are heavy so that dynamic behavior is even more critical.

Masonry

Most structural masonry for seismic-resistive buildings is produced with units of hollow, precast concrete and is reinforced both horizontally and vertically. Steel rods are inserted in vertically aligned voids, which are then filled with concrete to produce small reinforced concrete columns inside the wall. Horizontal rods are typically placed in horizontal rows of U-shaped blocks, which are filled with concrete to produce tie beams (see Fig. 3.3.)

Reinforced brick construction and various combinations of bricks and concrete units may also be used, but the all-concrete structure is used most widely in regions of high seismic risk. Brick and stone finishes most commonly occur as facings, applied as veneers over some substructure, which may be wood, steel, concrete, or reinforced concrete block construction. Adequate fastening of these veneers is a critical problem in seismic design.

The integrity of any masonry construction depends on the quality of the mortar joints and the structural character of the masonry units. It also depends a lot on the skill and care of the masons, on the control of the moisture content of the units during

FIGURE 3.3. Reinforced masonry construction with hollow concrete blocks.

laying, on protection against rapid drying of the mortar, and on the prevention of shock to the laid-up units prior to hardening of the mortar. The less the amount of reinforcing, the more the skill and care of the workers become critical.

With the wall of reinforced hollow units, the concrete-filled voids with their steel rods constitute a sort of reinforced concrete rigid frame inside the wall. This frame interacts in a composite manner with the infill of masonry units that is bonded to it. Theoretical analysis of this construction is quite complex, but its behavior has been demonstrated by tests, as well as by the response of many buildings to actual earthquakes.

There are many similarities between reinforced masonry and ordinary reinforced concrete. Stiffness and excessive weight is a liability for both in seismic action, as has been discussed. Nevertheless, the reinforced concrete unit masonry shear wall is widely used to brace low-rise buildings of various types of construction.

3.2 EFFECTS OF JOINTS

Buildings ordinarily consist of a great number of individual elements that are joined together to create the whole construction. Elements of the structure (decks, beams, columns, etc.) are connected, roof and floor planes intersect, and the building proper is achored to its foundation. The structural response of the building as a whole involves the actions of a vast number of joints.

Joints exist in endless variety, varying with the size, shape, and material of the connected elements, with the loads being transferred, and with the need for various nonstructural functions. The latter may include concerns for weather seal, water tightness, expansion and contraction, demountability, and so on. Of major concern is the need for the use of jointing methods that are technically and economically feasible, with emphasis on those that use standard, tested products and permit relatively simple and rapid assemblage.

With regard to seismic response there are a number of particular concerns for joints. Of prime concern is the ability to rely on a quantified load capacity for a specific joint function. For ordinary jointing methods such as nailing, welding, and bolting, this capacity is generally determinable by engineering computations. For special fasteners, capacities are mostly established by testing, using methods that are acceptable to the building codes.

For the dynamic effects of jolting and shaking due to earthquakes, joints must have resistances that are not usually required for gravity and wind loads. Fracture due to stress concentrations, progressive deformations, and loosening of joints are of particular concern in seismic response. Reliance on nail withdrawal in wood, unfinished bolts in steel, and simple doweling alone for anchorage in concrete is usually not acceptable in seismic design.

The complete and successful design and construction detailing of joints is a major part of the seismic design for a building. The most accurate and complete computations for the investigation of structural behavior and careful design of the major elements of the bracing system will be for naught if the load transfers through the system are not thoroughly developed by study of the jointing along the load paths.

While jointing of the elements of the lateral bracing system is of prime concern, the complete seismic design must also consider the problems of connection of the many nonstructural elements. Proper attachment of wall finish elements, light fixtures, doors, and so on is essential to prevent the objects from becoming flying missles during an earthquake.

In many situations jointing techniques must allow for some sort of movement while simultaneously providing support or attachment. Window glass must be securely held in a metal frame, but must have some degree of freedom to allow for thermal changes and to prevent structural deformations of the frame from transferring load to the glass. Structural joints must sometimes allow for a similar condition. A seismic control joint may provide for vertical support but allow horizontal movement. The specific requirements for any joint must be fully studied for a successful design.

3.3 DEAD WEIGHT OF CONSTRUCTION

The weight of the building must be determined for both gravity and seismic design. For most buildings this requires considerable approximation due to the complexity in terms of the number and form of elements comprising the complete construction. Since hair-splitting accuracy is not to be expected, the results of any investigation using the dead weight of the building should not be considered to have any disproportionate level of precision.

For determination of the weight of the building construction some information is required. This includes the following:

1. Knowledge of the exact details and dimensions of the construction. This is known at the *end* of the building design process but is usually somewhat sketchy during the early stages of design, which is when a major portion of the structural design is usually done.
2. The properties of the materials and elements of the construction. This includes the densities of the materials and the usual sizes, thicknesses, and so on. For simple, ordinarily elements, such as concrete slabs or plywood panels, this may be relatively easy. For complex or custom-designed elements, such as glass and metal curtain wall systems, it may involve a lot of guesswork.
3. The allowance necessary for indefinite elements, such as electrical fixtures, air ducts, plumbing lines, and so on, are typically accounted for by adding some amount to the uniformly distributed live or dead load. This amount is sometimes specified as a minimum value by building codes.

For the final structural design the dead load of the construction should be as accurately accounted for as is reasonably possible. By the time the final answers are required, it may well be that the final details of the construction are sufficiently developed to make such a determination possible. For preliminary design, however, it is usually necessary to use some very approximate values based on past experience with similar construction.

Information is provided in Chapter 32 for both the approximation of weights for preliminary investigations and the detailed determination of weights to the extent that may be desirable for the final design. Applications of this data are illustrated in the examples in Part 4.

3.4 Rehabilitation and Repair

For various reasons it is sometimes necessary to consider the problem of repairing or rehabilitating existing buildings to improve their seismic resistance. One such situation occurs when a building has sustained a severe earthquake shock and has suffered some damage that does not necessarily make it unsafe for ordinary conditions, but does make it unsafe with regard to another major seismic shock.

Work of this type is generally more feasible for structures with frames of wood or steel than for those with construction of concrete or masonry. Where extensive cracking and yielding of steel reinforcement has occurred in masonry or concrete structures, repair is often not economically feasible. However, when the damage is isolated, it may be possible to remove some parts of the structure and to either replace them with new construction or simply use the remaining, undamaged structure.

A problem sometimes encountered in such design work is that the design of the entire structure must often be brought into compliance with existing codes, whereas the original design may most likely have been done with less stringent code requirements. This is especially true for seismic design, as the general evolution of the codes has been toward increasingly severe requirements. In extreme cases this may often rule against the feasibility of restoration or rehabilitation of existing buildings.

Rehabilitation may be necessary because of a proposed remodeling or change of occupancy. It may also be required, or simply desired, to improve the seismic resistance of a building with a type of construction that has been shown to be unsafe. A classic situation of the latter type is that of the thousands of existing older buildings in regions of high seismic risk that have structures of unreinforced masonry. Due both to its weight and its lack of shock resistance, this type of construction has been responsible for a great amount of property loss and injury in many earthquakes. Its use has been generally outlawed in high-risk areas with some building code jurisdiction.

While unreinforced masonry is a major culprit, there are many other undesirable types of construction for earthquake-prone areas. These continue to exist primarily because they have not yet sustained a major shock, although the potential risk may be high. They are *permitted* to exist mainly because new requirements of building codes normally apply only to new construction or to remodeling or changes in occupancy.

Most buildings tend to age less than gracefully with regard to their seismic resistance. As we learn more, we tend to require more and *do* more about seismic design. Yesterday's designs become rapidly obsolete by today's standards. This is certainly a matter of degree for any specific building, but it is nevertheless a general trend. Thus just about *any* existing building more than a few years old may be viewed as a candidate for some rehabilitation, if the most stringent seismic resistance, as we are presently able to determine it and design for it, is desirable.

CHAPTER FOUR

Scale and Dimensional Factors

The actual size of a building—both in terms of its overall size and the size of its various parts—is important in seismic design, as well as in architectural planning. This chapter discusses some considerations of size that are critical to seismic response.

4.1 BUILDING SIZE AND SEISMIC RESPONSE

The size of a building relates to various concerns in seismic design. One concern is the degree of hazard in terms of potential injury or loss of life. Size often relates to the potential number of building occupants, which is partly reflected in the *I* factor presently used to modify the value for design force on the building due to seismic shock.

Size is an important factor in determining the fundamental period of the structure—or of a part of the structure for buildings of complex, multimassed forms. The fundamental period effects the amount of force received directly from seismic shock and also effects the building–site interaction during seismic movements. The implications of these factors are discussed in Chapter 11, and the effects on design are illustrated in the examples in Part 4.

For building–site interaction the critical quantity is the ratio of the fundamental periods of the building and the site rather than the specific value of either. Thus buildings of a particular size may be less safe on one site and more on another.

Building size also affects requirements for fire-resistive construction. Certain materials and types of construction are generally required when size is of a particular range. There are thus certain aspects of seismic response that may be anticipated for large buildings on the basis of the typical construction that may be expected.

As building size increases, the feasible limits for various types of lateral bracing systems may be exceeded. Based on current limits of construction materials and processes, on code requirements, and on the inherent limits of various basic systems (shear wall, braced frame, moment-resistive frame, etc.), the feasibility of system choice may be established for various situations. This applies also to the use of individual elements of systems: plywood diaphragms, sheet metal anchors, concrete block shear walls, and so on.

Low-rise buildings that are large in plan dimension often consist of structural units that are repeated. The division between units may be established by a fire wall, a thermal expansion joint, or a seismic separation joint—or possibly by all of these. Development of the logical size for the units, and the planning of the building in general, must reflect all the critical concerns.

4.2 LIMITING DIMENSIONAL FACTORS

Size limits for structural elements derive from a number of concerns. Based on general considerations of the construction, some limits are:

Available sizes of commercial products: nails, plywood sheets, reinforcing rods, and so on.

Minimum sizes based on slenderness ratios (L/r, h/t), span–depth ratios, deflections, and so on.

Minimum sizes based on practical construction limits: steel column plus fireproofing jacket, stud wall plus surfacing, and so on.

Maximum sizes of fabricated units based on limits of transportation, handling for erection, and so on.

Feasible limits exist for various types of spanning systems for roof and floor structures. Deck span limits affect the spacing of supporting beam or truss elements. Planning of bearing walls and columns relates to the spacing and span limits for the supported systems. The basic form of the structure, the dimension for the module of repeated elements, and the general geometric relations of the separate elements of the structural system derive from the various limitations mentioned. In addition, of course, the structure must accommodate and relate to the dimensions of any architectural modules.

When it is a critical concern, seismic behavior must also be considered in establishing building dimensions. Some specific size limits that derive from concern for seismic design are the following:

1. *Dimensional Ratio Limits for Diaphragms.* Limits for length-to-width for horizontal diaphragms and height-to-length of shear walls are established by design codes. Maximum load capacity or deflection limits may also produce limitations. These are discussed in Chapters 15 and 16.
2. *Capacity Requirements for Diaphragms.* The magnitude of loads and resulting stresses will influence the choice of overall dimensions and the size of the various parts of diaphragms. When load magnitude exceeds the upper limit for one type of construction, another—possibly with different dimensions—must be considered.
3. *Deflection Limits for Moment-resistive Frames.* In order to control lateral drift the column widths and beam depths in rigid frames must usually be greater than in frames designed only for gravity loading.
4. *Minimum Usable Plan Length for Shear Walls.* Based on dimensional ratios, load capacity, load sharing, deflection, and problems of anchorage and overturn, there are practical limits on the plan length of shear walls. These limits must be recognized at the earliest stages of architectural planning in order to develop a building layout that can be made to work for seismic resistance.

CHAPTER FIVE

Structural Versus Nonstructural Damage

Damage to the structure represents a direct loss of safety for the building. Damage to the building as a whole must be more broadly viewed. This chapter discusses issues relating to structural damage as well as problems of damage to the building in general.

5.1 SAFETY AND FAILURE AS RELATIVE VALUES

The concept of safety is essentially simple. It means that there is some margin—some unused strength—that cushions against failure. In structural design the fundamental definition of the safety factor is

$$\text{safety factor} = \frac{\text{limiting capacity of structure}}{\text{anticipated demand on structure}}$$

While this is a simple concept, design for real safety requires considerable judgment. This begins with a decision about the quantity for the safety factor. What level of margin (redundancy) is desirable or acceptable? If the redundancy is unreasonably high, the structure will be excessive in size and in cost. If it is too low, it infers a level of precision in the work—the ability to clearly define the loads and to precisely analyze for responses—that is unrealistic in most cases.

The average safety factor used in structural design work is 2. That is to say, structures are designed to be twice as strong as they need to be for the real situations anticipated. There may be some logical basis for this, but it is more likely that it is simply a nice round number and has become a long-standing habit.

In addition to the establishment of the numeric value for the safety factor, judgment must be exercised as to what constitutes failure. Collapse of the structure is a significant failure, easy to notice and reasonably capable of prediction with present experience and computational abilities. There are, however, many other ways in which a structure can fail to do its job. Some of these are:

1. *Failure of Individual Parts.* One element of the structure may fail. If it is not indispensible to the structure as a whole, it may be of less concern.

2. *Deformation Failure.* While retaining its strength in terms of load capacity, the structure may experience so much movement as to create problems for occupants or for attached nonstructural elements.
3. *Initial Failure.* Complex structures often fail in stages. If the building response does not progress beyond initial stages of failure, a margin of safety may be preserved. However, the exact nature of the initial failure as related to continued use of the building must be considered.

For seismic design it is sometimes necessary to define various levels (or stages) of failure and to match them to various intensities and probabilities of seismic activity. We may, for example, ask the following questions:

What can be tolerated in the form of response to a minor earthquake? (Low intensity; high probability.)

What response can be accepted for a moderate quake? (Moderate intensity; reasonable probability.) Can we tolerate minor, potentially repairable, damage?

What response, short of total collapse, must we consider in the event of a major quake? (High intensity; very remote possibility.) Can we tolerate major nonstructural damage? Can we use up the one-time early-stage response that may render the structure unrepairable and unusable?

5.2 ATTACHMENTS

While the major concern in structural design is for the behavior of the structure itself, attention must be given to the effects of structural response on the rest of the building. Finish materials, roofing, suspended ceilings, nonstructural walls, doors, and other elements are supported by the structure. So are the various elements of the building electrical, plumbing, heating and ventilating, and vertical transportation systems. When the structure moves, these elements will move with it, and concern must be given to the effects on the supported elements and on their attachment to the structure. Some specific concerns are the following:

Load Transfer to Nonstructural Elements. When the structure moves, it will often produce deformations in elements that are attached to it. If these elements have some resistance to deformation, they will develop stresses and literally carry some of the load in many cases.

Stress on Attachments. Due to the possibility of load transfers, there will be some stress on the materials or devices used to affect attachments.

Need for Controlled Support. It is often necessary to provide structural support in a manner that permits some types of load transfers, while preventing others. A pin-type connection is often used in structures when it is desired to resist direct forces or shears but to avoid the transfer of twisting or bending.

A common problem occurs when a relatively flexible structure is used to support some nonstructural elements that are quite rigid. Wall surfacing of plaster, brick veneer, or tile constitute such rigid, planar surfaces. When attached to walls of heavy masonry or concrete, these finishes will experience little stress, due to the minimal deformations of the supporting structure. When attached to light frameworks of wood or steel, however, the deformations of the structure are likely to transfer load to the rigid surface planes. In this situation it may be necessary to consider one of the following approaches:

1. Accept the eventuality of the load transfers and the failure of the nonstructural materials. This may be the practical course, if the repair or replacement of the materials is reasonably easy and the danger of falling materials is not a serious concern for the building occupants.
2. Use resilient or cushioned attachments that will control load transfer. This is sometimes easily and commonly done, as in the case of window glazing and suspended ceilings. In other cases it is not practical, as in the case of plaster or stucco attached to a wood frame.
3. Use control joints in the rigid surfacing. This is a means of controlling the degree of flexibility of the rigid surface and of channeling the cracks into predetermined patterns. These joints may also function to relieve cracking due to shrinkage, temperature variation, moisture expansion, and so on.
4. Develop the rigid supported materials to have some quantified stress capability. Masonry veneer may be reinforced; tile can be installed over plywood or a reinforced plaster base. Stucco, plaster, and gypsum drywall are now rated for a quantified resistance, when properly installed for this development.

There are, of course, many ways in which damage to nonstructural elements may be reduced or eliminated. The planning of the building and development of details of the construction can be done with this in mind. Selection of materials should be done with a sensitivity to their seismic vulnerability. Structural solutions should not be visualized as the best way to handle all of these problems.

CHAPTER SIX

Seismic Resistance of Ordinary Construction

Even when buildings are built with no consideration given to design for seismic response, they will have some natural capacity for lateral force resistance. This chapter presents some of the most ordinary types of construction and discusses the normal capabilities and vulnerabilities that they represent with regard to earthquake effects. Development of enhanced levels of response are discussed in later chapters.

6.1 WOOD FRAME CONSTRUCTION

Wood structures can be categorized broadly as light frame or heavy timber. Light frames—using mostly 2× dimension lumber for wall studs, floor joists, and roof rafters—account for the vast majority of small, low-rise buildings in the United States. In most cases the frames are covered on both surfaces by some type of surfacing material. Many of these surfacing systems have a usable, quantifiable, capacity for diaphragm resistance. Thus, without any major alteration of the basic structure, most light wood frames can be made resistive to lateral forces through the use of a combination of horizontal and vertical diaphragms.

Many of the ordinary elements of light wood frames can be utilized as parts of the lateral resistive system, serving as diaphragm chords, collectors, and edge transfer members. Wall studs, posts, sills and plates, and roof and floor framing members, occurring naturally in the structure, are often able to be utilized for these functions. Alterations necessary to make them more functional are often limited to moderate increases in sizes or to the use of some additional fastening or anchoring. When members are long, and not able to be installed as a single piece (as with the top plate on a long wall), it may be necessary to use stronger splicing than is ordinarily required for gravity resistance alone.

Common practices of carpentry result in a considerable amount of fastening between members of a light wood frame. Building codes often specify minimum requirements for such fastening. (See reprint of Table 25−Q from the *Uniform*

Building Code in Chapter 33.) Before undertaking to design such a structure for seismic loads, it well behooves the designer to become thoroughly familiar with these requirements, as well as current local practices of contractors and workmen. References with suggested details for construction become rapidly out of date as codes and construction practices and the availability of materials and equipment change.

In recent times there has been a trend toward the extensive use of sheet metal fastening devices for the assemblage of light wood structures. In general, these tend to increase resistances to lateral loads because the continuity of the frame and the anchorage of members is greater.

There is a considerable range in the diaphragm shear capability of various surfacing materials. The following are some widely used products:

Plywood. This may be used as the structural backup for a variety of finishes or may be used with a special facing as the complete surfacing. Most plywoods offer considerable potential for shear resistance. With an increase in structural quality, greater thickness than the minimum required for other functions, and a greater number of nails, it is possible to develop considerable strength for either horizontal or vertical diaphragm use.

Board Sheathing. Boards of 1 in. nominal thickness, with shiplap or tongue-and-groove edges, were once quite commonly used for sheathing. When applied in a position diagonal to the frame, they can produce some diaphragm action. Capacities are nowhere near that of plywood, however, so that for this and other reasons this type of structural covering is not much used at present.

Plaster. Portland cement plaster, applied over wire-reinforced backing and adequately secured to the framing, produces a very stiff diaphragm with load capacities equal to that of the thinner plywoods. On the exterior it is called stucco, which due to its popularity is the definitive vertical bracing material for light structures in southern and western United States.

Miscellaneous Surfacing. Gypsum drywall, gypsum plaster, nonstructural plywood, and fiberboard can develop some diaphragm capacity that may be sufficient for low-stress situations. Stiffness is minimal, thus narrow diaphragms should be avoided.

When the same material is applied to both sides of a wall, the code permits the use of the sum of the resistances of the two surfaces. For interior walls this is quite common and permits the utilization of low-strength surfacing. However, for exterior walls the two surfaces are seldom the same, thus the stronger (usually the exterior) must be used alone.

In the past a widely used method for bracing light wood frames consisted of diagonal bracing, typically in the form of 1-in. nominal thickness boards with their outside surfaces made flush with the face of the framing by cutting notches in the framing members (called let-in bracing). The acceptance of rated load capacities for a wider range of popular surfacing materials has made this practice largely redundant. When diagonal bracing is used today, it often consists of thin steel straps applied to the faces of the framing members.

Some of the problems encountered in developing seismic resistance with light wood frame construction follow.

1. Lack of adequate solid walls to serve as shear walls. This may be due to the building planning, with walls insufficient at certain locations or in a particular direction (see Fig. 2.2). Walls

may also simply be too broken up in short lengths by doors and windows. For multistory buildings there may be a problem where upper-level walls do not occur above walls in lower levels.

2. Lack of adequate diaphragm surfacing. Many types of surfacing have rated capacities for shear. Each, however, has its limits and some materials are not rated for code-acceptable loadings.
3. Lack of continuity of framing. Because it is often not required for gravity load conditions, members that could function as chords or collectors may consist of separate pieces that are not spliced for tension or compression continuity.
4. Lack of adequate connections. Load transfers—most notably those from horizontal to vertical diaphragms—may not be possible without modification of the construction details, involving additional framing, increased nailing, or use of special anchorage devices. Ordinary code-required minimum wall sill bolting is not acceptable for overturn resistance and is often not adequate for sliding resistance.

Wood post and beam structures can sometimes be made to function as braced frames or moment-resistive frames, the latter usually being somewhat more difficult to achieve. Often, however, these structures occur in combination with wall construction of wood frame plus surfacing or reinforced masonry so that the post and beam frame need not function for lateral bracing. Floor and roof decks are often the same as for light frames and thus are equally functional as horizontal diaphragms.

A problem with post and beam construction is often the lack of ability for load transfers between members required for resistance to lateral loads. At present this is less the case due to the increasing use of metal framing devices for beam seats, post caps and bases, and so on.

When heavy timber frames are exposed to view, a popular choice for roof or floor decks is a timber deck of 1.5 in. or greater thickness. Although such a deck has a minimal shear capacity, it is usually not adequate for diaphragm development, except for small buildings. The most common and economical means for providing the necessary diaphragm resistance is to simply nail a continuous plywood surface to the top of the timber deck.

6.2 MASONRY WALLS

For seismic zones 3 and 4 the only masonry structural construction permitted is *reinforced masonry*. These comments are confined to structural masonry walls constructed with hollow concrete units (concrete blocks), with the voids partly or wholly reinforced and filled with grout (see Fig. 3.3).

Structural masonry walls have considerable potential for utilization as shear walls. There are, however, a number of problems that must be considered.

1. *Increased Load.* Due to their weight, stiffness, and brittleness, masonry walls must be designed for higher lateral seismic forces. The *UBC* requires an additional increase of 50% in the load for stress analysis of the wall.
2. *Limited Stress Capacity.* The unit strength and the mortar strength must be adequate for the required stress resistances. In addition, both vertical and horizontal reinforcing is required for major shear wall functions.
3. *Cracks and Bonding Failures.* Walls not built to the specifications usually used for seismic-resistive construction often have weakened mortar joints

and cracking. These reduce seismic resistance, especially in walls with minimal reinforcing.

Code specifications for concrete block walls result in a typical minimum construction that has a particular limit of shear wall capacity. This limit is beyond the limit for the strongest of the wood-framed walls with plywood on a single side, so the change to a masonry wall is a significant step. Beyond the minimum value the load capacity is increased by adding additional reinforcing and filling more of the block voids. At its upper limits the reinforced masonry wall approaches the capacity of a reinforced concrete wall.

Anchorage of masonry walls to their supports is usually simply achieved. Resistance to vertical uplift and horizontal sliding can typically be developed by the usual doweling of the vertical wall reinforcing. The anchorage of horizontal diaphragms to masonry walls is another matter and typically requires the use of more "positive" anchoring methods than are ordinarily used when seismic risk is low.

6.3 REINFORCED CONCRETE CONSTRUCTION

Poured concrete elements for most structures are ordinarily quite extensively reinforced, thus providing significant compensation for the vulnerability of the tension-weak material. Even where structural demands are not severe, minimum two-way reinforcing is required for walls and slabs to absorb effects of shrinkage and fluctuation of temperature. This form of construction has considerable natural potential for lateral force resistance.

Subgrade building construction most often consists of thick concrete walls, in many cases joined to horizontal concrete structures with solid poured-in-place slabs. The typical result is a highly rigid, strong boxlike structure. Shears in the planes of the walls and slabs can be developed to considerable stress levels with minimum required reinforcing. Special attention must be given to the maintaining of continuity through pour joints and control joints and to the anchorage of reinforcing at wall corners and intersections and at the joints between slabs and walls. This does not always result in an increase in the amount of reinforcing but may alter some details of its installation.

Structures consisting of poured concrete columns used in combination with various concrete spanning systems require careful study for the development of seismic resistance as rigid frame structures. The following are some potential problems:

1. *Weight of the Structure.* This is ordinarily considerably greater than that of wood or steel construction, with the resulting increase in the total seismic force.
2. *Adequate Reinforcing for Seismic Effects.* Of particular concern are the shears and torsions developed in framing elements and the need for continuity of the reinforcing or anchorage at the intersections of elements. A special problem is that of vertical shears developed by vertical accelerations, most notably punching shear in slab structures.
3. *Ductile Yielding of Reinforcing.* This is the desirable first mode of failure, even for gravity load resistance. With proper design it is a means for developing a yield character in the otherwise brittle, tension-weak structure.
4. *Detailing of Reinforcing.* Continuity at splices and adequate anchorage at member intersections must be assured

by careful layout of reinforcing installation.

When poured concrete walls are used in conjunction with concrete frames, the result is often similar to that of the plywood-braced wood frame, the walls functioning to absorb the major portion of the lateral loads due to their relative stiffness. At the least, however, the frame members function as chords, collectors, drag struts, and so on. The forces at the intersections of the walls and the frame members must be carefully studied to assure proper development of necessary force transfers. This is even more critical if there is an actual interaction of the wall and frame systems.

As with masonry structures, considerable cracking is normal in poured concrete structures; much of it due to shrinkage, temperature expansion and contraction, settlement or deflection of supports, and the normal development of internal tension forces. In addition, built-in cracks of a sort are created at the cold joints that are unavoidable between successive, separate pours. Under the back-and-fourth swaying actions of an earthquake, these cracks will be magnified, and a grinding action may occur as stresses reverse. The grinding action can be a major source of energy absorption but can also result in progressive failures or simply a lot of pulverizing and flaking off of the concrete. If reinforcing is adequate, the structure may remain safe, but the appearance is sure to be affected.

It is virtually impossible to completely eliminate cracking from masonry and poured-in-place concrete buildings. Good design, careful construction detailing, and quality construction can reduce the amount of cracking and possibly eliminate some types of cracking. However, the combination of shrinkage, temperature expansion, settlement of supports, creep, and flexural stress is a formidable foe.

6.4 STEEL FRAME CONSTRUCTION

Structures with frames of steel can often quite readily be made resistive to lateral loads, usually by producing either a braced (trussed) frame or a moment-resistive frame. Steel has the advantage of having a high level of resistance to all types of stress and is thus not often sensitive to multidirectional stresses or to rapid stress reversals. In addition, the ductility of ordinary structural steel provides a toughness and a high level of energy absorption in the plastic behavior mode of failure.

The high levels of stress obtained in steel structures are accompanied by high levels of strain, resulting often in considerable deformation. The actual magnitudes of the deformations may affect the building occupants or contents or may have undesirable results in terms of damage to nonstructural elements of the building construction. Deformation analysis is often a critical part of the design of steel structures, especially for moment-resistive frames.

The ordinary post and beam steel frame is essentially unstable under lateral loading. Typical framing connections have some minor stiffness and moment resistance but are not effective for development of the rigid joints required for a moment-resistive frame. Frames must therefore either be made self-stable with diagonal bracing or with specially designed moment-resistive connections or be braced by shear walls.

Steel frames in low-rise buildings are often braced by walls, with the steel structure serving only as the horizontal spanning structure and vertical gravity load resisting structure. Walls may consist of masonry or of wood or metal frames with various shear-resisting surfacing. For the wall-braced structure, building planning must incorporate the necessary solid wall construction for the usual shear wall braced building. In addition, the frame will usually be used for chord and collector actions; therefore, the connections between the decks, the walls, and the frame must be designed for the seismic load transfers.

The trussed steel structure is typically quite stiff, in a class with the wall-braced structure. This is an advantage in terms of reduction of building movements under load, but it does mean that the structure must be designed for as much as twice the total lateral force as a moment-resistive frame. Incorporating the diagonal members in vertical planes of the frame is often a problem for architectural planning, essentially similar to that of incorporating the necessary solid walls for a shear wall braced structure.

In the past steel frames were mostly used in combination with decks of concrete or formed sheet steel. A popular construction for low-rise buildings at present—where fire-resistance requirements permit its use—is one that utilizes a wood infill structure of joists or trusses with a plywood deck. A critical concern for all decks is the adequate attachment of the deck to the steel beams for load transfers to the vertical bracing system. Where seismic design has not been a factor, typical attachments are often not sufficient for these load transfers.

Buildings of complex unsymmetrical form sometimes present problems for the development of braced or moment-resistive frames. Of particular concern is the alignment of the framing to produce the necessary vertical planar bents. Randomly arranged columns and discontinuities due to openings or voids can make bent alignment or continuity a difficult problem.

6.5 PRECAST CONCRETE

Precast concrete structures present unique problems in terms of lateral bracing. Although they share many characteristics with poured-in-place concrete structures, they lack the natural member-to-member continuity that provides considerable lateral stability. Precast structures must therefore be dealt with in a manner similar to that for post and beam structures of wood or steel. This problem is further magnified by the increased dead weight of the structure, which results in additional lateral force.

Separate precast concrete members are usually attached to each other by means of steel devices that are cast into the members. The assemblage of the structure thus becomes a steel-to-steel connection problem. Where load transfer for gravity resistance is limited to simple bearing, connections may have no real stress functions, serving primarily to hold the members in position. Under lateral load, however, all connections will likely be required to transfer shear, tension, bending, and torsion. Thus for seismic resistance many of the typical connections used for gravity resistance alone will be inadequate.

Because of their weight, precast concrete spanning members may experience special problems due to vertical accelerations. When not sufficiently held down against upward movement, members may be bounced off their supports (a failure described as dancing.)

Precast concrete spanning members are often also prestressed, rather than simply utilizing ordinary steel reinforcing. This presents a possible concern for the effects of the combined loading of gravity and lateral forces or for upward movements due to vertical acceleration. Multiple loading conditions and stress reversals tend to greatly complicate the design of prestressing.

As with frames of wood or steel, those of precast concrete must be made stable with trussing, moment connections, or infill walls. If walls are used, they must be limited to masonry or concrete. Connections between the frame and any bracing walls must be carefully developed to assure the proper load transfers.

6.6 MISCELLANEOUS CONSTRUCTION

The preceding sections in this chapter have discussed the major types of construction for building structures. We now discuss some concerns for other aspects of the building and its construction.

Foundations

Where considerable below-grade construction occurs—with heavy basement walls, large bearing footings, basement or sublevel floor construction of reinforced concrete, and so on—the below-grade structure as a whole usually furnishes a solid base for the above-grade building. Not many extra details or elements are required to provide for seismic actions. Of principal concern is the tying together of the base of the building, which is where the seismic movements are transmitted to the building. If the base does not hold together as a monolithic unit, the result will be disasterous for the supported building. Buildings without basements or those supported on piles or piers may not ordinarily be sufficiently tied together for this purpose, requiring some additional construction. Foundation design problems are discussed more thoroughly in Chapter 20.

Freestanding Structures

Freestanding structures include exterior walls used as fences as well as large signs, water towers, and detached stair towers. The principal problem is usually the large overturning effect. Rocking and permanent soil deformations that result in vertical tilting must be considered. It is generally advisable to be quite conservative in the design for soil pressure due to the overturning effects. When weight is concentrated at the top—as in the case of signs or water towers—the dynamic rotational effect is further increased. These concerns apply also to the elements that may be placed on the roof of a building.

CHAPTER SEVEN

Special Considerations for Seismic Resistance

When the need to develop resistance to seismic forces is kept in mind throughout the entire process of the building design, it will have bearing on many areas of the design development. This chapter deals with various considerations that may influence the general planning as well as choices for materials, systems, and construction details.

7.1 BUILDING PLANNING

When seismic design is dealt with as an afterthought, rather than being borne in mind in the earliest decisions on form and planning of the building, it is quite likely that optimal conditions will not be developed. Various relations of building form to seismic response are discussed in Chapter 2, and the general seismic response of typical building construction materials and systems is treated in Chapter 6. Some of the major issues that should be kept in mind in the early planning stages follow.

1. The need for *some kind* of lateral bracing system. In some cases, because of the building form or size or the decision to use a particular structural material or system, the choice may be highly limited. In other situations there may be several options, with each having different required features (alignment of columns, incorporation of solid walls, etc.). The particular system to be used should be established early, although it may require considerable exploration and development of the options in order to make an informed decision.

2. Implications of architectural design decisions. When certain features are desired, it should be clearly understood that there are consequences in the form of problems with regard to seismic design. Some typical situations that commonly cause problems are the following:

General complexity and lack of symmetry in the building form.

Random arrangement of vertical elements (walls and columns), resulting in a haphazard framing system in general.

Lack of continuity in the horizontal

structure due to openings, multiplane roofs, split-level floors, or open spaces within the building.

Buildings consisting of aggregates of multiple, semidetached units; requiring considerations for linking or separation for seismic interaction.

Special forms (curved walls, sloping floors, etc.) that limit the performance of the structure.

Large spans, heights, or wall openings that limit placement of structural elements and result in high concentrations of load.

Use of nonstructural materials and construction details that result in high vulnerability to damage caused by seismic movements.

3. Allowance for seismic design work. Consideration should be given to the time, cost, and scheduling for the seismic investigation and design development. This is most critical when the building is complex or when an extensive dynamic analysis is required. Sufficient time should be allowed for a preliminary investigation of possible alternatives for the lateral bracing system, since a shift to another system at late stages of the architectural design work will undoubtedly cause problems.

4. Design styles not developed with seismic effects in mind. In many situations popular architectural design styles or features are initially developed in areas where seismic effects are not of concern. When these are imported to regions with high risk of seismic activity, a mismatch often occurs. Early European colonizers of Central and South America and the West Coast of North America learned this the hard way. The learning goes on.

Various special problems related to architectural design choices are discussed in Chapter 8. Some general relations between building form and seismic response are discussed in Chapter 9. These matters are also discussed in the presentation of many of the design examples in Part 4.

An excellent presentation of the general relation of building form to seismic response is contained in *Building Configuration and Seismic Design* (Ref. 29).

7.2 LIMITS OF BASIC ELEMENTS AND SYSTEMS

Although the range of possibilities for development of lateral resistive structures is considerable, it is necessary to be aware of the limitations that exist. Limitations may be real, established by stress level capacities of materials or the available sizes of ordinary construction elements. Limits may also be somewhat artificial or ambiguous, established by code requirements or rules of thumb in design practice. The latter are not developed arbitrarily but rather are evolved from the collective experience and judgments of generations of professional researchers and designers. However, in time, changes in construction methods, design practices, and use of materials and products make old rules lose touch with reality.

The limits of basic types of construction are discussed in Chapter 6. The limits of typical elements of lateral resistive structural systems are discussed in Part 3. The following are some basic types of limits that should be recognized, although the specific quantification of data and the relative status of particular materials or systems tend to exist in a constant state of flux.

1. *Limits of Materials.* Common materials have specific stress limits; variation is typical over some range of quality or type (stress grade of wood, f'_c of concrete, etc.).

2. *Available Sizes.* Concrete is unique in its ability to form elements of virtually any size; most other materials and products have practical limits in terms of commonly available shapes and sizes.
3. *Functional Limits.* For various reasons particular materials, products, construction techniques, or details of construction are often limited to specific uses, or excluded from specific uses. Unfinished bolts, which are mostly used only for temporary or minor connections, are excluded from use in braced frames. Wood-framed shear walls should not be used to brace a concrete frame. For the bolts the limits are due to their inability to maintain a tight (nonslipping) joint. For the wood-framed wall it is simply not reasonable to brace the heavy, rigid, system with a light, deformable one.
4. *Size Limits.* Upper and lower limits of size exist—sometimes for actual reasons of the availability of products, sometimes merely for practicality, as in the case of the limits of minimum size for concrete members (wall thickness, beam width, column diameter, etc.).
5. *Aspect Ratios.* Examples of aspect ratios are the practical, or in some cases the code-specified, limits for the h/t of columns or L/d of beams. A major type of limitation is the length-to-width ratio of wood diaphragms (expressed as span-width for horizontal diaphragms and as height-width for vertical diaphragms by the *UBC* in Table 25–I of the code.)
6. *Structural Behavior.* Performance of the structure is mostly limited by the assigning of values for allowable stresses (in the working stress method) or load factors (for the ultimate strength method). Except for vertical deflection of beams and lateral drift (horizontal deflection) of individual stories, codes do not provide much criteria for limitation of deformations.

7.3 LOAD PROPAGATION AND STRUCTURAL CONTINUITY

Most buildings contain elements of the construction that can be called upon to resist some amount of lateral force action. Solid, continuous roof and floor decks, solid exterior and interior walls, and frames of wood, steel, or concrete occur in most buildings. Individual elements may need to be increased in size or strength grade in order to function for resistance to lateral effects, but it is not a matter of introducing wholly missing elements. What is often missing in a building that has not been consciously designed with seismic resistance in mind is the necessary connection of elements for the transfer of the loads through the system.

Connections for resistance to seismic loads must have the usual resistance to direct forces, just as in the case of gravity loads. However, for seismic actions resistance to dynamic jolting and back-and-forth shaking actions is equally critical, and is often where ordinary connections fall short.

Of particular concern are connections between major elements of the bracing system. For gravity resistance roofs and floors need often merely sit on their supports, with only minor attachments to hold the parts together. These attachments must often be enhanced—or even replaced by other means—in order to develop the load transfers required for seismic design.

Ordinary elements of wood and steel construction that must function as chords, collectors, or drag struts must be developed with splices, ties, or other devices to assure their continuity. Choice of sizes and shapes and location of joints for these members may be affected by concerns for the seismic load transfers.

Anchorage of the building to its foundations—and in general to the ground—must be studied in view of the multidirectional movements during an earthquake. This is a much more complicated problem than that of gravity load resistance, which is essentially limited to resolution of downward vertical forces.

CHAPTER EIGHT

Special Problems

The material in this chapter deals with a number of matters of special concern in developing resistance to the lateral force effects of seismic actions.

8.1 VULNERABLE ELEMENTS

There are many commonly used elements of buildings that are especially vulnerable to damage due to earthquakes. Most of these are nonstructural, that is, not parts of the structural system for resistance of gravity or lateral loads. Because of their nonstructural character, they do not routinely receive thorough design study by the structural designer; thus, in earthquake country, they constitute major areas of vulnerability. Some typical situations are the following:

1. *Suspended Ceilings.* These are subject to horizontal movement. If not restrained at their edges, or hung with elements that resist horizontal movement, they will swing and bump other parts of the construction. Another common failure consists of the dropping of the ceiling due to downward acceleration if the supports are not resistive to a jolting action.
2. *Cantilevered Elements.* Balconies, canopies, parapets, and cornices should be designed for significant seismic force in a direction perpendicular to the cantilever. In most cases, codes provide criteria for consideration of these forces.
3. *Miscellaneous Suspended Objects.* Lighting fixtures, signs, HVAC equipment, loud speakers, catwalks, and other items that are supported by hanging should be studied for the effects of pendulumlike movements. Supports should tolerate the movement or should be designed to restrain it.
4. *Piping.* Building movements during seismic activity can cause the rupture of piping that is installed in a conventional manner. In addition to the usual allowances for thermal expansion, provisions should be made for the flexing of the piping or for isolation from the structure that is sufficient to prevent any damage. This is obviously most critical for piping that is pressurized.
5. *Stiff, Weak Elements.* Any parts of the building construction that are stiff but not strong are usually vulnerable to

damage. This includes window glazing, plastered surfaces, wall and floor tile (especially of ceramic or cast materials), and any masonry (especially veneers of brick, tile, precast concrete, or stone). Reduction of damage and of hazard to occupants requires careful study of installation details for attachment. In most cases extensive use of control joints is advised to permit movements without fractures.

8.2 THE SOFT STORY

Any discontinuity that constitutes an abrupt change in the structure is usually a source of some exceptional distress. This is true for static load conditions as well, but is especially critical for dynamic loading conditions. Any abrupt increase or decrease in stiffness will result in some magnification of deformation and stress in a structure subjected to energy loading. Openings, notches, necking-down points, and other form variations produce these abrupt changes in either the horizontal or vertical structure. An especially critical situation is the so-called soft-story, as shown in Fig. 8.1.

The soft story could—and indeed sometimes does—occur at an upper level. However, it is more common at the ground-floor level, between a rigid foundation system and some relatively much stiffer upper-level system. The tall, open ground floor has both historical precedent and current popularity as an architectural feature. This is not always strictly a matter of design style, as it is often required for functional reasons.

Several failures of such structures in recent years have focused attention on their vulnerability and on the inadequacy of the equivalent static method for their design. The *UBC* presently disclaims the use of static methods for buildings with this sort of discontinuity (see Section 2312(e)3 of the 1982 edition of the *UBC*, reprinted in Chapter 33). The soft story presents a strong case for a true dynamic analysis, or at the least a very conservative design with the static load method.

If the tall, relatively open ground floor is necessary, Fig. 8.2 presents some possibilities for having this feature with a reduction of the soft story effect. The methods shown consist of the following:

1. Bracing some of the open bays (Fig. 8.2*a*). If designed adequately for the forces, the braced frame (truss) should have a class of stiffness closer to a rigid shear wall, which is the usual upper structure in these situations. However, the soft story effect can also

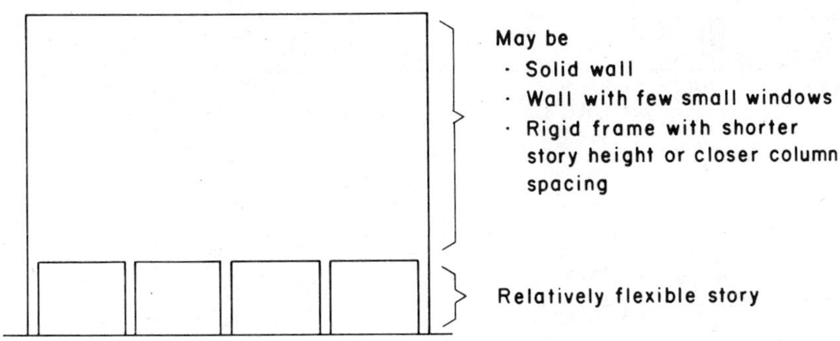

FIGURE 8.1. The soft story.

occur in rigid frames, where the "soft" story is simply significantly less stiff.

2. Keeping the building plan periphery open, while providing a rigidly braced interior (Fig. 8.2*b*).
3. Increasing the number and/or stiffness of the ground-floor columns for an all-rigid frame structure (Fig. 8.2*c*).
4. Using tapered or arched forms for the ground-floor columns to increase their stiffness (Fig. 8.2*d*).
5. Developing a rigid first story as an upward extension of a heavy foundation structure (Fig. 8.2*e*).

Figure 8.3 shows the exterior of a building with a first-story development that appears to be similar to that shown in Fig. 8.2*e*. However, the tapered first-story base elements could as well be primarily decorative (or used for recreation), and the true situation for the lateral bracing system be that shown in Fig. 8.2*b*.

The soft story is actually a method for providing critical damping or major energy absorption, which could be a *positive* factor in some situations. However, the major stress concentrations and deformations must be carefully provided for, and a true dynamic analysis is certainly indicated.

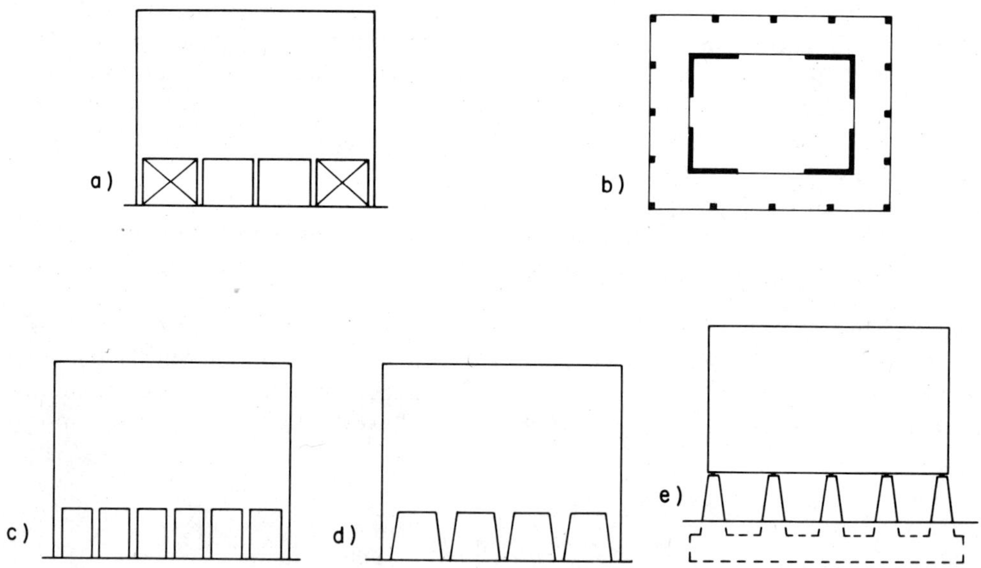

FIGURE 8.2. Some remedies for the soft story.

FIGURE 8.3. Transitional support elements used to raise the building free of the ground and create an open ground floor. CN Park Place office building, Los Angeles. Architects: Langdon and Wilson, Los Angeles.

CHAPTER NINE

Design for Optimal Seismic Resistance

The building design process involves a long series of compromises between disparate and sometimes conflicting requirements. Designing exclusively for seismic performance is unrealistic, just as is designing exclusively for construction cost, fire safety, energy use, or ease of maintenance. However, it is useful to be aware of the idealistic optimal situation for any requirement so that compromising can be done with some point of reference. This chapter presents some ideas that may help this situation in design.

9.1 DESIRABLE CHARACTERISTICS

The following are some attributes of a building, or of its lateral resistive structure, that will usually result in improvement of response to seismic activity.

1. *Simplicity, Regularity, and Symmetry of Building Form.* Investigation of response will be simpler and more reliable; structural layout, design, and detailing will be easier, with less opportunity for error; discontinuities, torsional effects, and complex load transfers may be avoided.
2. *Simplicity of Construction.* Control of both design and construction will be easier and more assured; detailing can be more carefully developed with the same available design time and effort; workmanship will probably be better if confusion is less.
3. *Light Weight.* Whatever the other factors may be, the impelling seismic forces are directly proportional to the building mass. Reduce the mass (weight) and the load on the structure will be less.
4. *Tough Materials.* For the structure this means a lack of sudden failure and a greater margin of safety. For the building construction in general it means less damage and lower risk of injury to occupants.
5. *Positive Connections.* This means connections that are not subject to loosening, slipping, brittle fracture, or progressive loss of strength, or are otherwise vulnerable to the jolting,

back-and-forth, multidirection, dynamic actions during an earthquake. This is of prime concern for the connections of all major elements of the lateral resistive structure, but it is also desirable for all connections in the building construction.

6. *Torsional Strength of the Building.* The elements of the lateral resistive structure (most notably the vertical elements) should be arranged so as to produce high resistance to torsional effects on the building. This is highly significant for the multidirection movements.

Individually considered, each of these is important. However, it is seldom possible to achieve them all, as improvements in one area are often at the expense of another. For example, the strongest shear walls are those of reinforced concrete; however, they are also the heaviest and have some aspect of brittle fracture in their failure.

9.2 SITUATIONS TO AVOID

A list of bad situations would begin with the absence or reverse of the characteristics enumerated in the preceding section. Because of competing criteria for design, it is common for some of these less desirable situations to occur, but the consequences in terms of seismic response should be understood. Some additional situations that can contribute to poor seismic response are the following:

1. Use of rigid nonstructural elements in combination with a flexible bracing system. This is not impossible, but care must be exercised in developing control joints or motion-isolating joints. Otherwise, the stiffer elements will absorb the load until they fail and others take over. The *load* is not aware of what is structural and what is intended to be nonstructural.
2. Concentration of mass at upper levels. You would not use a sledge hammer by grasping the head and swinging the handle; you swing the heavy head at the outer end of the handle to maximize the dynamic impact. The building will behave the same way if you put the mass mostly at the end of the handle (top of the building). Using the heaviest roof construction and roofing, choosing heavy materials for decorative cornices, and placing the heaviest items of the HVAC system on the roof will assure this response. Amazingly, it is common to use heavy clay or concrete roof tiles and rooftop HVAC systems on the West Coast of the United States. Clearly, there are other priorities.

PART TWO

ANALYSIS FOR SEISMIC EFFECTS

CHAPTER TEN

Lateral Load Resistance of Buildings

One of the most difficult aspects of learning to design for seismic effects on buildings is the development of the ability to visualize the nature of the forces and of the form of the building's responses to them. One reason for this is that it is necessary to visualize these actions in three dimensions. The material in this chapter is presented to aid in the visualization of the force actions and the various ways in which buildings of different form and construction respond to them.

10.1 GENERAL EFFECTS OF LATERAL LOADS

The typical building exists as a vertical projection from the horizontal ground surface, to which it is reasonably securely attached. As such, its general response to horizontal forces is analogous to that of a vertically cantilevered beam (see Fig. 10.1). The direct responses of such a structure may be visualized in the form of the typical shear, moment, and deflected shape diagrams, as shown in the illustration.

The shear and moment effects require the development of internal resistance, similar to that required for the beam at its various cross sections. Both of these effects, when considered as equivalent static forces, become most critical in magnitude at the base of the building. However, failure may occur at any point in the structure if a lack of sufficient resistance exists at that location.

Horizontal deflection (usually called *drift* for the building as a whole) must be considered for a number of reasons. Excessive drift will likely be objectionable to the building occupants and possibly cause damage to the building contents and to nonstructural portions of the construction. On the other hand movement is a means for dissipating some of the energy of the seismic force action and is thus a positive consideration for the structure in many cases.

The direct action of the horizontal forces may result in a building being slid off its foundation, as shown in Fig. 10.2*a*. If the building is adequately anchored to the foundation, the next concern is for adequate resistance of the foundation it-

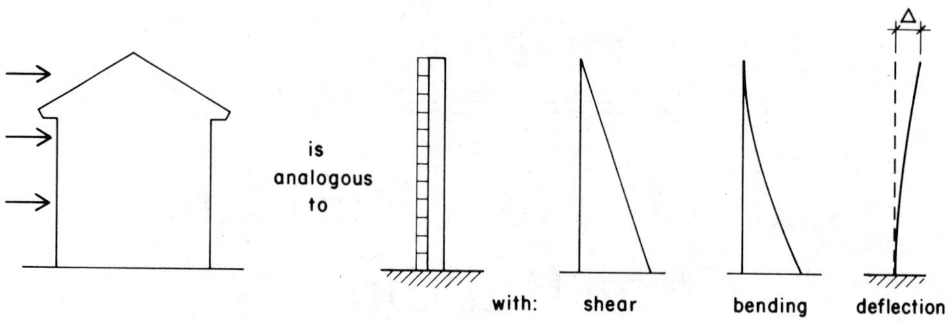

FIGURE 10.1. Cantilever functions of the lateral resistive structure.

self, in the form of some combination of horizontal sliding friction and lateral earth pressure (see Fig. 10.2*b*). Sliding failure can also occur within the building structure, a classic case being that of the dislocation of a lightly attached roof, as shown in Fig. 10.2*c*.

One form of the bending effect is that of the so-called overturning moment. This may result in the building being tipped over—with or without its foundation—as shown in Fig. 10.3*a*. For this effect the building mass has a dual role, being at once the source of the horizontal momentum force and a stabilizing resistance to the overturn tipping failure. The critical nature of the overturning effect has much to do with the form of the building's vertical profile. Buildings that are relatively squat in form (Fig. 10.3*b*) are quite unlikely to fail in this manner, while those with tall, slender forms (Fig. 10.3*c*) are highly vulnerable.

It must be borne in mind that the resistance to the horizontal forces is in many cases not actually developed by the building as a whole, but rather by some discrete elements of the building structure. Thus the concerns illustrated in Figs. 10.2 and 10.3 may have to do with the response of these discrete elements. Fig. 10.3*d* shows a building in which lateral resistance is developed by a shear wall that is considerably narrower than the building total width. The sliding and overturn effects

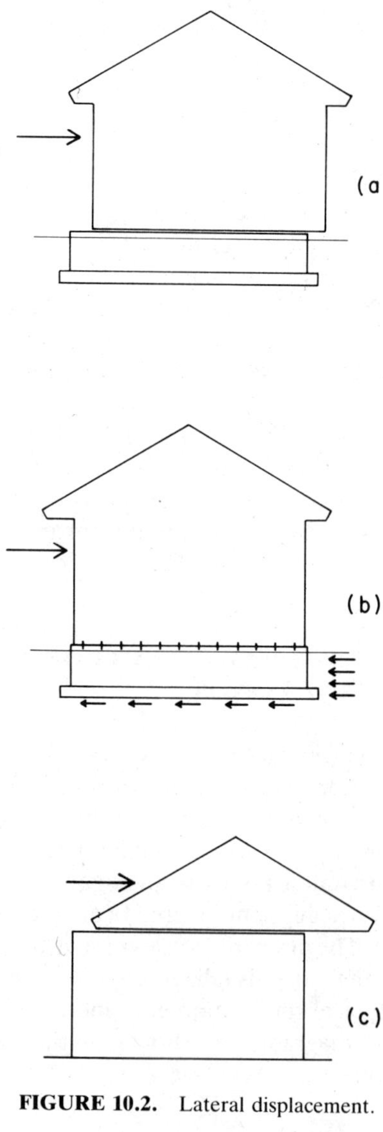

FIGURE 10.2. Lateral displacement.

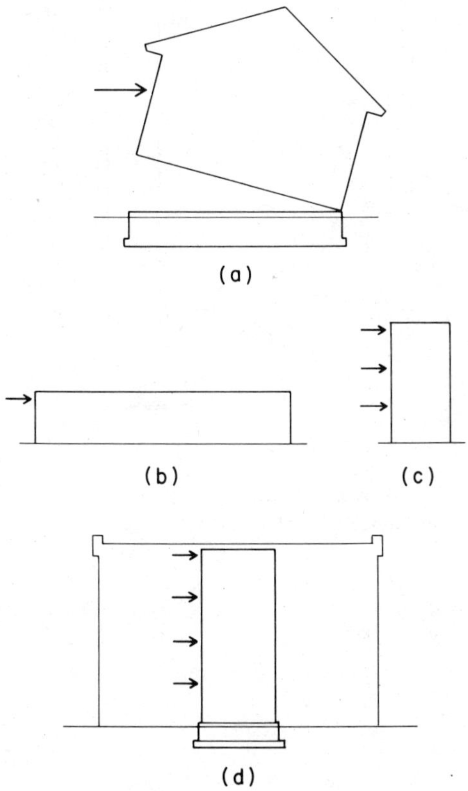

FIGURE 10.3. Overturning.

must be considered for this shear wall. In this case, while the building itself is relatively squat in profile, the shear wall is actually quite slender, and overturning is likely to be a critical concern.

It is also necessary to keep in mind that the seismic action does not produce a single loading of the structure, as in the case of the gravity-loaded beam. The shear, bending, and deflection effects are actually rapidly reversed and repeated. This reversal and repetition can have a number of effects, resulting in the loosening of connections, progressive cracking of brittle materials, or simply the shaking apart of the building in general. One type of effect of the repeated force action is that shown in Fig. 10.4. In this case the rocking back and forth of the tall structure produces permanent deformations of the soil, resulting in a progressive loss of resistance to the structure's motions.

As with all types of loads, it is necessary to consider the ways in which forces due to seismic action may combine with other load effects. The most general combination that must be studied is that of the seismic action plus effects of the gravity loads. Figure 10.5*a* illustrates the common situation of internal stress in a tall structure due to the vertical compression effect of the gravity loads and the bending induced by the lateral forces caused by seismic activity. If the structure is quite narrow in profile, it is possible that the net stress condition may be as shown in the illustration, requiring some tensile resistance at lower points in the structure. This tensile-resistive character may be required of the structure itself, making it necessary to design columns or walls for upward tensile resistance, or it may apply essentially to anchorage connections and splices. If this stress condition is potential at the bottom of the foundation, it is most critical since a bearing-type foundation cannot offer such resistance.

In the analysis for overturning effects, the basic investigation is for the net combination of the lateral load overturn moment versus the restoring moment due to the gravity loads (see Fig. 10.5*b*). If the overturning moment prevails, tipping will occur. For safety, it is desired that the ratio of the restoring moment to the overturning moment be something greater than 1—usually a minimum of 1.5. If this limit is not obtained, some kind of anchorage must be added to provide the necessary level of safety.

A point of difficulty arises in these investigations in that some of the gravity loads are permanent while others are transitory in nature. When performing the overturn investigation, it is essential that only the permanent loads (due to the dead weight of the permanent construction) be

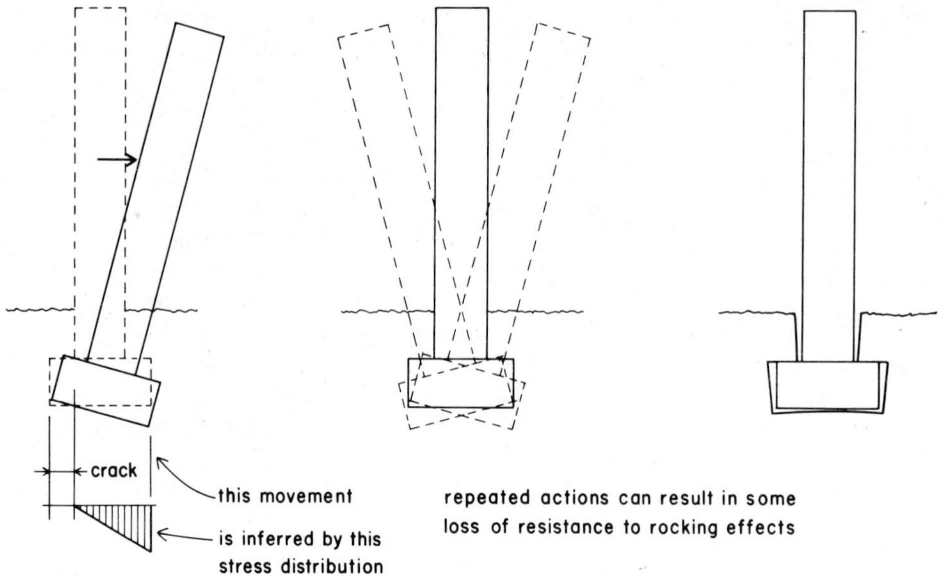

FIGURE 10.4

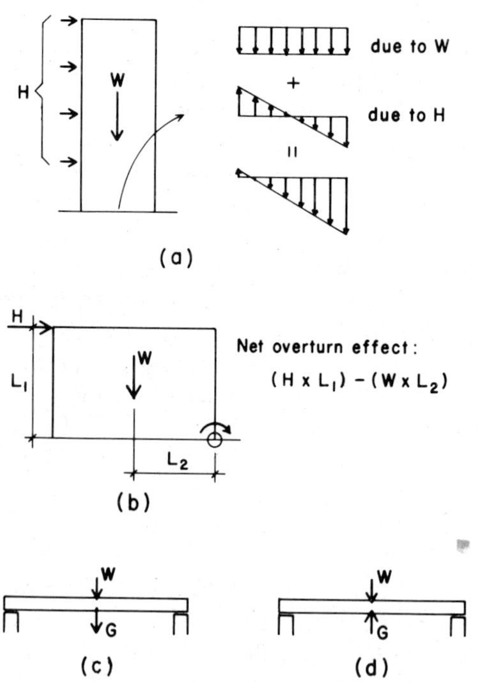

FIGURE 10.5

considered for the restoring moment. However, the lateral force generated by the building gravity mass will be greater if *all* existing gravity loads are included, such as those caused by the weight of movable partitions, optional plastered ceilings, heavy floor coverings, building equipment, and heavy furniture. In the analysis illustrated in Fig. 10.5*a*, the maximum tensile effect may occur with the least vertical compression, while the maximum combined compression stress will occur with the greatest vertical compression force.

The actions illustrated in Fig. 10.5*a* and *b* deal with combinations of the vertical gravity effects and the horizontal seismic effects. Since seismic activity also induces up-and-down motions, it is necessary to consider the possible combination of gravity and seismic forces when both operate in a vertical direction. When the building mass moves downward and then suddenly reverses direction, the effect is as shown in Fig. 10.5*c*, in which the downward gravity effect *W* is added to the inertial force *G*. Thus the normal static dead load forces and stresses are increased by some magnitude. For this action it is essential to consider all of the gravity load. When the building moves in the opposite direction, the two force actions are opposite in direction, in which case the critical concern is for the possibility that the *G* force may be greater than the *W* force, producing uplift or internal stress reversals. For the latter situation the worst case will occur when only the permanent gravity load is considered to be present.

Again, it is necessary to remember that the effects illustrated in Fig. 10.5 do not occur a single time but rather in rapidly reversing succession, as the building is shaken back and forth and bounced up and down. Failure may occur with a single, major jolt or as a result of the progressive weakening due to the repeated actions.

Most of our techniques of building construction—built up over centuries of experience—relate primarily to the direct resistance of gravity forces. This is a natural result of our own personal experience with the gravity forces acting on our bodies and our direct handling of the elements with which we build. Unless conscious thought is given to the potential actions of lateral forces, we tend to produce constructions that are vulnerable to the effects of horizontal forces. Figure 10.6*a* symbolizes such a structure, well oriented to gravity load resistance but vulnerable to the load condition shown in Fig. 10.6*b*.

Developing external stability for the gravity-resistant structure may consist of modifying the existing (gravity-resisting) supports or of adding some separate bracing system. Fig. 10.6*c* shows the simple solution used for fence posts and utility poles—burying them a sufficient distance in the ground to develop the required overturning moment. A more common solution for structures other than simple posts is shown in Fig. 10.6*d*, consisting of spreading the base to increase the moment arm for the gravity-developed stabilizing moment (see Fig. 10.5*b*). External bracing may be directly applied to individual structures, as symbolized in Fig. 10.6*e*, or may consist of linking parts of a complex structure to other, fully braced, parts.

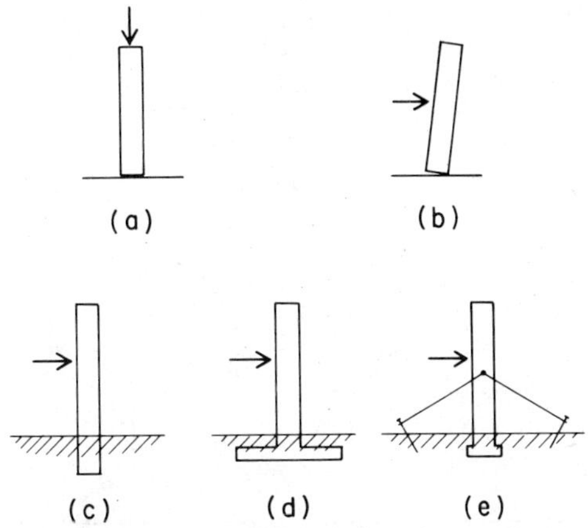

FIGURE 10.6. Resistance to overturn.

Building frameworks, assembled in the traditional post and beam system, will generally lack lateral force resistance when consisting of discrete elements, as shown in Fig. 10.7*a*. For a single, planar bay of such a system, stability is usually achieved by one of four means:

1. By making moment-resistive connections between the elements of the frame (Fig. 10.7*b*).
2. By trussing or X-bracing to achieve triangulation (Fig. 10.7*c*).
3. By independently stabilizing the columns (Fig. 10.7*d*).
4. By using rigid infilling wall surfaces (Fig. 10.7*e*).

Election of one or the other of these bracing techniques depends on the scale and complexity of the framework, on the

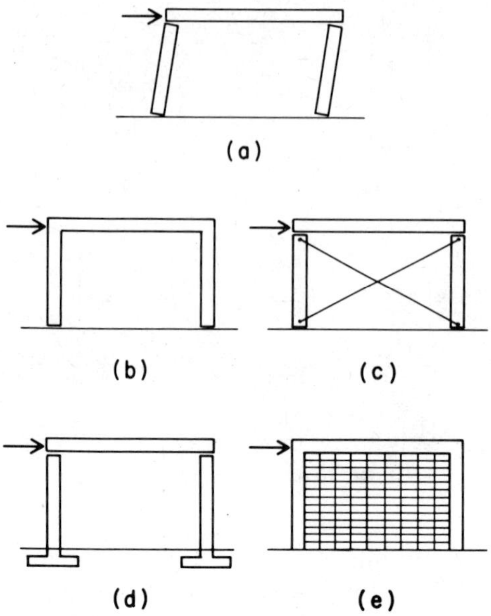

FIGURE 10.7. Bracing the post and beam structure.

materials and form of the frame elements, and on the various problems of architectural planning and construction detailing.

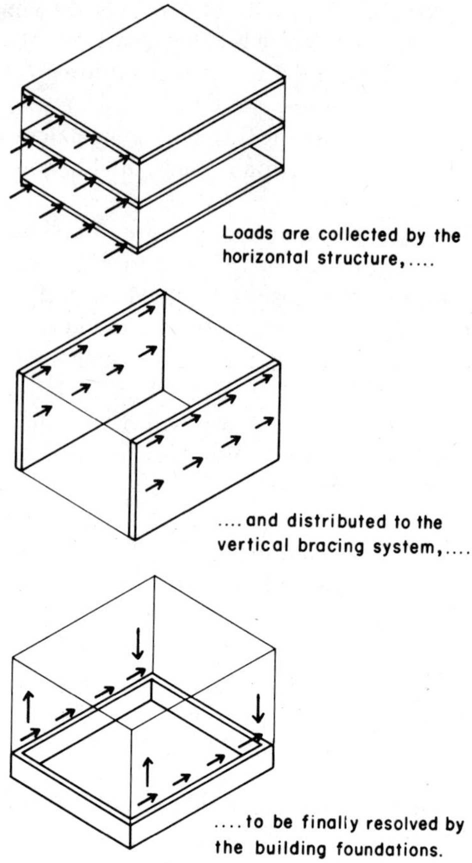

FIGURE 10.8. Load propagation in the box system.

10.2 PROPAGATION OF LATERAL LOADS

Seismic loads are generated by the dead weight of the building construction. In visualizing the application of seismic forces, we look at each part of the building and consider its weight as a horizontal force. The flow (or propagation) of these forces through the structure is symbolized in Fig. 10.8.

In most buildings the initial element of the lateral resistive system is the horizontal structure consisting of the spanning elements of the roof and framed floor systems. These elements are loaded by their own weight as well as the weight of objects resting on or hung from them. Most of the vertical elements of the construction (primarily walls) will deliver the loads generated by their weight to one or more of the horizontal structures.

The horizontal structures (above ground level) must be braced by some vertical bracing elements. In Fig. 10.8 these are symbolized by the exterior walls that are parallel to the lateral load direction. In resisting the lateral loads, the horizontal structures span between these bracing elements and transfer the loads to them.

The vertical bracing elements in turn transfer the loads to the building foundations. This is the end of the line for the internal force transfers. The foundations must then resolve the forces externally to the supporting ground.

Actions of individual elements of such a building system are shown in Fig. 10.9. Vertical walls that are perpendicular to the direction of the lateral loads tend to span vertically, delivering part of their load to the structure below and part to that above. The horizontal structure will span between vertical bracing elements, delivering its loads as concentrated forces to the supporting structure. The vertical bracing elements will be loaded by a combination of the gravity loads and the horizontal forces delivered by the horizontal structure.

Again, it must be recalled that these actions occur in rapidly reversing directions (back and forth) and in all possible separate directions (north–south, east–west, etc.). In complex buildings of large size different parts of the building are likely to be moving in opposite directions

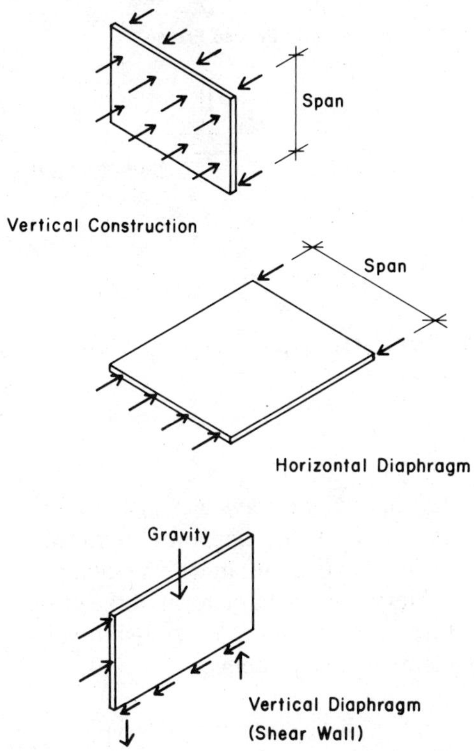

FIGURE 10.9. Functions of the horizontal and vertical construction.

at the same instant. Because of their relative stiffness in different directions, solid walls may function differently, depending on the direction of the lateral load with respect to the plane of their surface. Perpendicular to their planes, walls are usually relatively flexible. In their own planes, however, they tend to be quite stiff. If this is the case, the actions of the vertical walls in a one-story building may be assumed to be as shown in Fig. 10.10.

10.3 TYPES OF LATERAL RESISTIVE SYSTEMS

Although the form, materials, and details of building construction are subject to endless variation, there are only a few basic types of lateral resistive building structures. As illustrated in Fig. 10.11, the most common systems consist of the following:

The Box, or Panelized, System. The box, or panelized, system is usually of the type shown in the previous example, consisting of some combination of horizontal and vertical planar elements. Actually, most buildings use horizontal diaphragms, simply because the existence of roof and floor construction provides them as a matter of course. The other types of systems usually consist of variations of the vertical bracing elements. An occasional exception is a roof structure that must be braced by trussing or other means when there are a large number of roof openings or a roof deck with little or no diaphragm strength.

Internally Braced Frames. The typical assemblage of post and beam elements is not inherently stable under lateral loading unless the frame is braced in some manner. Shear wall panels may be used to achieve this bracing, in which case the system functions as a box, even though there is a frame structure. It

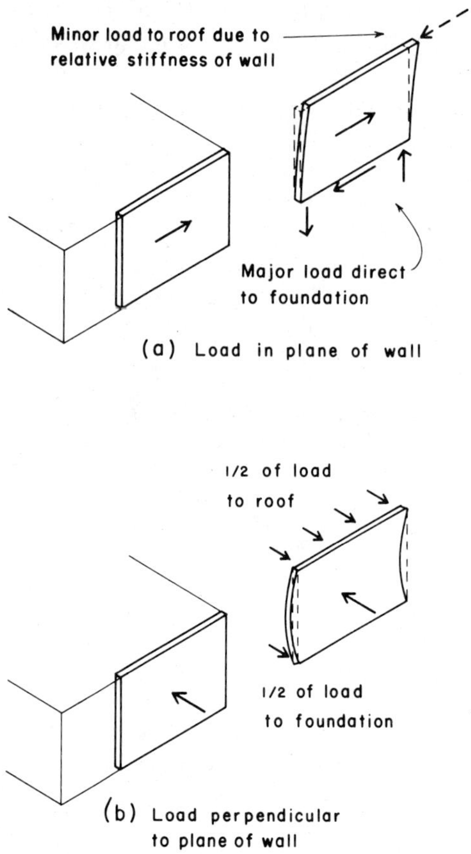

FIGURE 10.10. Lateral loads caused by wall weight.

is also possible, however, to use diagonal members, X-bracing, knee braces, struts, and so on to achieve the necessary stability of the rectangular frame. The term *braced frame* usually refers to these techniques.

Rigid Frames. Although the term *rigid frame* is a misnomer, since this technique usually produces the most *flexible* lateral resistive system, the term refers to the use of moment-resistive joints between the elements of the frame.

Externally Braced Frames. The use of guys, struts, buttresses, and so on that are applied externally to the structure or the building results in externally braced frames.

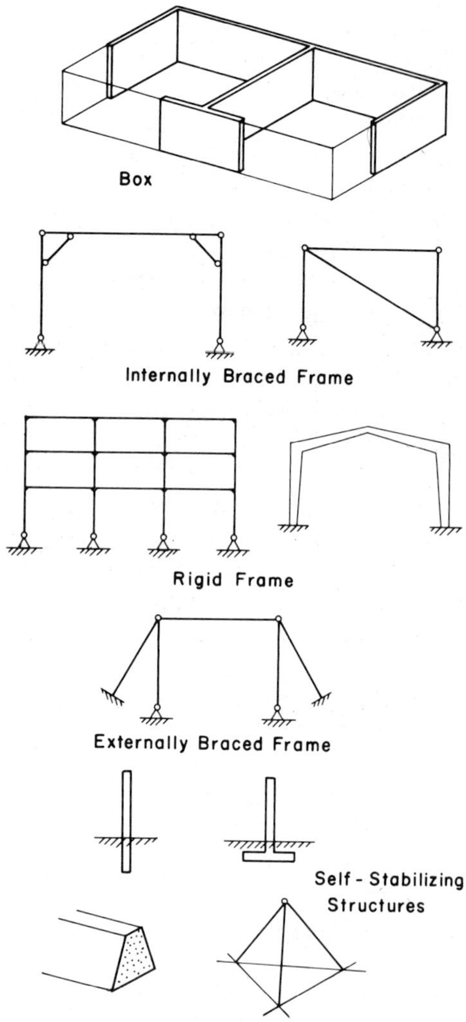

FIGURE 10.11. Types of lateral load resistive systems.

Self-Stabilizing Elements and Systems. Retaining walls, flagpoles, pyramids, tripods, and so on, in which stability is achieved by the basic form of the structure are examples of self-stabilizing elements and systems.

Each of these systems has variations in terms of materials, form of the parts, details of construction, and so on. These variations may result in different behavior characteristics, although each of the basic types has some particular properties. An important property is the relative stiffness or resistance to deformation, which is of particular concern in evaluating energy effects, especially for response to seismic loads. A box system with diaphragms of poured-in-place concrete is usually very rigid, having little deformation and a short fundamental period. A multistory rigid frame of steel, on the other hand, is usually quite flexible and will experience considerable deformation and have a relatively long fundamental period. In seismic analysis these properties are used to modify the percentage of the dead weight that is used as the equivalent static load to simulate the seismic effect.

Elements of the building construction developed for the gravity load design, or for the general architectural design, may become natural elements of the lateral resistive system. Walls of the proper size and in appropriate locations may be theoretically functional as shear walls. Whether they can actually serve as such will depend on their construction details, on the materials used, on their length-to-height ratio, and on the manner in which they are attached to the other elements of the system for load transfer. It is also possible, of course, that the building construction developed only for gravity load resistance and architectural planning considerations may *not* have the necessary attributes for lateral load resistance, thus requiring some replanning or the addition of structural elements.

In many cases it is neither necessary nor desirable to use every wall as a shear wall or to brace every bay of the building frame. The illustrations in Fig. 10.12 show various situations in which the lateral bracing of the building is achieved by partial bracing of the system. This procedure does require that there be some load-distributing elements, such as the roof and floor diaphragms, horizontal struts, and so on, that serve to tie the unstabilized portions of the building to the lateral resistive elements.

Many buildings consist of mixtures of the basic types of lateral resistive systems (see Fig. 10.13). Walls existing with a frame structure, although possibly not used for gravity loads, can still be used to brace the frame for lateral loads. Shear walls may be used to brace a building in one direction whereas a braced frame or rigid frame is used in the perpendicular direction. Multistory buildings occasionally have one type of system, such as a rigid frame, for the upper stories and a different system, such as a box system or braced frame, for the lower stories to reduce deformation and take the greater loads in the lower portion of the structure.

There is a possibility that some of the elements of the building construction that are not intended to function as bracing elements may actually end up taking some of the lateral load. In frame construction surfacing materials of plaster, drywall, wood paneling, masonry veneer, and so on may take some lateral load even though the frame is braced by other means. This is essentially a matter of relative stiffness, although connection for load transfer is also a consideration. What can happen in these cases is that the stiffer finish materials take the load first, and if they are not strong enough, they fail and the intended bracing system then goes to work. Although collapse may not occur, there can be considerable damage to the building construction as a result of the failure of the supposed nonstructural elements.

The choice of the type of lateral resistive system must be related to the loading conditions and to the behavior characteristics required. It must also, however, be coordinated with the design for gravity loads and with the architectural planning considerations. Many design situations allow for alternatives, although the choice may be limited by the size of the building, by code restrictions, by the magnitude of lateral loads, by the desire for limited deformtaion, and so on.

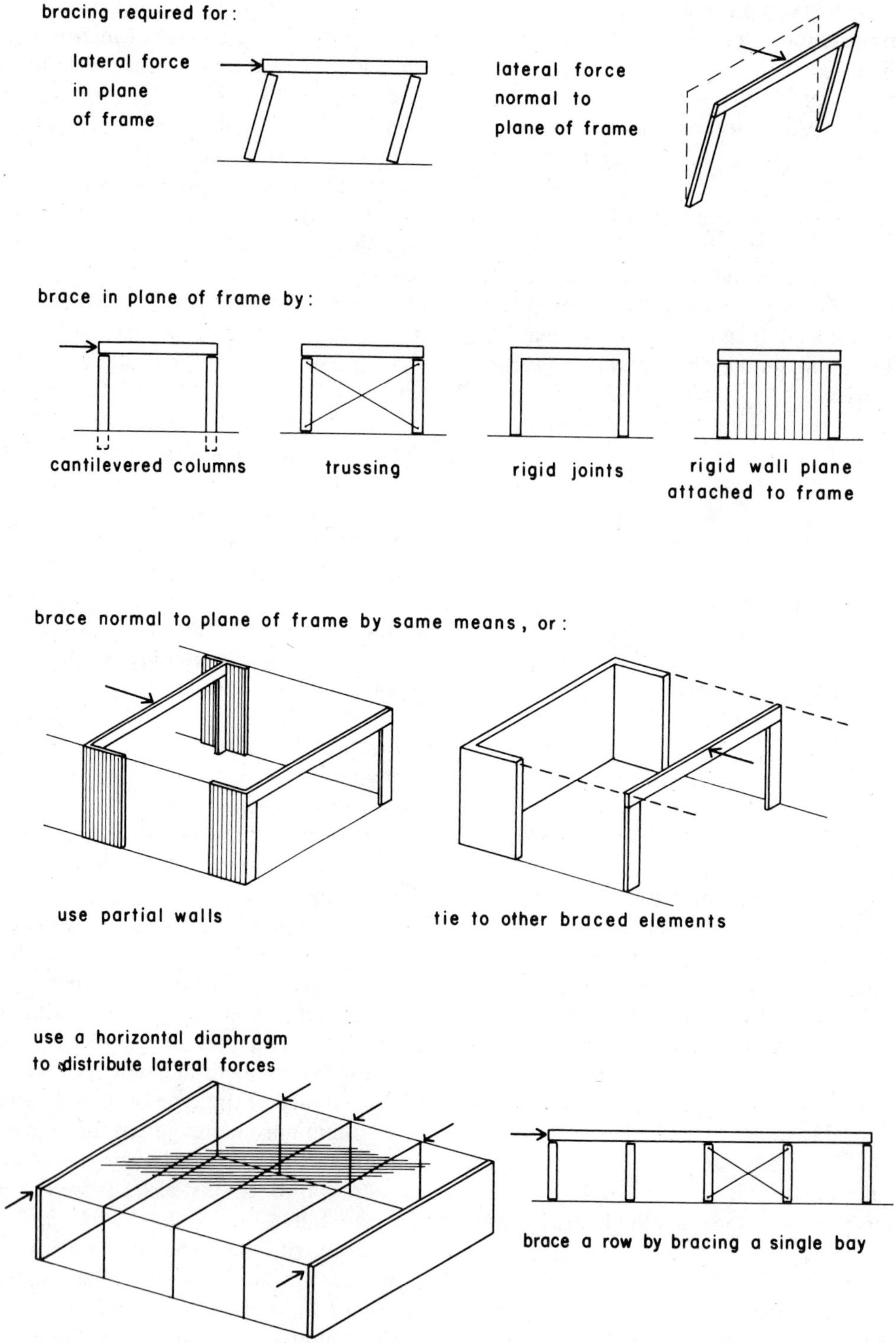

FIGURE 10.12. Bracing of framed structures.

FIGURE 10.13. Large, multimassed building employing several systems of bracing: rigid frame, braced frame, and shear wall.

CHAPTER ELEVEN

Earthquake Effects on Building Structures

Previous chapters have dealt with various issues that are involved in the response of buildings to seismic effects. This chapter focuses on the particular problems of investigation and design of the lateral resistive structure.

11.1 GENERAL EFFECTS

There are a number of potential problems that must be considered in evaluating a structure with respect to its response to seismic movements. The following are some of the main considerations:

1. *Maximum Direct Force.* At some point during an earthquake the structure will receive the quake's hardest punch. Theoretically, this will occur at the point where the horizontal acceleration of the ground reaches its highest peak magnitude. However, since the building is also moving, it may occur at some other time. It is this maximum jolt that is translated into a major force on the structure. This force will produce stresses and deformations as well as destabilizing effects (overturn, sliding, etc.). Since the time duration of this force is only a fraction of a second, it is common to allow slightly higher stress levels or to use slightly lower safety factors when designing.

2. *Harmonic Effects.* Since the building structure will respond to rapid deformation or movement in the manner of an elastic spring, its properties that relate to harmonic motion must be considered. This is complicated by the fact that the earthquake is also harmonic in nature and that the site in the vicinity of the building may also have some harmonic response. These issues are discussed more fully in later sections.

3. *Progressive Failures.* Even though it is able to sustain the single maximum jolt, the structure may fail due to the repeated shaking action of the quake. Connections may be steadily loosened. Minor cracks that may not be critical in response to static loads may be built into major ones by repeated

stress and reversal of the load directions.

4. *Three-Dimensional Movement.* The fact that the building moves in all directions may result in unique problems for different structures. Members or connections may sustain forces of a direction or sense that they do not normally need to resist. Tension members may have to sustain compression; bending-resistive elements may have to sustain moments of opposite rotational sense. Types of force actions not anticipated for elements may occur, such as torsion in beams, shear in columns, bending in walls, and so on.
5. *Effects of Deformations.* These are generally more critical with regard to nonstructural parts of the construction. However, there can be problems with regard to deformation of connections, distribution of loads due to relative stiffness, *P*-delta effects in columns, and so on.
6. *Energy Absorption.* For the structure, its most significant strength in terms of seismic response is its energy-absorbing capacity. This is the most meaningful measure of its dynamic strength, versus its static strength which can be measured in simpler direct force units (pounds, tons, etc.). Seismic effects are most often dealt with in terms of equivalent static effects, but the true dynamic character of the actions must not be forgotten.

11.2 CONSIDERATIONS OF DYNAMIC BEHAVIOR

A good lab course in physics should provide a reasonable understanding of the basic ideas and relationships involved in dynamic behavior. A better preparation is a course in engineering dynamics that focuses on the topics in an applied fashion, dealing directly with their applications in various engineering problems. The material in this section consists of a brief summary of basic concepts in dynamics that will be useful to those with a limited background and that will serve as a refresher for those who have studied the topics before.

The general field of dynamics may be divided into the areas of *kinetics* and *kinematics*. *Kinematics* deals exclusively with motion, that is, with time–displacement relationships and the geometry of movements. *Kinetics* adds the consideration of the forces that produce or resist motion.

Motion can be visualized in terms of a moving point, or in terms of the motion of a related set of points that constitute a body. The motion can be qualified geometrically and quantified dimensionally. In Fig. 11.1 the point is seen to move along a path (its geometric character) a particular distance. The distance traveled by the point between any two separate locations on its path is called *displacement*. The idea of motion is that this displacement occurs over time, and the general mathematical expression for the time–displacement function is

$$s = f(t)$$

Velocity is defined as the rate of change of the displacement with respect to time. As an instantaneous value, the velocity is expressed as the ratio of an increment of displacement (*ds*) divided by the increment of time (*dt*) elapsed during the displacement. Using the calculus, the velocity is thus defined as

$$v = \frac{ds}{dt}$$

That is, the velocity is the first derivative of the displacement.

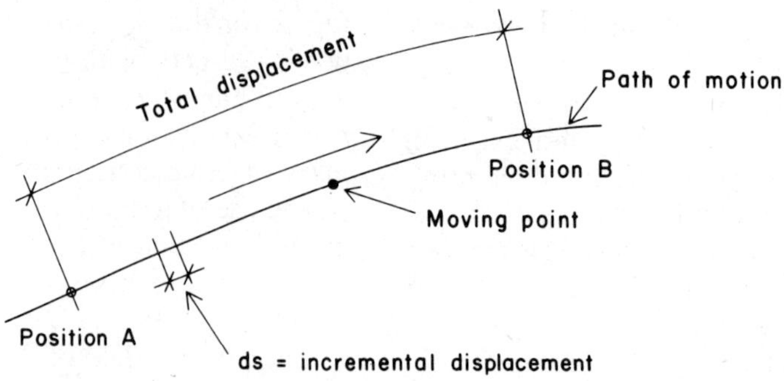

FIGURE 11.1. Motion of a point.

If the displacement occurs at a constant rate with respect to time, it is said to have *constant velocity*. In this case the velocity may be expressed more simply without the calculus as

$$v = \frac{\text{total displacement}}{\text{total elapsed time}}$$

When the velocity changes over time, its rate of change is called the *acceleration* (a). Thus, as an instantaneous change

$$a = \frac{dv}{dt} = \frac{d^2s}{dt^2}$$

That is, the acceleration is the first derivative of the velocity or the second derivative of the displacement with respect to time.

Except for the simplest cases, the derivation of the equations of motion for an object generally require the use of the calculus in the operation of these basic relationships. Once derived, however, motion equations are generally in algebraic form and can be used without the calculus for application to problems. An example is the set of equations that describe the motion of a free-falling object acted on by the earth's gravity field. Under idealized conditions (ignoring air friction, etc.) the distance of fall from a rest position will be

$$s = f(t) = 16.1t^2 \qquad (s \text{ in ft, } t \text{ in sec})$$

This equation indicates that the rate of fall (the velocity) is not a constant but increases with the elapsed time, so that the velocity at any instant of time may be expressed as

$$v = \frac{ds}{dt} = \frac{d(16.1t^2)}{dt} = 32.2t \qquad (v \text{ in ft/sec})$$

and the acceleration as

$$a = \frac{dv}{dt} = \frac{d(32.2t)}{dt} = 32.2 \text{ ft/sec}^2$$

which is the acceleration of gravity.

Kinematics also includes the study of the various forms of motion: translation, rotation, plane motion, motion of deformable bodies, and so on. A study of the mechanics of motion is very useful in the visualization of the deformation of a structure by static as well as dynamic forces.

As stated previously, kinetics includes the additional consideration of the forces that cause motion. This means that in addition to the variables of displacement and time, we must consider the mass of the moving objects. From Newtonian physics the simple definition of mechanical force is

$$F = ma = \text{mass} \times \text{acceleration}$$

Mass is the measure of the property of inertia, which is what causes an object to resist change in its state of motion. The more common term for dealing with mass is *weight*, which is a force defined as

$$W = mg$$

where g is the constant acceleration of gravity (32.2 ft/sec^2).

Weight is literally a dynamic force, although it is the standard means of measurement of force in statics, when the velocity is assumed to be zero. Thus, in static analysis we express forces simply as

$$F = W$$

and in dynamic analysis, when using weight as the measure of mass, we express force as

$$F = ma = \frac{W}{g}a$$

If a force moves an object, work is done. *Work* is defined as the product of the force multiplied by the displacement (distance traveled). If the force is constant during the displacement, work may be simply expressed as

$$w = Fs = \text{force} \times \text{total distance traveled}$$

If the force varies with time, the relationship is more generally expressed with the calculus as

$$w = \int_{s_2}^{s_1} F_t ds$$

indicating that the displacement is from position s_1 to position s_2, and the force varies in some manner with respect to time.

Figure 11.2 illustrates these basic relationships. In dynamic analysis of structures the dynamic "load" is often translated into work units in which the distance traveled is actually the deformation of the structure.

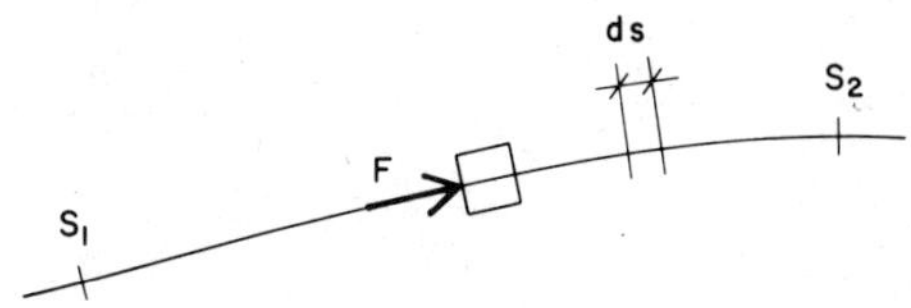

$$\text{Work} = \int_{S_1}^{S_2} F_t \, ds \quad \text{(F variable with time)}$$

$$= F(S_2 - S_1) \quad \text{(F constant with time)}$$

FIGURE 11.2. Kinetics of a moving object.

Energy may be defined as the capacity to do work. Energy exists in various forms: heat, mechanical, chemical, and so on. For structural analysis the concern is with mechanical energy, which occurs in one of two forms. *Potential energy* is stored energy, such as that in a compressed spring or an elevated weight. Work is done when the spring is released or the weight is dropped. *Kinetic energy* is possessed by bodies in motion; work is required to change their state of motion, that is, to slow them down or speed them up (see Fig. 11.3).

In structural analysis energy is considered to be indestructible, that is, it cannot be destroyed, although it can be transferred or transformed. The potential energy in the compressed spring can be transferred into kinetic energy if the spring is used to propel an object. In a steam engine the chemical energy in the fuel is transformed into heat and then into pressure of the steam and finally into mechanical energy delivered as the engine's output.

An essential idea is that of the conservation of energy, which is a statement of its indestructibility in terms of input and output. This idea can be stated in terms of work by saying that the work done on an

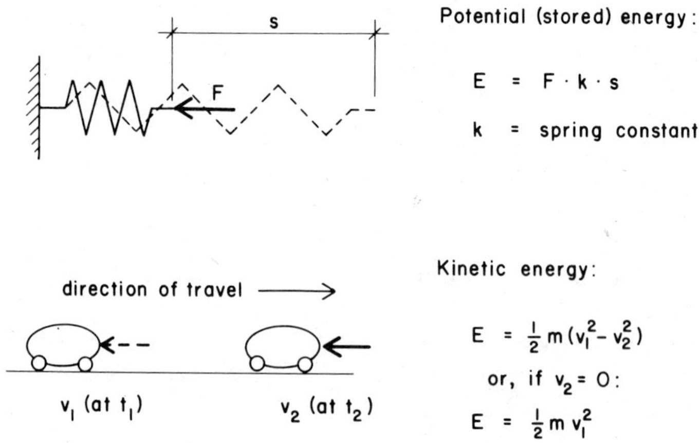

FIGURE 11.3. Forms of mechanical energy.

object is totally used and that it should therefore be equal to the work accomplished plus any losses due to heat, air friction, and so on. In structural analysis we make use of this concept by using a "work equilibrium" relationship similar to the static force equilibrium relationship. Just as all the forces must be in balance for static equilibrium, so the work input must equal the work output (plus losses) for "work equilibrium."

A special kinematic problem of major concern in structural analysis for dynamic effects is that of *harmonic motion.* The two elements generally used to illustrate this type of motion are the swinging pendulum and the bouncing spring. Both the pendulum and the spring have a neutral position where they will remain at rest in static equilibrium. If one displaces either of them from this neutral position, by pulling the pendulum sideways or compressing or stretching the spring, they will tend to move back to the neutral position. Instead of stopping at the neutral position, however, they will be carried past it by their momentum to a position of displacement in the opposite direction. This sets up a cyclic form of motion (swinging of the pendulum; bouncing of the spring) that has some basic characteristics.

Figure 11.4 illustrates the typical motion of a bouncing spring. Using the calculus and the basic motion and force equations, the displacement/time relationship may be derived as

$$s = A \cos Bt$$

The cosine function produces the basic form of the graph, as shown in Fig. 11.4. The maximum displacement from the neutral position is called the *amplitude.* The time elapsed for one full cycle is called the *period.* The number of full cycles in a given unit of time is called the *frequency* (usually expressed in cycles per second) and is equal to the inverse of the period. Every object subject to harmonic motion has a fundamental period (also called natural period), which is determined by its weight, stiffness, size, and so on.

Any influence that tends to reduce the amplitude in successive cycles is called a *damping effect.* Heat loss in friction, air resistance, and so on are natural damping effects. Shock absorbers, counterbalances, cushioning materials, and other devices can also be used to damp the amplitude. Figure 11.5 shows the form of a damped harmonic motion, which is the normal form of most such motions, because per-

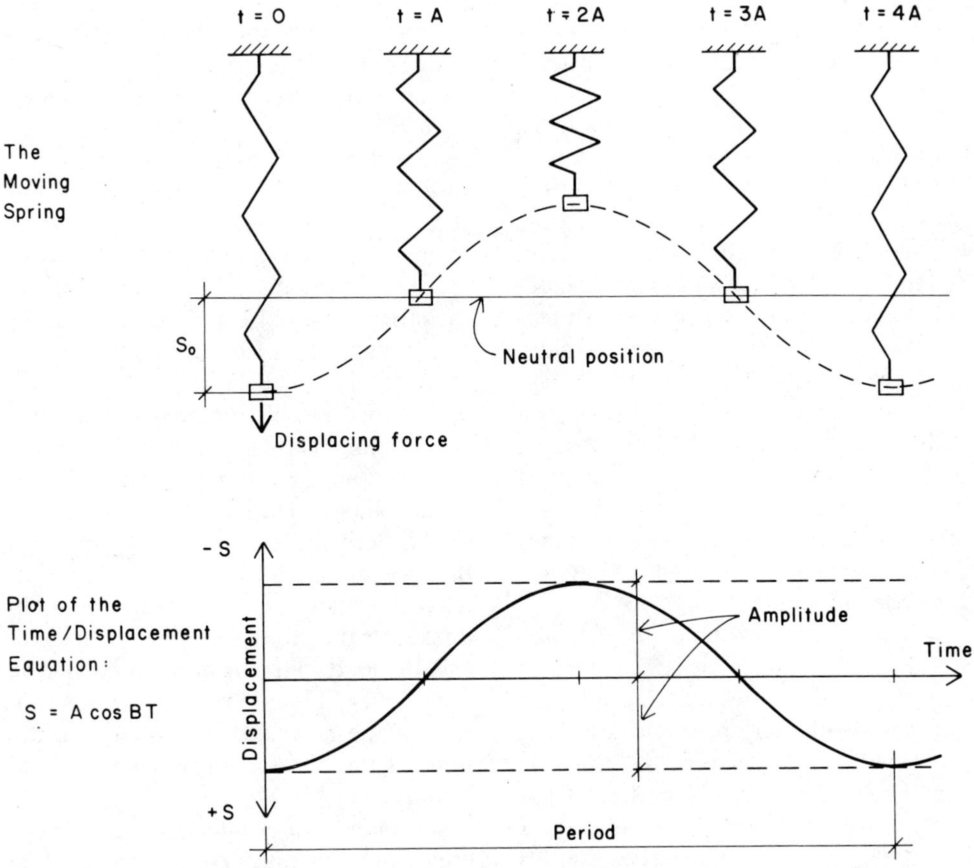

FIGURE 11.4. Harmonic motion of a bouncing spring.

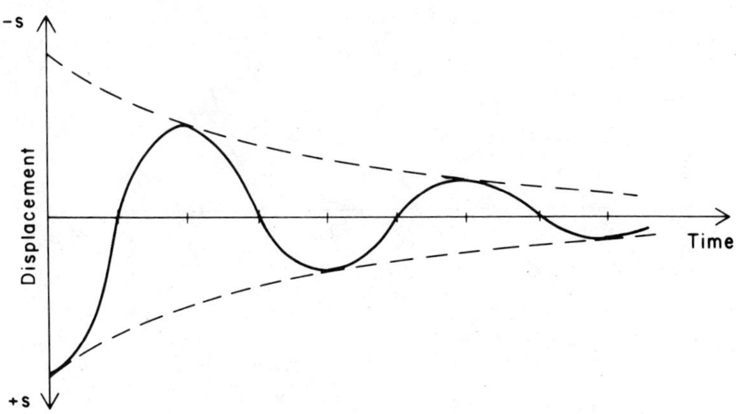

FIGURE 11.5. Form of a damped harmonic motion.

petual motion is not possible without a continuous reapplication of the original displacing force.

Resonance is the effect produced when the displacing effort is itself harmonic with a cyclic nature that corresponds with the period of the impelled object. An example is someone bouncing on a diving board in rhythm with the board's fundamental period, thus causing a reinforcement, or amplification, of the board's free motion. This form of motion is illustrated in Fig. 11.6. Unrestrained resonant effects can result in intolerable amplitudes, producing destruction or damage of the moving object or its supports. A balance of damping and resonant effects can sometimes produce a constant motion with a flat profile of the amplitude peaks.

Loaded structures tend to act like springs. Within the elastic stress range of the materials, they can be displaced from a neutral (unloaded) position and, when released, will go into a form of harmonic motion. The fundamental period of the structure as a whole, as well as the periods of its parts, are major properties that affect responses to dynamic loads.

Load sources that involve motion, such as wind, earthquakes, walking people, moving vehicles, and vibrating heavy machinery, have the potential to cause dynamic effects on structures. Analyzing for their effects requires consideration of essential dynamic properties of the structure. These properties are determined by the size, weight, relative stiffness, fundamental period, type of support, and degree of elasticity of the materials of the structure and by various damping influences that may be present.

Dynamic load sources deliver an energy load to the structure that may be in the form of an impact, such as that caused by the moving air bumping into the stationary building. In this case the energy load is derived from the kinetic energy of the moving air, which is a product of its mass and velocity. In the case of an earthquake, or the vibration of heavy machinery, the load source is not a force as such but, rather, something that induces motion of the structure, in which case the mass of the building is actually the load source.

An important point to note is that the effects of a dynamic load on a structure are determined by the structure's response as well as by the nature of the load. Thus the same dynamic load can produce different

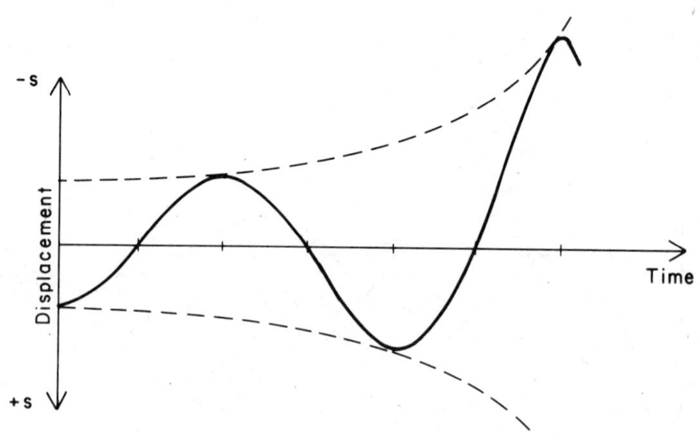

FIGURE 11.6. Form of a resonant harmonic motion.

effects in different structures. Two buildings standing side by side can have significantly different responses to the same earthquake shock if they have major differences in their dynamic properties.

Dynamic effects on structures may be of several types. Some of the principal effects are the following:

Total energy load is the balance between the peak magnitude of the load and the maximum work required by the structure and is known as the *work equilibrium concept*. Work done to the structure by the load equals the work done by the structure in resisting the load.

Unstabilizing effects occur if the dynamic load produces a stability failure of the structure. Thus a free-standing wall may topple over, an unbraced post and beam system may collapse sideways, and so forth, because of the combined effects of gravity and the dynamic load.

Harmonic effects of various types may be set up in the structure, especially if the load source is cyclic in nature, such as the footsteps of marching troops. Earthquake motions are basically cyclic, in the form of vibration or shaking of the surface of the ground. Relations between these motions and the harmonic properties of the structure can result in various effects, such as the flutter of objects at a particular wind velocity, the resonant bouncing of floors, and the resonant reinforcing of the swaying of buildings during an earthquake.

Using up of the structure's energy capacity can occur if the energy of the load exceeds the limit for the structure. Actually, there are several degrees, or stages, of energy capacity, instead of a single limit. Four significant stages are the following:

1. The *resilience limit*, or the limit beyond which some form of permanent damage—however slight—may occur.
2. The *minor damage limit*, the damage being either relatively insignificant or easily repairable.
3. The *major damage limit*, short of total destruction, but with loss of some minor elements. The structure as a whole remains intact, but some major repairs may be required to restore it to its original level of capacity.
4. The *toughness limit*, or the maximum, ultimate capacity represented by the destruction of the structure.

Failure under repeated loadings can result in some cases when structures progressively use up their dynamic resistance. The structures may successfully resist a single peak load of some dynamic effort, only to fail later under a similar, or even smaller, loading. This failure is usually due to the fact that the first loading used up some degree of structural failure, such as ductile yielding or brittle cracking, which absorbed enough energy to prevent total failure but was a one-time usable strength only.

A major consideration in design for dynamic loads is what the response of the structure means to the building as a whole. Thus, although the structure may remain intact, that may be only a minor accomplishment if there is significant damage to the building as a whole. A high-rise building may swing and sway in an earthquake without there being any significant damage to the structure, but if the occupants are tossed about, the ceilings fall, the windows shatter, the partitions and curtain walls collapse, the plumbing bursts, and the elevators derail, it can hardly be said that the building was adequately designed.

In many cases analysis and design for dynamic effects are not done by working directly with the dynamic relationships but simply by using recommendations and rules of thumb that have been established by experience. Some testing or theoretical analysis may have helped in deriving ideas or data, but much of what is used is based on the observations and records from previous disasters.Even when actual calculations are performed, they are mostly done with data and relationships that have been translated into simpler static terms—so-called equivalent static analysis and design. The reasons for this practise have to do frankly with the degree of complexity of dynamic analysis. Even with the use of programmable calculators or computers, the work is quite laborious in all but the simplest of situations.

11.3 EQUIVALENT STATIC EFFECTS

Use of equivalent static effects essentially permits simpler analysis and design by eliminating the complex procedures of dynamic analysis. To make this possible the load effects and the structure's responses must be translated into static terms.

For wind load the primary translation consists of converting the kinetic energy of the wind into an equivalent static pressure, which is then treated in a manner similar to that for a distributed gravity load. Additional considerations are made for various aerodynamic effects, such as ground surface drag, building shape, and suction, but these do not change the basic static nature of the work.

For earthquake effects the primary translation consists of establishing a hypothetical horizontal static force that is applied to the structure to simulate the effects of sideward motions during ground movements. This force is calculated as some percentage of the dead weight of the building, which is the actual source of the kinetic energy loading once the building is in motion—just as the weight of the pendulum and the spring keep them moving after the initial displacement and release. The specific percentage used is determined by a number of factors, including some of the dynamic response characteristics of the structure.

An apparently lower safety factor is used when designing for the effects of wind and earthquake because an increase of one-third is permitted in allowable stresses. This is actually not a matter of a less-safe design but is merely a way of compensating for the fact that one is actually adding static (gravity) effects and *equivalent* static effects. The total stresses thus calculated are really quite hypothetical because in reality one is adding static strength effects to dynamic strength effects, in which case 2 + 2 does not necessarily make 4.

Regardless of the number of modifying factors and translations, there are some limits to the ability of an equivalent static analysis to account for dynamic behavior. Many effects of damping and resonance cannot be accounted for. The true energy capacity of the structure cannot be accurately measured in terms of the magnitudes of stresses and strains. There are some situations, therefore, in which a true dynamic analysis is desirable, whether it is performed by mathematics or by physical testing. These situations are actually quite rare, however. The vast majority of building designs present situations for which a great deal of experience exists. This experience permits generalizations on most occasions that the potential dynamic effects are really insignificant or that they will be adequately accounted for by design for gravity alone or with use of the equivalent static techniques.

Use of the seismic design criteria of the

Uniform Building Code (Ref. 2)—most of which is keyed to an analysis for equivalent static effects—is explained in detail in Chapter 12.

11.4 MOVEMENTS AND DEFORMATIONS

Earthquake effects can be visualized most directly in terms of movement. Stresses and strains in the building structure are induced by the deformations of the structure caused by the relative movement of parts of the structure with respect to other parts. Just as the visualization of the deformed structure is an essential tool in analysis for static force effects, the visualization of the dynamically moving structure is necessary for a full understanding of earthquake effects. In this section we will summarize the various types of movements and their significance to the building structure and the general problem of seismic design.

It is the ground of course that does the initial moving. For purposes of design the earthquake effect is abstractly translated into a hypothetical horizontal force applied externally to the building. It is well to remember, however, that what really happens is that the building is being shaken, as if it were held in the hand of some giant (the ground) who is angry at it.

The ground movements have their first effects on the ground itself. If the movements are intense and prolonged, there are likely to be various types of failure of the ground. Loose surface materials may be dislodged. Poorly consolidated or weakly bonded soil masses may be subject to subsidence or slides. Potentially quick soils may develop liquefaction, producing a situation for the building somewhat like that for a person trying to walk on a smooth, lubricated surface. For seismic design it is necessary to consider the dynamic, as well as static, properties of foundation materials, as these may present unique difficulties for the building.

For the building, it is actually quite useful to consider the idea of the angry giant shaking it in his hand. In fact, the single most useful study tool is a shaking investigation—performed on a shaking table or by computer simulation. In this investigation a physical or mathematical model of the building is subjected to the actual movements of recorded severe earthquakes. Watching a model on a shaking table during such a test is an experience without comparison. While this type of investigation is not feasible for most building design work, the experience from many such tests has greatly influenced the procedures and data used for simpler design methods in current practice.

Movements within the structure—that is, the movement of one part of the structure with respect to other parts—is what causes the major structural actions during an earthquake. Thus, as shown in Fig. 11.7, the horizontal movement of the roof level of the one-story building with respect to the floor (or ground) level is what induces stress in the rigid frame or shear walls. This single distortion is important for visualization of the nature of internal forces (bending, shear, etc.) in the members of the structure. Again, however, it is necessary to remember that the motion is vibratory in nature, with the distortions rapidly reversing between the two forms shown in Fig. 11.7*a* and *b*.

Failure in the structure may occur due to a single, major distortion—producing a maximum magnitude of stress. Failure may also occur due to the rapid reversal of stresses, especially in structures with materials that are subject to fatigue or have specific weaknesses, such as tension failure in concrete, shear failure in wood, or tearing or brittle fracture in welded steel. Connection failures are often due to repeated reversal effects, rather than to a single major jolt.

As discussed in Section 10.1, most buildings exist in the form of vertical projections (or cantilevers upward) from the ground. Their single direction distortions (such as that shown in Fig. 11.7*a*) may be visualized as being produced by a single static force. However, because the distortion is actually dynamically produced, there are various differences in response from that produced by an actual static force. One of these is the rapid reversal effect just discussed. Another has to do with the so-called center of percussion effect. The effect of the gravity-induced weight of a cantilever, as a constant, static force, is such that the moment at the support may be determined using the weight as a single force considered to be concentrated at the centroid (center of gravity) of the cantilever. If the weight is dynamically impelled, however, there is an increase in the effect of the momentum of the mass at the outer end, which may be visualized by considering the mass to be concentrated at some point slightly farther out than the actual center of gravity. This hypothetical point is called the center of percussion.

When an investigation is made for the effect of horizontal movements on the building, the force effects are visualized as being produced by the momentum of the building mass, and thus are proportional to the mass that exists at various locations (notably at various vertical positions) in the building. Because of the center of percussion effect, however, it is necessary to make some correction if the forces are actually so determined and located. This type of correction has only recently been added to seismic design criteria, and is discussed in Chapter 12.

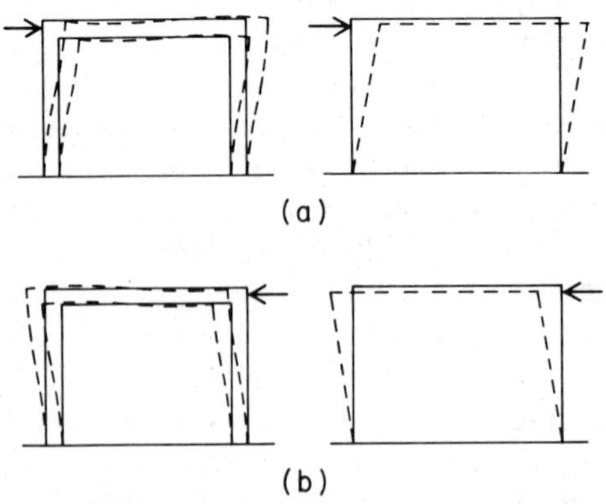

FIGURE 11.7. Reversal effects of seismic motions.

11.5 ENERGY ABSORPTION

Resistance to seismic effects generally has two factors. One is the resistance to movement, which requires stability as well as sufficient stiffness to reduce drift, swaying, and bouncing. Movement failures include overturning, sliding, torsional twisting, and dancing.

The second type of resistance is that of absorption of the mechanical energy vested in the inertial momentum of the moving building. This requires work by the structure, since energy is essentially the capacity to do work. Similar to the concept of static equilibrium, there is a work equilibrium relationship: the structure must do the same amount of work as the load. Work done by the structure—and the whole building in some cases—includes the following:

1. *Stress Development in the Structure.* In work terms this means the product of stress and strain. The total amount of work produced depends on the magnitude of the stress, the magnitude of the strain (deformation), and the total amount of material involved. Increasing the energy capacity of the structure in these terms requires increasing one or more of these: stress, strain, or the amount of material.
2. *Inelastic Deformations.* Plastic yielding of ductile steel is a notable example of this. This is a useful contribution to resistance to ultimate collapse,

but is generally not desirable under minor loading since the permanent deformations involved may likely render the building at least partly unusable.

3. *Other Structural Actions.* Various types of structural behavior mechanisms that do not necessarily constitute significant unsafe conditions may contribute to energy absorption. Cracking of concrete and masonry and the grinding back and forth along crack lines is such a case, possibly resulting in unsightly conditions, but not necessarily causing unsafe situations. Slipping in bolted joints, elastic buckling of redundant truss members, and shear splitting of wood beams and columns are other examples.

4. *Structural Actions of Nonstructural Elements.* If a plywood shear wall is faced with securely attached plaster or stucco, the rigid facing will initially resist the shear distortion of the wall. Failure of this material will absorb considerable energy, before the plywood becomes significantly involved. Rigid partitions and ceilings applied directly to framing, ceramic tile flooring, insulative concrete fill on roof decks, and concrete fireproofing jackets on steel columns are some additional examples.

5. *Actions of Shock-Absorbing Devices.* Special devices may be contained in the load transfer path for the purpose of absorbing energy. These may serve primarily to cushion the structure, or may provide damping for vibratory movements (bouncing). Such devices may function in a resilient manner or may fail and provide energy absorption through their failure mechanism. The inclusion of one-time usable structural elements that are replaceable and not essential to the stability of the system is a recent innovation in seismic design.

In most cases the ultimate resistance of a building to seismic shock is significantly greater than that of the structure alone, especially if the structure's resistance is measured in terms of static strength. While this may be true and significant to the safety margin against total collapse, it may not be of use in reduction of damage. Design for energy absorption should be done with a recognition of the contributions that occur at various stages of loading and an understanding of any failures that are irreversible and damage producing.

CHAPTER TWELVE

Determination of Earthquake Effects

This chapter presents a discussion of current practices for the determination of earthquake effects on building structures. While some other references are used, the emphasis is on the use of the requirements set forth in the 1982 edition of the *Uniform Building Code* (Ref. 2), which was the edition in force at the time of this writing. Direct reproduction of significant parts of this code are given in Chapter 33.

12.1 BUILDING CODE REQUIREMENTS

The following is a discussion of the various requirements of the 1982 edition of the *UBC* for seismic design. The code for seismic design is usually quite up-to-date, a new edition being published every three years. The main body of material on seismic design is in Section 2312. We begin with a discussion of this section and proceed to a discussion of some materials to be found in other sections of the code. See Chapter 33 for reprints of the code.

Section 2312: General Design Requirements

In the beginning of this section are definitions of various terms and symbols used in the code. Reference should be made to this portion whenever any term or symbol is not clear to the reader.

The basic formula for determination of the lateral load on the structure is *UBC* formula 12-1:

$$V = ZIKCSW$$

in which:

Z varies from $\frac{3}{16}$ to 1, depending on the seismic zone as identified in the maps in Figures 1, 2, and 3 in *UBC* Chapter 23.

I is a factor relating to the occupancy or use of the building, as given in *UBC* Table 23-K.

K is based on the type of lateral resistive structural system and is given in *UBC* Table 23-I.

C is a factor empirically derived from the relationship of the building period to the average earthquake effect.

S is a factor that accounts for the effect of local ground conditions, which may dampen or amplify the shock.

W is the dead weight of the building.

C is determined from *UBC* formula 12–2 as

$$C = \frac{1}{15\sqrt{T}}$$

in which *T* is the fundamental period of the building in seconds.

In the words of the code, *T* is to be established by a "properly substantiated analysis." A recommended formula for this determination is given in *UBC* formula 12-3, which incorporates considerations of the seismic load distribution and the deformation characteristics of the structure. This is a somewhat complex analysis, so the code permits alternative determinations by the simpler formulas, 12-3A and 12-3B.

Formula 12-3A expresses *T* as a function of the height of the structure and its depth in the direction of the lateral load.

$$T = \frac{0.05h_n}{\sqrt{D}}$$

Formula 12-3B expresses *T* simply as a function of the number of stories of the structure.

$$T = 0.10N$$

Formula 12-3B is to be used only for buildings in which the lateral resistive structure consists of a ductile rigid frame.

The *UBC* stipulates that the value of *C* need not exceed 0.12. Entering this value in the formula for *C*, we can derive a lower limit for *T*. Thus

If $$C = \frac{1}{15\sqrt{T}} = 0.12$$

Then $$T = \left[\frac{1}{15(0.12)}\right]^2 = 0.309 \text{ sec}$$

Figure 12.1 shows a plot of *C* as a function of *T*, using *UBC* formula 12-2. The form may be seen to be that of the typical spectrum curve. The maximum value for *C* is seen to correspond to the derived value of 0.309 sec for *T*. Below the graph are shown some interpretations of *UBC* formulas 12-3A and 12-3B that relate to the values for *T* in the graph.

Calculation of the *S* factor requires a determination of the fundamental period of the site (T_s) as well as of the structure. Determining T_s requires rather extensive geological information of the site conditions. In the absence of this information, the *UBC* permits the use of a maximum value for the product of *C* and *S* of 0.14. Because the code also requires a minimum value for *S* of 1.0, if the maximum value of *C* of 0.12 is used, the bracketed range for the product of *C* and *S* is between 0.12 and 0.14. The net result of this is that, unless the building project cost is high, the true *C* value is less than 0.12, or if the necessary geological information is available from previous studies in the immediate vicinity, the maximum *CS* value of 0.14 is used for most designs.

The use of the *S* factor is a fairly recent addition to the code, and there is some controversy regarding its validity as a requirement for all structures. In time the use of this factor may be more rationally developed in the code criteria so as not to amount to a de facto general increase in the required lateral load.

Subsection (e) of Section 2312 deals with the distribution of seismic loads on the structure. Formula 12-5 requires that the total load on the structure be

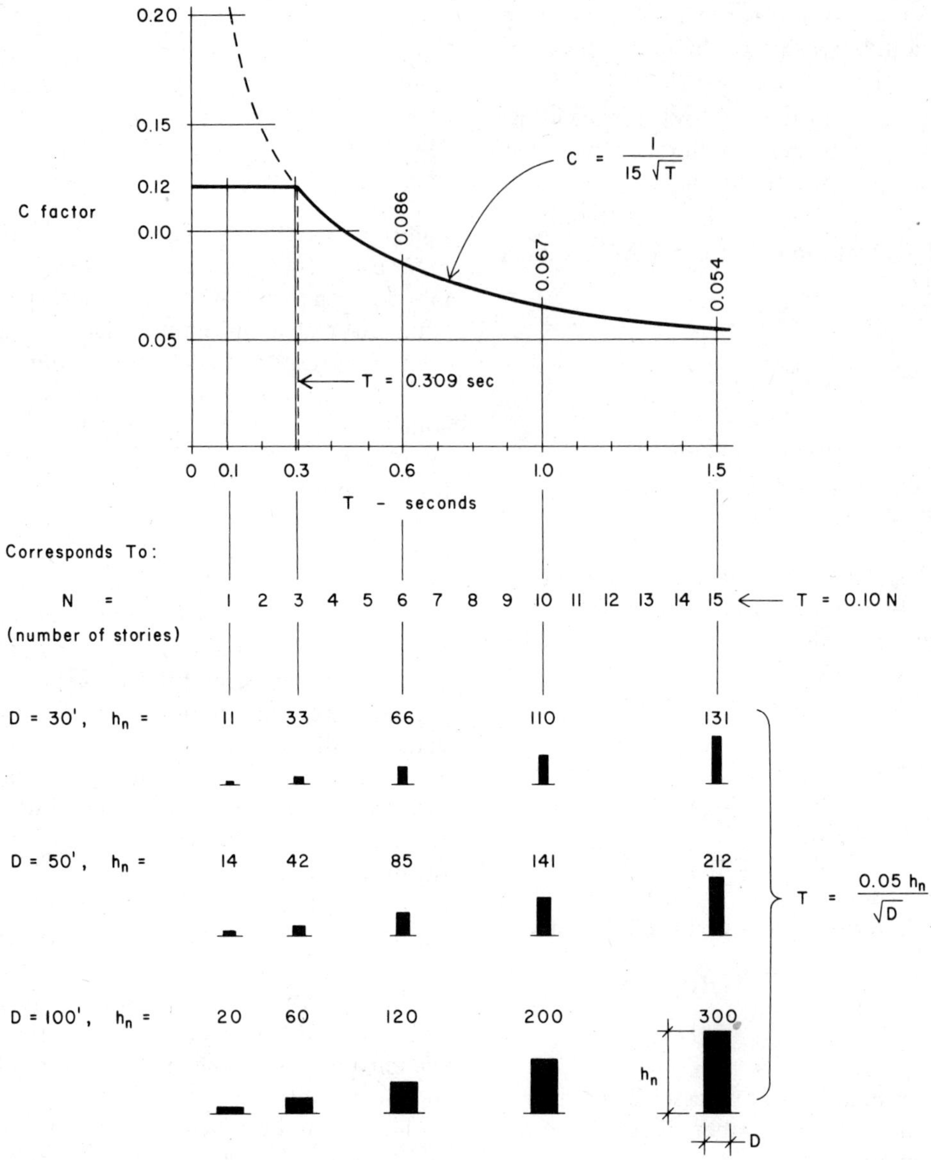

FIGURE 12.1. Relation of *C* and *T* factors.

$$V = F_t + \sum_{i=1}^{n} F_i$$

The summation portion of this formula consists of the loads as normally calculated and applied at the various levels of the structure. F_t is an added force that must be applied at the top of the structure as well as being included in the total, and it is found using formula 12-6.

$$F_t = 0.07TV$$

The maximum required value for F_t is $0.25V$ and F_t may be considered to be zero where T is 0.7 sec or less. As shown in Fig. 12.1, a T value this high generally occurs only with relatively tall buildings.

The value of the lateral shear force at each level of a multilevel structure (F_x) is determined from formula 12-7. The use of this formula for the analysis of multistory structures is illustrated in the examples in Part 4.

The code formulas are generally applicable only to buildings that are reasonably simple in shape and have little change in their basic structural systems or their plans from top to bottom. Changes in these characteristics can result in behaviors more complex than those assumed for the general formulas and their limits. Subsection 2312(e)3 of the *UBC* requires that these more complex structures be analyzed by dynamic methods, rather than by the simpler equivalent static load method. An exception is made in Subsection 2312(e)2 for setbacks of less than 25% in multistory buildings.

Subsection 2312(e)5 requires that the structure be analyzed for torsion when the center of rigidity of the structure does not coincide with the centroid of the load. Even when this eccentricity is not present, the code requires design for a minimum assumed eccentricity of 5% of the maximum plan dimension of the building when horizontal diaphragms are used to distribute the shear.

Overturning of the building or the structure is provided for in Subsection 2312(f). When the lateral resistive system has a number of isolated elements (such as individual shear walls), the overturn effect is distributed to them in the same manner as for the distribution of the lateral shear forces.

Subsection 2312(g) provides that parts of the structure be designed for individual lateral forces that result from the parts' own weight. This force is calculated from formula 12-8 as

$$F_p = ZIC_pW_p$$

The C_p factor in this equation replaces the C, S, and K factors as used in the total load for the building in formula 12-1. Values for C_p are given in *UBC* Table 23-J for various types of elements and situations.

Subsection 2312(h) limits the relative deflection of one story with respect to another (called the *drift*) to 0.5% of the story height, unless an analysis can demonstrate that more deflection is tolerable. A multiplying factor of $1/K$ must be used for the deflection calculation when K is less than 1 for the structure.

Subsection 2312(j) contains a number of requirements for the design of specific structural systems and elements. Reference is made to these requirements in the examples in Part 4.

Subsection 2312(k) defines essential facilities as they are listed in Table 23-K for establishment of the I factor in the lateral load formula, 12-1.

Lateral Load Requirements in Other Chapters

The following is a summary of some of the requirements for lateral load deign to be found in other portions of the *UBC*. In some cases these requirements apply to both wind and seismic loads.

1. Stress increase: Section 2303(d). This section contains the provision for a one-third increase in allowable stresses when load combinations include wind or seismic force.
2. Load combinations: Section 2303(f). The highly remote possibility of a simultaneous earthquake and wind storm loading is excluded by this section.
3. Anchorage of masonry and concrete walls: Section 2310. This section requires that the connections of

masonry and concrete walls to roofs and floors that provide lateral support for the walls be designed for a minimum load of 200 lb/ft.

4. Increase of force on masonry shear walls: Table 24-H. The third footnote to this table requires that the seismic force on reinforced masonry shear walls be increased by 50% over that required by Section 2312 when investigating shear stress in the wall.
5. Wood diaphragms: Section 2513. This is an extensive section with data and requirements for plywood and board-sheathed diaphragms.
6. Fiberboard-sheathed diaphragms: Section 2514. This is a special section with design requirements for vertical diaphragms of fiberboard on a wood frame. The fiberboard referred to must meet the requirements of one of the *UBC* standards. Allowable loads for these diaphragms are given in Table 25-P.
7. Bracing of stud walls: Section 2517(g)3. This section requires that every exterior wall and main interior wall be braced at or near its ends and at least every 25 ft throughout its length. The approved methods of bracing include diagonal, let-in braces, diagonal board sheathing, plywood sheathing, fiberboard sheathing, gypsum sheathing, particleboard sheathing, gypsum drywall, and cement plaster. Reference is made to various code tables for the load capacities of these various constructions. Although all these wall constructions produce potential shear walls, the details for anchorage and load transfer must be developed in order to use them.
8. Maximum diaphragm dimension ratios for wood diaphragms: Table 25-I. This table gives the limits for the span-to-width ratios of simple-span horizontal diaphragms and the height-to-width ratios for vertical diaphragms with plywood or diagonal board sheathing.
9. Allowable loads on horizontal plywood diaphragms: Table 25-J. This table gives the allowable shear load per foot of width for horizontal plywood diaphragms. Variables include the type and thickness of plywood, size and spacing of nails, and details of the framing and panel layout.
10. Allowable loads on plywood shear walls: Table 25-K. This table is similar to Table 25-J, giving the same information for shear walls with plywood sheathing.
11. Minimum nailing requirements: Table 25-Q. This table gives minimum nailing requirements for the various elements of light wood frame construction. Reference should be made to this table when developing the details for structural connections to assure that calculated nailing is not used when it falls below the requirements of this table.
12. Concrete rigid frames: Sections 2625 and 2626. These sections give the requirements for ductile frames of reinforced concrete.
13. Concrete shear walls and trussed frames: Section 2627. This section gives requirements for concrete shear walls and "braced frames" (trussed frames, as defined by *UBC* 2312.)
14. Steel rigid frames: Sections 2722 and 2723. These sections give the requirements for steel rigid frames used to resist seismic load.
15. Wood-framed walls with miscellaneous sheathing: Section 4713 and Table 47-I. This material gives requirements and data for wood-framed shear walls faced with vari-

ous materials other than plywood, diagonal boards, or fiberboard.

Although the *UBC* is reasonably well organized and indexed, it takes some time to gain a familiarity with all the material that is pertinent to structural design for wind and earthquakes. The use of much of the material just described is illustrated in the examples in Part 4, but not all possible situations can be covered in a limited number of examples.

12.2 DETERMINATION OF LATERAL LOADS

For the equivalent static load method, the determination of lateral loads consists of the following:

1. *Visualization of the Loading Pattern.* This consists of determining the manner in which the mass of various parts of the buildings become horizontal forces that are applied to the lateral resistive structure. This varies with the form of the building and with the type and disposition of the structure.
2. *Determination of the Building Mass.* This is a matter of figuring out how much the various parts weigh, and how they aggregate as combined quantities for the loading analysis. (Example: the total mass that constitutes the lateral load applied to the roof diaphragm.)
3. *Conversion of Mass to Force.* Whether for the load on a single element, or the total base shear for the building, this is a matter of using the code formulas with the various appropriate modifying factors.
4. *Redistribution (if any) of Forces.* This refers primarily to the adjustments for multistory buildings, as described in Section 2312(e) of the 1982 edition of the *UBC* (see Chapter 33).
5. *Determination of Design Values for Individual Elements.* Once the load magnitudes and their patterns of application are found, this is a matter of analyzing the behavior of the particular structure—box, truss, frame, and so on.

Procedures for various types of structures are illustrated in the examples in Part 4. Data for the determination of the building weight is provided in Chapter 32.

12.3 CALCULATION OF DEFORMATIONS

There are a few situations in seismic design in which determination of structural deformations must be made. In some cases the actual dimension of the movement is required, while in other cases it is mostly a matter of establishing relative stiffnesses. Some common situations are the following:

1. *Lateral Drift of Frames.* This is quoted in terms of the relative horizontal movements of the levels of the frame. The code establishes limits as a percent of the story height. (See Section 2312(h) of the 1982 edition of the *UBC*, reprinted in Chapter 33.) Of concern in some situations is the effect of the frame distortion on walls in the plane of the frame. The calculation of drift for braced frames is discussed in Section 17.4, where the primary contribution is the change in length of the diagonals. Drift of rigid frames is discussed in Section 18.1.
2. *Deflection of Shear Walls.* This is mostly of concern for wood-framed walls. Empirical formulas are used for plywood walls, incorporating the various contributions of chord length change, shear distortion of the panels, and bending of the nails. Proce-

dures for calculations are discussed in Section 16.4.

3. *Deflection of Horizontal Diaphragms.* These are mostly of concern for diaphragms that are wood framed or those that use light steel deck. Empirical formulas and available data can be used for both situations, as discussed in Section 15.4.

Actual dimensions of movement may be of concern for various reasons. Satisfying code limits for story drift or determining effects on nonstrucutral elements—as mentioned previously—are examples. Another problem is that of providing a basis for the detailing and dimensioning of structural separation joints. In truth, however, movements calculated by the static load method are quite fictitious, and considerable judgment is needed if they are used for design.

12.4 DYNAMIC ANALYSIS

For the equivalent static load method, the energy loads are translated into static forces, and the analysis then proceeds in the usual manner as for real static loads. Loads are quantified in static units (pounds or kips) and the response of the structure is evaluated in static terms (stress). A more realistic dynamic analysis requires that the load be dealt with in dynamic units (energy, not force) and the structure be evaluated in dynamic terms (work done or energy absorbed).

In fact, the current application of the static load method—as specified in the *UBC*—is considerably modified by dynamic considerations. Determination of the loading is modified by dynamic responses of the structure through use of the K, C, and S factors. Distribution of loads for tall structures is modified to approximate the true distribution of dynamic shear. Stress magnification is required for shear in masonry walls and for forces in the members and connections of braced frames.

Nevertheless, true dynamic responses can only be determined using dynamic relationships, dynamic properties of the structure, and dynamic units for loads. This type of analysis is being increasingly done in engineering practice as the operating facilities (mostly computer software) become more readily available and usable. In time, dynamic analysis will probably be routinely done for most buildings. At present, however, it is mostly only done as a check or design adjustment after a preliminary design has been done by conventional methods. And then mostly only for buildings of unusual configuration, complex structural form, considerable size and cost, or where there is a heightened concern for safety (e.g., nuclear power plants).

A brief discussion of the use of currently available computer-aided methods for seismic design is presented in Chapter 34. For most of the building design examples presented in this book, work based on the static load method is currently used for the general determination of the structural system.

CHAPTER THIRTEEN

Interactions in the Lateral Resistive System

Failures of buildings during seismic actions are often due to the inability of the parts of the building structure to work together in resisting lateral forces. This is true whether or not the building was designed for resistance to earthquakes. The visualization and determination of the required internal force transfers and the development of the construction details to facilitate them is a major part of seismic design.

13.1 LOAD TRANSFERS BETWEEN ELEMENTS

The most difficult aspect of visualizing the propagation of loads due to seismic activity is the fact that it is essentially a three-dimensional problem. The seismic actions, the resulting forces, and the building structure are all three dimensional. Add this to the fact that the actions are dynamic and the problem is quite enough.

The difficulty of the design and investigative work may be a contributory reason for the number of failures of connections between elements of lateral resistive systems. Neither architects nor civil engineers are, by training, well prepared to deal with three-dimensional problems of force resolution.

Load propagation in general is discussed in Section 10.2, and the actions of various elements and systems are discussed and illustrated in Part 3. More direct and practical applications are presented in the building design examples in Part 4.

Interactions in the lateral resistive system are commonly illustrated as single static load conditions for individual elements (the classic free-body diagram). Since the separate elements are usually two dimensional in character (single shear wall, single rigid bent, etc.), illustrations are simplified as two-dimensional force resolutions. The concept of the three-dimensional actions fades away when one works continuously with the relatively simple, two-dimensional problem sets.

Individual elements are not loaded individually, but rather are loaded simultaneously with all other parts of the system. Load transfers are thus a matter of load distributions (or load sharing) in an interactive, multiunit system. This is similar to the action of a continuous rigid frame, in which the action of any one individual member is conditioned by the response of all the members in the system.

13.2 ANCHORING AND SPLICING

Two functions that occur repeatedly in lateral resistive systems are those of anchoring and splicing. Anchoring is the general term for the securing of some structural entity at its boundaries. Thus a horizontal diaphragm is anchored to the vertical system, shear walls are anchored at their bases, and so on. Splicing occurs in all structures in one form or another, being generally necessary whenever continuity is required in long elements that are assembled from short individual parts.

For either anchoring or splicing there is a need for so-called positive connection. This is a somewhat nebulous term but generally means a connecting method that is relatively secure against the jolting, repetitive, multidirectional actions during an earthquake. Many connecting devices and methods that are acceptable for resistance to static gravity forces are not so qualified. The seismic behavior of ordinary connection methods is discussed in Chapter 6 and specific issues relating to typical elements of lateral resistive systems are presented in Part 3.

The positive character of anchorage is often a matter of degree. In some cases it may be achieved by simple redundancy, much like wearing both a belt and suspenders, or using a shotgun to kill a flea.

Figure 13.1 shows the situation of anchoring the top reinforcing for a beam in a supporting column or spandrel girder. The positive character of this connection may be increased by degrees as follows:

1. Anchorage can be achieved by simple embedment by providing a sufficient length (L in Figs. 13.1*a* and *b*) for development of bond between the bar surface and the surrounding concrete.
2. Additional anchorage is provided by adding a hook to the end of the bar, a common procedure when the length L is not sufficient. The hook substantially increases the positive character of the connection.
3. Finally, the hook may be made more effective if it is not merely made to grasp the concrete mass but is wrapped around a transverse reinforcing bar, as shown in Figs. 13.1*c* and *d*.

The following are some additional considerations regarding the positive character of connections:

Screws or bolts are more positive than nails.

Nails loaded laterally are more positive than those subjected to withdrawal.

Welding of steel is more positive than bolting; high-strength friction bolts are more positive than unfinished bolts.

Lapped splices of reinforcing bars that comply with the requirements of the ACI code are usually adequate for either tension or compression anchorage. However, some designers prefer the more positive action of welded or mechanically achieved anchorage at splice points. The latter, however, must be done by methods approved by local codes and may require special inspection during construction.

13.3 DISCONTINUITIES

In general, seismic response will be better when the flow of forces in the structure is simple, direct, and uninterrupted. The existence of various discontinuities will work to reduce all of these qualities so that the force flow becomes complex and ineffi-

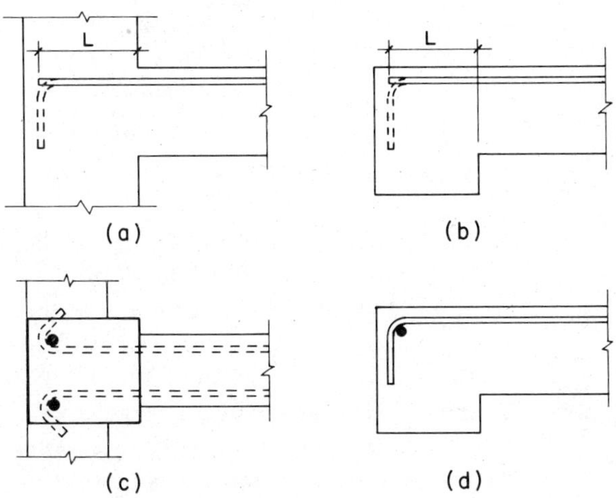

FIGURE 13.1. Progressive development of bar anchorage.

cient. This is not a blank indictment for complexity, as the primary concern is for the lateral resistive structure. It is possible to have richness of form and detail in the building, while still maintaining a simple, continuous, well-ordered bracing system.

Certain discontinuities in the building plan or general form, however, will often be reflected in the form of the lateral bracing system. Some of these situations are as follows:

1. *Openings.* Windows, doors, skylights, stairs, ducts, and elevators require openings. These may result in holes in horizontal or vertical diaphragms or may produce misalignment or discontinuity of rigid frame bents.
2. *Multiplane Roofs.* The farther the roof is from a single, flat, continuous plane, the less effective it will be for use as a horizontal diaphragm.
3. *Split-level Floors.* This is the same problem as for the roof. Where this is required, the locations of level changes should be carefully worked out with the layout of the vertical elements of the bracing system so that the horizontal diaphragm does not have to try to maintain its continuity across level changes.
4. *Interrupted Shear Walls or Offset Columns.* While not uncommon, these represent major discontinuities and should be avoided if possible.
5. *Nonlinear Edges.* If the building edge is curved, saw-toothed, zigzagged, or otherwise not straight, the development of diaphragm chords and collectors may be difficult. A useful technique may be to use a line somewhat inside the actual edge as the boundary of the diaphragm and simply cantilever the edge for horizontal force.

13.4 SITE–STRUCTURE INTERACTION

Building site problems are usually of one of the following type:

1. Subsurface conditions that complicate the building foundation design problem: weak, compressible soil deposits; high groundwater levels; highly organic materials.
2. Topographic situations: hillsides, surface water.

3. Instability conditions: collapsing soils, highly erodable soils, soils subject to dynamic liquefaction.
4. Structure–site interaction during motion that increases seismic effects on the building. (See the 1982 edition of the UBC for use of the *S* factor.)

In effect, the ground must be considered to be an element of the lateral resistive system. Once the building is set in motion, the forces induced by its momentum must be resolved into the ground, just as must be done for gravity forces. Vertical soil pressure, horizontal sliding and lateral soil pressure, pressure concentrations due to overturning, and resistance to uplift must all be given consideration in the design of the foundations. The general problems of foundation design for seismic effects are discussed in Chapter 20.

Concern for site–structure interaction in the form of vibratory relationships has only relatively recently been incorporated in code criteria. It is likely that this criterion will be further refined in future editions of the *UBC*. With more experience in the use of design criteria and the data in the form of geological studies required for its implementation, it is likely that this area of investigation will have more influence. To an extent this is a trend of concern for geological factors beyond the scale of the immediate site as traditionally reflected in ordinary shallow soil borings at the location of the building.

At present, for most small building projects, the minimum factors provided for in the code are used in lieu of obtaining the necessary geological studies to support any real investigation in this area. If the obtaining of necessary geological data becomes as routine as soil borings are now for most buildings, the code criteria is likely to become more detailed and more routinely used. As in the case of real dynamic analysis of seismic effects in general, computer-aided methods are also likely to become routine.

13.5 LOAD SHARING

A principal concern in the design of the vertical elements of the lateral bracing system is the distribution of loads to the individual elements of the system. The portion of the total horizontal force on the building that is carried by a single element will be affected by several factors. The principal considerations are as follows:

1. *Location of the Vertical Elements in the Building Plan.* This may be of significance primarily with respect to the locations of other vertical elements. If loads are distributed on a peripheral basis (also called tributary load basis), the periphery for a single element is defined by the location of adjacent elements. If the building is subject to significant torsion, the location of an element will determine its contribution to torsional resistance.
2. *Relative Stiffness.* Where elements exist in rows, such as a series of shear wall piers in a single wall surface, load sharing is on the basis of the relative stiffness of the elements in the linked series. If the distributing element (usually the horizontal diaphragm) is quite stiff, distribution to all members may be on a stiffness basis, rather than on a peripheral basis. Illustrations and discussion of these situations are presented in the design examples in Part 4.
3. *Control Devices.* Load distribution may be controlled in part through the use of various devices such as expansion joints, seismic separation joints, pinned joints, collectors, drag struts, and ties. Natural force paths may thus be diverted or redirected to some extent.

Loads may be redistributed or shifted when staged responses occur. Initial failures, such as yielding of steel reinforcing and plastic hinging in frames, will change the relative resistance of elements so that loads may be shifted to other members in the linked system. From initial response to final failure of the entire system, load distributions will most likely shift considerably among the vertical elements.

In many cases determination of the design load for a single element of the vertical system requires some amount of judgment involving many qualifying assumptions. When a rational process seems elusive, it is not uncommon to consider a range of possibilities, to investigate each, and to provide for all the worst eventualities.

13.6 COUPLING AND DECOUPLING

It is generally desirable that the building move as a single unit in response to seismic actions. However, in the typical multimassed building, with many parts of different mass and stiffness, it is not reasonable to think of movements as simple as those of a tuning fork or a single bouncing spring. Nevertheless, it is usually a first preference to try to tie the separate elements of the building together for a single joined response if possible. This is particularly critical for the building base, thus other options are seldom considered for the foundations.

As in other situations elements of the construction may serve useful purposes for the tying together of the building. Horizontal diaphragms do the most work in this regard. Also effective are continuous spandrel beams and the continuous top plates of stud walls. If tying is a specific need, these elements often need positive splicing, although they may already be developed for use as chords or collectors.

Building corners and wall intersections are critical locations for interactions. If the presence of framing does not offer potential for tying at these locations, the wall joints themselves should be studied for this purpose.

There are times, of course, when tying elements is not practical or not even desirable. When this is the case, the construction must be designed to permit independent actions by use of flexible attachments or structural separation joints. Detachment of rigid finishes from flexible supports, decoupling of tall shear walls, and separation of separate parts of a multimassed building are examples of these cases. The judgment as to the need for separation, and the development of the construction details to facilitate the degree of required independence, is one of the most difficult parts of seismic design of buildings.

CHAPTER FOURTEEN

Miscellaneous Problems

This chapter contains discussions of some issues that relate to the problem of determination of seismic effects on buildings.

14.1 DETERMINATION OF BUILDING WEIGHT

Translation of seismic motions into force effects involves the use of many theoretical relationships, most of which are empirical, complex, judgmental, and to say the least, mystical. One evident piece of hard data used in the process is that of the building weight. Here is a touchpoint with reality—something believable and logically determinable. However, the physical fabric of most buildings is quite complex, and the computation of building weight is typically a process that employs considerable approximation. Some of the factors that contribute to this situation are as follows:

1. The densities of many building materials vary over some range, requiring the use of *average* values, unless precise material specifications are available. "Plaster" is a description that permits considerable range of interpretation—from the very low density of sprayed-on acoustic coating of ceilings to the high density of exterior stucco.
2. Installation details vary. Thicknesses vary; required auxiliary devices, such as attachments, reinforcing, bracing, and so on, may vary; designer judgment affects many relatively common details. A simple exterior wood stud wall with drywall on the inside and plywood plus exterior finish on the outside can range over a considerable spread in terms of weight per square foot of wall surface. Variables include the size, spacing, and type of wood of the studs, the need for blocking between studs, the thickness of the plywood, the thickness of the drywall, and the type of exterior finish—from light wood siding to real brick veneer.
3. Structural design calculations are often done with information of a vague nature regarding the materials and details of the building construction. This is simply because the definitive structural design must often be done quite early in the design process since the architectural designer needs to know basic facts about the structure quite early. It is hoped that major

changes in the process of development of the construction will be reflected in later adjustments to the structural design. Nevertheless, the preliminary structural design—which often affects major decisions—is usually done with unavoidably sketchy information.

4. Discontinuities in the construction make determination of material quantities difficult. Doors, windows, stairs, openings for ducts, and other building features make precise quantity takeoffs quite laborious.

In short, the determination of the building weight is generally accepted to be an imprecise process that involves considerable judgment and some sheer guesswork. This does not excuse the structural designer from trying to do the best job possible, but the pursuit of precision is doomed to frustration.

Data is provided in Chapter 32 to assist in the determination of the weight of building construction, once some amount of detail is known about the choices of materials and type of construction. If the exact nature of the construction is established (working drawings and specifications complete), the densities of materials provided in Table 32.1 may be used to find weights. If the construction is only approximately determined—at some midpoint of the design work—Table 32.2 may be used for average values of typical construction elements. For very early calculations—used for quick judgments concerning basic system choices and determination of approximate structural sizes—the material in Table 32.3 may prove helpful.

14.2 DISTRIBUTION OF LATERAL LOADS

To a degree the design requirements and the formulas and data presented in codes are based on buildings of simple shape with lateral resistive structures that are well ordered and free of major discontinuities. Most real buildings are not of this type, and the degree to which they do not conform to simple conditions determines the relative validity of the code criteria.

A major point of consideration in this regard is the relationship between the disposition of the building mass (as generater of lateral forces) and the arrangement of the lateral resistive system. The idea is to try to get the catcher positioned to catch the ball. This may be imaged as a relationship between the entire mass and the entire resistive system, as illustrated in Fig. 2.5. In most real situations the problem is both more complex in nature and less subject to simple computation.

Figure 14.1*a* shows an irregular building plan with an arrangement of shear walls that are disposed for resistance to lateral forces in one direction. The distribution of the lateral loads to these walls on a peripheral basis is illustrated by the zones defined by the dashed lines on the plan. On this basis the total building weight between the lines labeled 1 and 2 would be used to find the load on the horizontal diaphragm in this area, and this load would be distributed to the three shear walls labeled H, I, and J. Even though these three walls are not symmetrically disposed in the zone area, the stiffness of the diaphragm and the torsional stabilizing effect of the walls and horizontal diaphragm outside the zone would be considered adequate to prevent any significant torsion on the affected walls.

The situation described in Fig. 14.1 is one that is subject to considerable judgmental interpretation. Some factors that could allow for different interpretations are as follows:

1. Is this a one-story or a multistory building? If it is multistory, it may be advisable to use a seismic separation joint at one side or the other of the

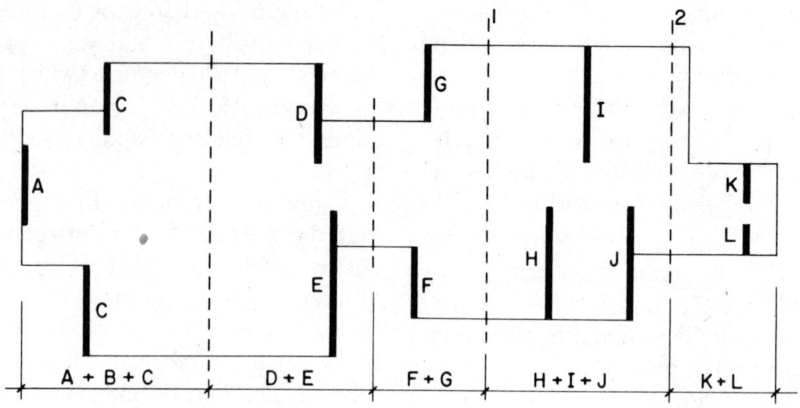

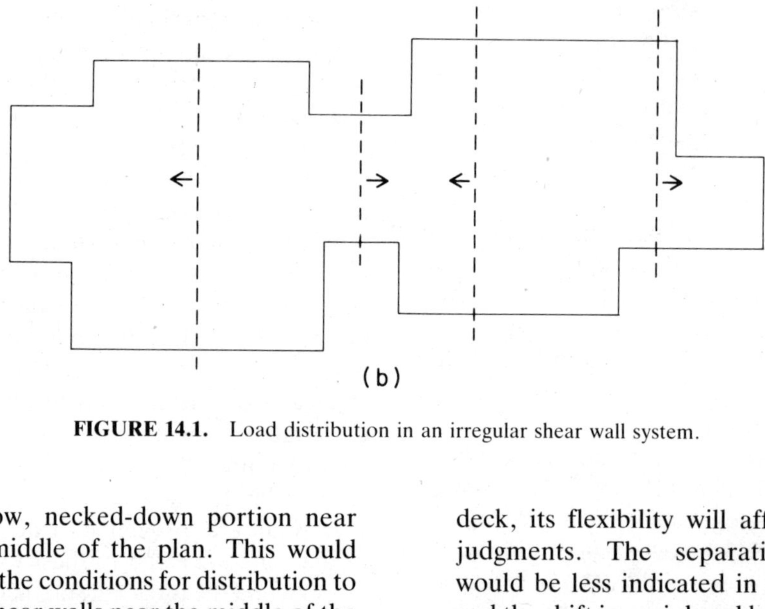

FIGURE 14.1. Load distribution in an irregular shear wall system.

narrow, necked-down portion near the middle of the plan. This would alter the conditions for distribution to the shear walls near the middle of the building.

2. Are the shear walls of wood frame or of masonry or concrete construction? This relates primarily to the load distribution to individual walls. However, if the walls are of masonry or concrete, some designers would move the peripheral zone boundaries slightly to acknowledge the relative stiffness of the longer walls. Possibilities for such shifting are shown in Fig. 14.1*b*.
3. What is the construction of the horizontal diaphragm? If it is wood frame with plywood, or a light-gauge metal deck, its flexibility will affect some judgments. The separation joint would be less indicated in this case, and the shift in peripheral boundaries more questionable. If the diaphragm is a poured-in-place concrete deck, the situation is reversed.
4. Should the system be designed for torsion? If the deck is flexible, and the separation joint is used, probably not. If the deck is concrete, and the separation joint *is* used, each independent half of the building should probably be investigated for torsion.

The use of the peripheral load distribution method and the procedures for consideration of torsional effects are illustrated in the design examples in Part 4.

PART THREE

ANALYSIS AND DESIGN OF ELEMENTS AND SYSTEMS

CHAPTER FIFTEEN

Horizontal Diaphragms

Most lateral resistive structural systems for buildings consist of some combination of vertical elements and horizontal elements. The most common vertical elements are shear walls, braced (trussed) frames, and rigid frames. The horizontal elements are most often the roof and floor framing and decks. When the deck is of sufficient strength and stiffness to be developed as a rigid plane, it is called a *horizontal diaphragm*. Decks most commonly used for this purpose are constructed of plywood, diagonally placed wood boards, formed sheet steel, poured-in-place concrete, and units of precast concrete. This chapter deals with various general concerns for horizontal diaphragms. Design applications of various types of diaphragms in different situations are illustrated in the examples in Part 4. Data pertaining to design is given in Chapter 28 for wood framed diaphragms and in Chapter 29 for formed sheet steel diaphragms.

15.1 GENERAL BEHAVIOR

A horizontal diaphragm typically functions by collecting the lateral forces at a particular level of the building and then distributing them to the vertical elements of the lateral resistive system. For wind forces the lateral loading of the horizontal diaphragm is usually through the attachment of the exterior walls to its edges. For seismic forces the loading is partly a result of the weight of the deck itself and partly a result of the weights of other parts of the building that are attached to it.

The particular structural behavior of the horizontal diaphragm and the manner in which loads are distributed to vertical elements depend on a number of considerations that are best illustrated by various example cases in Part 4. Some of the general issues of concern are discussed in the following sections.

Relative Stiffness of the Horizontal Diaphragm

If the horizontal diaphragm is relatively flexible, it may deflect so much that its continuity is negligible and the distribution of load to the relatively stiff vertical elements is essentially on a peripheral basis. If the deck is quite rigid, on the other hand, the distribution to vertical elements will be essentially in proportion to their relative stiffness with respect to each other. The possibility of these two situations is illustrated for a simple box system in Fig. 15.1.

Torsional Effects

If the centroid of the lateral forces in the horizontal diaphragm does not coincide with the centroid of the stiffness of the vertical elements, there will be a twisting action (called *rotation effect* or *torsional effect*) on the structure as well as the direct force effect. Figure 15.2 shows a structure in which this effect occurs because of a lack of symmetry of the structure. This effect is usually of significance only if the horizontal diagphragm is relatively stiff. This stiffness is a matter of the materials of the construction as well as the depth-to-span ratio of the horizontal diaphragm. In general, wood and metal decks are quite flexible, whereas poured concrete decks are very stiff.

Relative Stiffness of the Vertical Elements

When vertical elements share load from a rigid horizontal diaphragm, as shown in the lower figure in Fig. 15.1, their relative

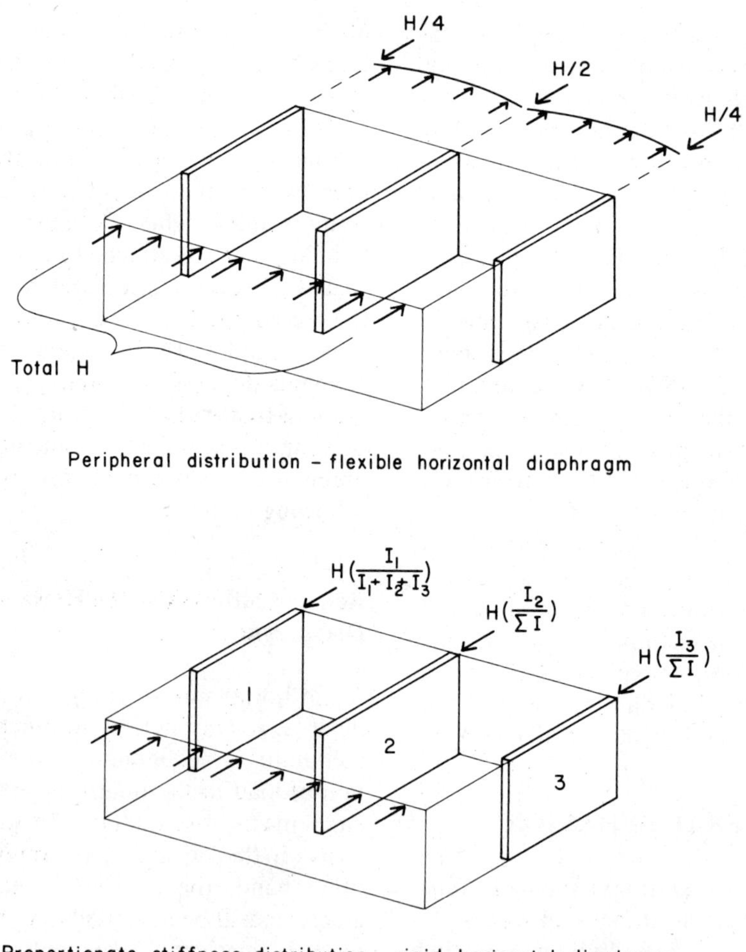

FIGURE 15.1. Peripheral distribution versus proportionate stiffness distribution.

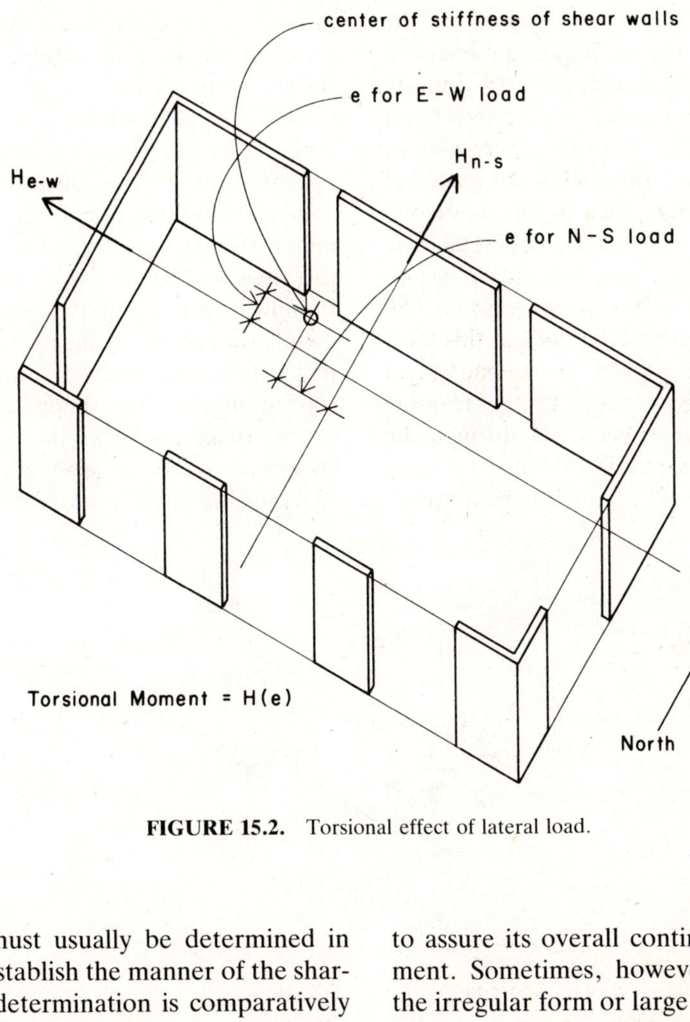

FIGURE 15.2. Torsional effect of lateral load.

stiffness must usually be determined in order to establish the manner of the sharing. The determination is comparatively simple when the elements are similar in type and materials, such as all-plywood shear walls. When the vertical elements are different, such as a mix of plywood and masonry shear walls or of some shear walls and some braced frames, their actual deflections must be calculated in order to establish the distribution, and this may require laborious calculations. A general discussion of deflections due to lateral loads is given in Section 15.4.

Use of Control Joints

The general approach in design for lateral loads is to tie the whole structure together to assure its overall continuity of movement. Sometimes, however, because of the irregular form or large size of a building, it may be desirable to control its behavior under lateral loads by the use of structural separation joints. In some cases these joints function to create total separation, allowing for completely independent motion of the separate parts of the building. In other cases the joints may control movements in a single direction while achieving connection for load transfer in other directions. A general discussion of separation joints is given in Section 21.5.

In performing their basic tasks, horizontal diaphragms have a number of potential stress problems. A major consideration is that of the shear stress in the plane of the diaphragm, caused by the spanning action of the diaphragm as shown in Fig. 15.3. This spanning action results in shear stress in the material as well as a force that must be transferred across joints in the deck when the deck is composed of separate elements such as sheets of plywood or units of formed sheet metal. The sketch in Fig. 15.4 shows a typical plywood framing detail at the joint between two sheets. The stress in the deck at this location must be passed from one sheet through the edge nails to the framing member and then back out through the other nails to the adjacent sheet.

As is the usual case with shear stress, both diagonal tension and diagonal compression are induced simultaneously with the shear stress. The diagonal tension becomes critical in materials such as concrete. The diagonal compression is a potential source of buckling in decks composed of thin sheets of plywood or metal. In plywood decks the thickness of the plywood relative to the spacing of framing members must be considered, and it is also a reason why the plywood must be nailed to intermediate framing members (not at edges of the sheets) as well as at edges. In metal decks the gauge of the sheet metal and the spacing of stiffening ribs must be considered. Tables of allow-

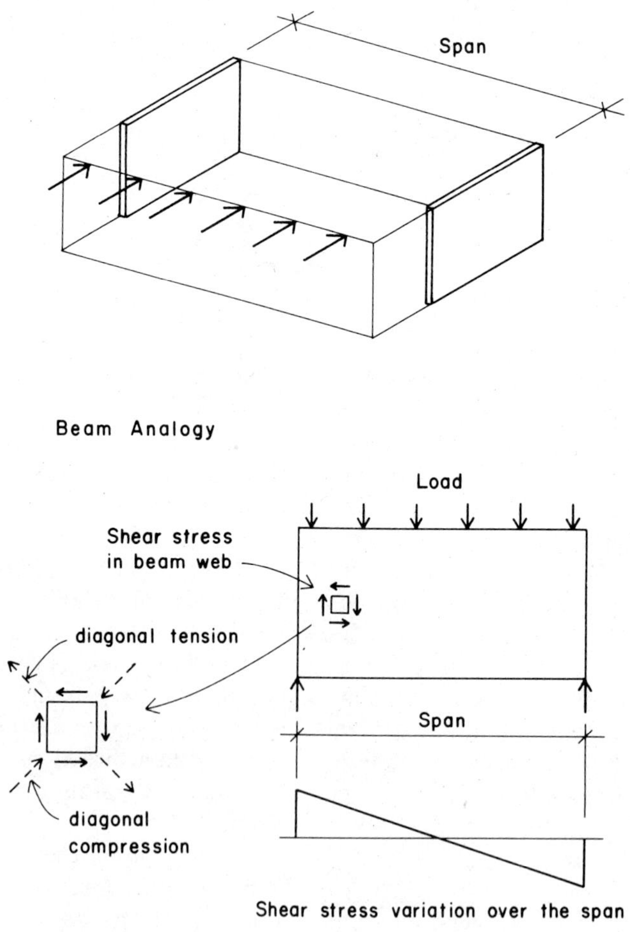

FIGURE 15.3. Function of a horizontal diaphragm.

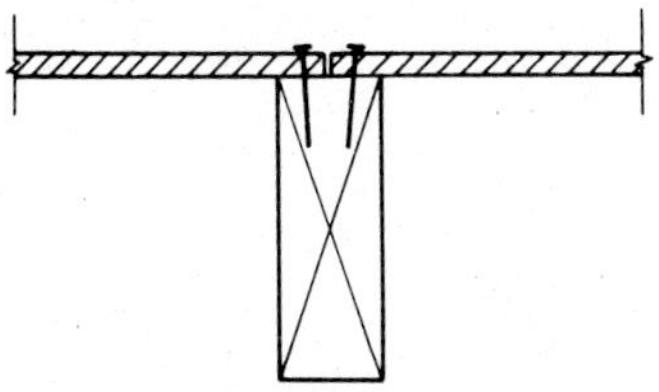

FIGURE 15.4

able loads for various deck elements usually incorporate some limits for these considerations.

Diaphragms with continuous deck surfaces are usually designed in a manner similar to that for webbed steel beams. The web (deck) is designed for the shear, and the flanges (edge-framing elements) are designed to take the moment, as shown in Fig. 15.5. The edge members are called *chords,* and they must be designed for the tension and compression forces at the edges. With diaphragm edges of some length, the latter function usually requires that the edge members be spliced for some continuity of the forces. In many cases there are ordinary elements of the framing system, such as spandrel beams or top plates of stud walls, that have the potential to function as chords for the diaphragm.

In some cases the collection of forces into the diaphragm or the distribution of loads to vertical elements may induce a stress beyond the capacity of the deck alone. Figure 15.6 shows a building in which a continuous roof diaphragm is connected to a series of shear walls. Load collection and force transfers require that some force be dragged along the dotted lines shown in the figure. For the outside walls it is possible that the edge framing used for chords can do double service for this purpose. For the interior shear wall, and possibly for the edges if the roof is cantilevered past the walls, some other framing elements may be necessary to reinforce the deck.

The diaphragm shear capacities for commonly used decks of various materials are available from the codes or from load tables prepared by deck manufacturers. Loads for plywood decks are given in the *UBC* (see Chapter 28). An example of a lateral load table for steel deck is given in Chapter 29. Other tabulations are available from product manufacturers, although care should be exercised in their use to be certain that they are acceptable to the building code of jurisdiction.

A special situation is that of a horizontal system that consists partly or wholly of a braced frame. Care may be required when there are a large number of openings in the roof deck, or when the diaphragm shear stress is simply beyond the capacity of the deck. In the event of a deck with no code-accepted rating for shear, the braced

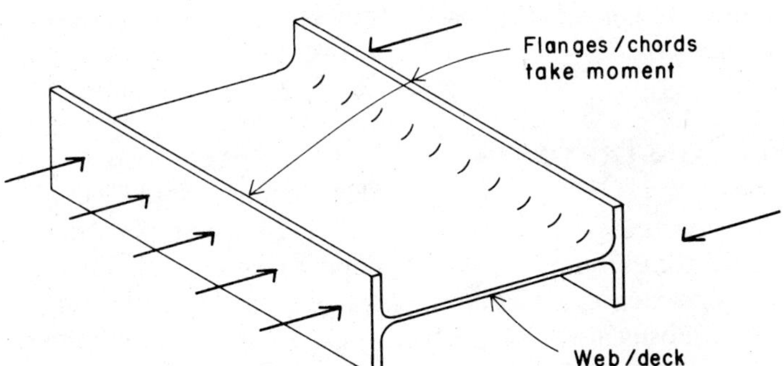

FIGURE 15.5. Horizontal diaphragm functions— beam analogy.

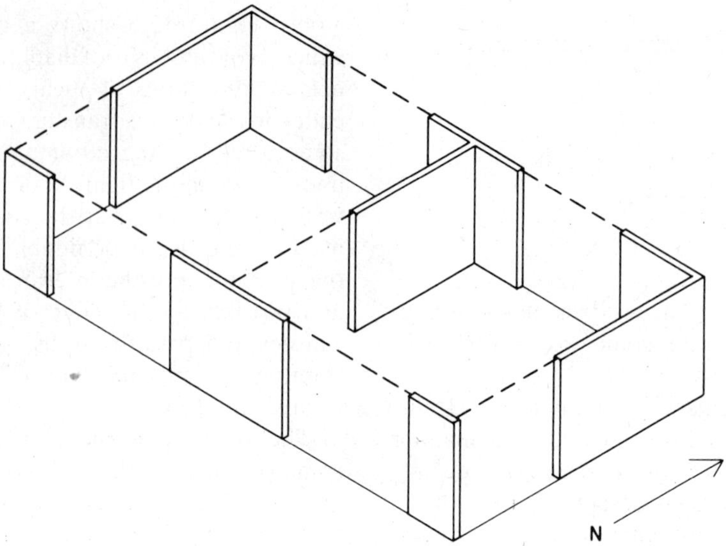

FIGURE 15.6. Collector funcions in the box system.

frame may have to be used for the entire horizontal system.

The horizontal deflection of flexible decks, especially those with high span-to-depth ratios, may be a critical factor in their design. Calculation of actual deflection dimensions may be required to determine the effect on vertical elements of the building construction or to establish positively whether the deck must be considered as essentially flexible or rigid, as discussed previously.

The use of subdiaphragms may also be required in some cases, necessitating the design of part of the whole system as a separate diaphragm, even though the deck may be continuous.

15.2 TYPICAL CONSTRUCTION

The most common horizontal diaphragm is the plywood deck, for the simple reason that wood frame construction is so popular and plywood is mostly used for roof and floor decks. For roofs the deck may be as thin as $\frac{3}{8}$ in., but for flat roofs with waterproof membranes decks are usually $\frac{1}{2}$ in. or more. Attachment is typically by nailing, although glued floor decks are used for their added stiffness and to avoid nail popping and squeaking. Mechanical devices for nailing may eventually become so common that shear capacities will be based on some other fastener; at present the common wire nail is still the basis for load rating.

Attachment of plywood to chords and collectors and load transfers to vertical shear walls are also mostly achieved by nailing. Code-acceptable shear ratings are based on the plywood type and thickness, the nail size and spacing, and features such as size and spacing of framing, use of blocking, and so on. Load capacities for plywood decks are given in *UBC* Table 25−J, which is reprinted as Table 28.1 in Chapter 28.

In general, plywood decks are quite flexible and should be investigated for deflection when spans are large or span−depth ratios are high. Deflection problems are discussed in Section 15.4.

Decks of boards or timber, usually with tongue-and-groove joints, were once popular but are given low rating for shear capacity at present. Where the exposed plank-type deck is desired, it is not uncommon to use a thin plywood deck on top of it for lateral force development.

Steel decks offer possibilities for use as diaphragms for either roof or floors. Acceptable shear capacities should be obtained from the supplier for any particular product, as capacities vary considerably and code approval is not consistent. Stiffnesses are generally comparable with those of plywood decks. Floor decks receive concrete fill, which significantly increases the stiffness of the deck. In some situations it is practical to use the concrete as the basic shear-resisting element. An example of diaphragm design information supplied by a deck manufacturer is given in Chapter 29.

Poured-in-place concrete decks provide the strongest and stiffest diaphragms. Precast concrete deck units, as well as the slab portions of precast systems, can be used for diaphragms. Precast units must be adequately attached to each other and to supporting members for diaphragm actions; if designed only for gravity, ordinary attachments will usually not be adequate for lateral forces. As with steel deck, concrete fill is sometimes placed on top of precast units and both stiffens and strengthens the system.

There are many other types of roof deck construction that may function adequately for diaphragm action, especially when the required unit shear resistance is low. Acceptability by local building code administration agencies should be determined if any construction other than those described are to be used.

15.3 LOAD CAPACITY

Load capacities for wood-framed decks are given in the *UBC*. Load tables from the 1982 edition of the *UBC* are reprinted in this book in Chapter 28. Ratings for other types of construction are generally a matter of local code approvals. Manufacturers and suppliers often provide data for their products, some of which have secured code approval. When contemplating any type of roof or floor deck construction, other than wood framed, the designer is advised to investigate the products and systems used locally and to determine which are approved and rated by local code-enforcing agencies. A sample of data supplied by one manufacturer of steel roof deck is given in Chapter 29 for illustration only.

15.4 STIFFNESS AND DEFLECTION

Horizontal Diaphragms

As spanning elements, the relative stiffness and actual dimensions of deformation of horizontal diaphragms depends on a number of factors, such as:

- The materials of the construction.
- The continuity of the spanning diaphragm over a number of supports.
- The span-to-depth ratio of the diaphragm.
- The effect of various special conditions, such as chord length changes, yielding of connections, influence of large openings, and so on.

In general, wood and light-gauge metal decks tend to produce quite flexible diaphragms whereas poured concrete decks tend to produce the most rigid diaphragms. Ranging between these extremes are decks of lightweight concrete, gypsum concrete, and composite constructions of lightweight concrete fill on metal deck. For true dynamic analysis the variations are more complex because the weight and degree of elasticity of the materials must also be considered.

With respect to their span-to-depth ratios, most horizontal diaphragms approach the classification of deep beams. As shown in Fig. 15.7, even the shallowest of diaphragms, such as the maximum 4 to 1 case allowed for a plywood deck by the *UBC*,

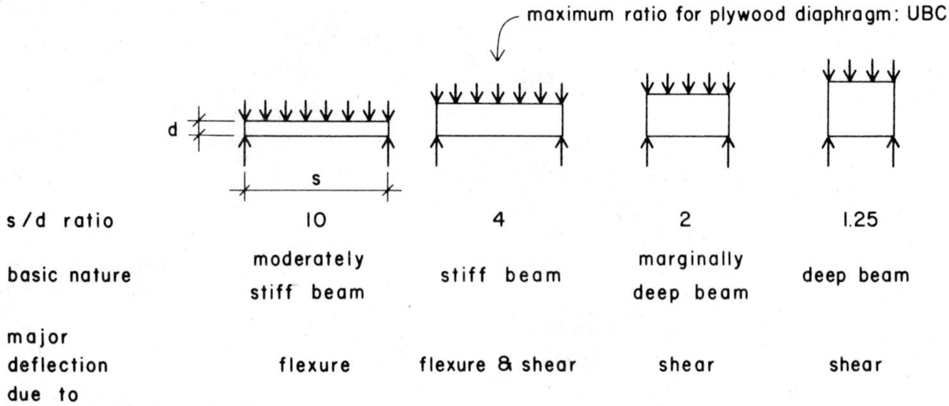

FIGURE 15.7. Behavior of horizontal diaphragms related to depth-to-span ratios.

tends to present a fairly stiff flexural member. As the span-to-depth ratio falls below about 2, the deformation characteristic of the diaphragm approaches that of a deep beam, in which the deflection is primarily caused by shear strain rather than by flexural strain. Thus the usual formulas for deflection caused by flexural strain become of limited use.

The following formula (see *Western Woods Use Book*, Ref. 19) is used for the calculation of deflection of simple span plywood diaphragms:

$$\Delta = \frac{5vL^3}{8EAb} + \frac{vL}{4Gt} + 0.094Le_n + \Sigma\frac{(\Delta_c X)}{2b}$$

in which the four terms account for four different contributions to the deflection, as follows:

Term 1 accounts for the length change of the chords.

Term 2 accounts for the shear strain in the plywood panels.

Term 3 accounts for the lateral bending of the nails.

Term 4 accounts for additional change in the chord lengths caused by slip in the chord splices.

The reader is referred to the Ref. 19 or to other sources on wood design for data and explanations of the factors in this equation and for examples of its use.

The deflection of steel deck diaphragms is discussed and illustrated in the *Inryco Lateral Diaphragm Data Manual 20-2* published by Inryco, Inc. The formula used for calculating the deflection of a simple span deck is

$$\Delta_t = \frac{5\ W L_s^4 \times 1728}{384\ E \times I} + q \times F \times \frac{L_s}{2} \times 10^{-6}$$

in which the first term accounts for flexural deflection caused by the length change of the chords and the second term for shear strain and panel distortion in the diaphragm web.

The quantities q and F vary as a function of the type and gauge of the deck, the fastening patterns and methods used, and the possible inclusion of concrete fill.

As with deck load capacities, deflections and the relative stiffness of deck systems are generally based on materials presented in the *Diaphragm Design Manual*, published by the Steel Deck Institute (Ref. 17). For a specific product, however, designers should obtain information from the product supplier and verify that any data or procedures used are acceptable to the building-permit-approving agency for the work.

CHAPTER SIXTEEN

Vertical Diaphragms

Vertical diaphragms are usually the walls of buildings. As such, in addition to their shear wall function, they must fulfill various architectural functions and may also be required to serve as bearing walls for the gravity loads. The location of walls, the materials used, and some of the details of their construction must be developed with all these functions in mind.

The most common shear wall constructions are those of poured concrete, masonry, and wood frames of studs with surfacing elements. Wood frames may be made rigid in the wall plane by the use of diagonal bracing or by the use of surfacing materials that have sufficient strength and stiffness. Choice of the type of construction may be limited by the magnitude of shear caused by the lateral loads but will also be influenced by fire code requirements and the satisfaction of the various other wall functions, as described previously.

16.1 GENERAL BEHAVIOR

Some of the structural functions usually required of vertical diaphragms are the following (see Fig. 16.1):

1. *Direct shear resistance* usually consists of the transfer of a lateral force in the plane of the wall from some upper level of the wall to a lower level or to the bottom of the wall. This results in the typical situation of shear stress and the accompanying diagonal tension and compression stresses, as discussed for horizontal diaphragms.
2. *Cantilever moment resistance.* Shear walls generally work like vertical cantilevers, developing compression on one edge and tension on the opposite edge, and transferring an overturning moment (M) to the base of the wall.
3. *Horizontal sliding resistance.* The direct transfer of the lateral load at the base of the wall produces the tendency for the wall to slip horizontally off of its supports.

The shear stress function is usually considered independently of other structural functions of the wall. The maximum shear stress that derives from lateral loads is compared to some rated capacity of the wall construction, with the usual increase of one-third in allowable stresses because the lateral load is most often a result of wind or earthquake forces. For concrete

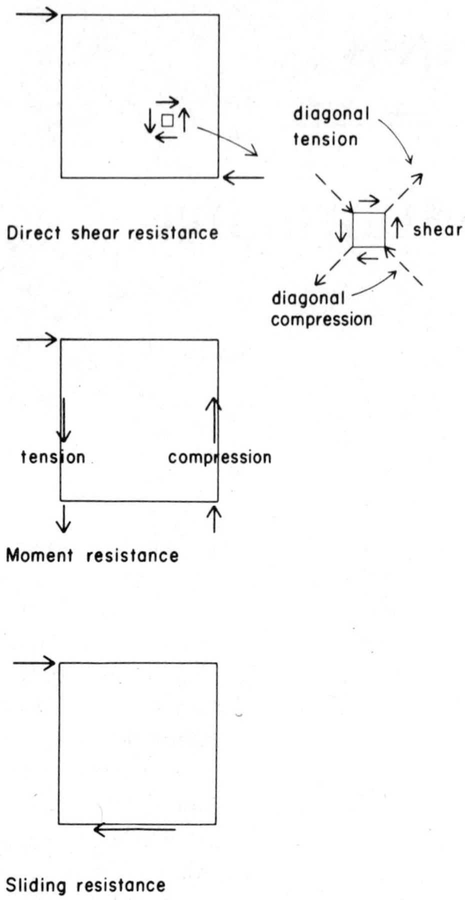

FIGURE 16.1. Functions of a shear wall.

and masonry walls the actual stress in the material is calculated and compared with the allowable stress for the material. For structurally surfaced wood frames the construction as a whole is generally rated for its total resistance in pounds per foot of the wall length in plan. For a plywood-surfaced wall this capacity depends on the type and thickness of the plywood; the size, wood species, and spacing of the studs; the size and spacing of the plywood nails; and the inclusion or omission of blocking at horizontal plywood joints.

The analysis and design of shear walls of concrete and reinforced masonry are illustrated in the examples in Part 4. For walls of concrete the procedures are based on the requirements of the American Concrete Institute (ACI) code (Ref. 14). For walls of reinforced hollow concrete block construction the procedures are based on the requirements of Chapter 24 of the *UBC*.

For wood stud walls the *UBC* provides tables of rated load capacities for several types of surfacing, including plywood, diagonal wood boards, plaster, gypsum drywall, and fiberboard. This material is included in the *UBC* reprints in Chapter 33.

Although the possibility exists for the buckling of walls as a result of the diagonal compression effect, this is usually not critical because other limitations exist to constrain wall slenderness. The thickness of masonry walls is limited by maximum values for the ratio of unsupported wall height or length-to-wall thickness. Concrete thickness is usually limited by forming and pouring considerations, so that thin walls are not common except with precast construction. Slenderness of wood studs is limited by gravity design and by the code limits as a function of the stud size. Because stud walls are usually surfaced on both sides, the resulting sandwich-panel effect is usually sufficient to provide a reasonable stiffness.

As in the case of horizontal diaphragms, the moment effect on the wall is usually considered to be resisted by the two vertical edges of the wall acting as flanges or chords. In the concrete or masonry wall this results in a consideration of the ends of the wall as columns, sometimes actually produced as such by thickening of the wall at the ends. In wood-framed walls the end framing members are considered to fulfill this function. These edge members must be investigated for possible critical combinations of loading because of gravity and the lateral effects.

The overturn effect of the lateral loads must be resisted with the safety factor of 1.5 that is required by the *UBC*. The form of the analysis for the overturn effect is as shown in Fig. 16.2. If the tiedown force is actually required, it is developed by the anchorage of the edge-framing elements of the wall.

Resistance to horizontal sliding at the base of a shear wall is usually at least partly resisted by friction caused by the dead loads. For masonry and concrete walls with dead loads that are usually quite high, the frictional resistance may be more than sufficient. If it is not, shear keys must be provided. For woodframed walls the friction is usually ignored and the sill bolts are designed for the entire load.

An important judgment that must often be made in designing for lateral loads is that of the manner of distribution of lateral force between a number of shear walls that share the load from a single horizontal diaphragm. In some cases the existence of symmetry or of a flexible horizontal diaphragm may simplify this consideration. In many cases, however, the relative stiffnesses of the walls must be determined for this calculation.

If considered in terms of static force and elastic stress–strain conditions, the relative stiffness of a wall is inversely proportionate to its deflection under a unit load. Figure 16.3 shows the manner of

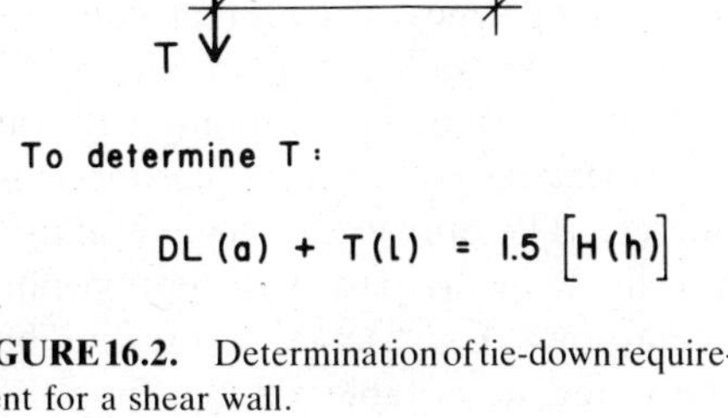

To determine T:

DL (a) + T(L) = 1.5 [H(h)]

FIGURE 16.2. Determination of tie-down requirement for a shear wall.

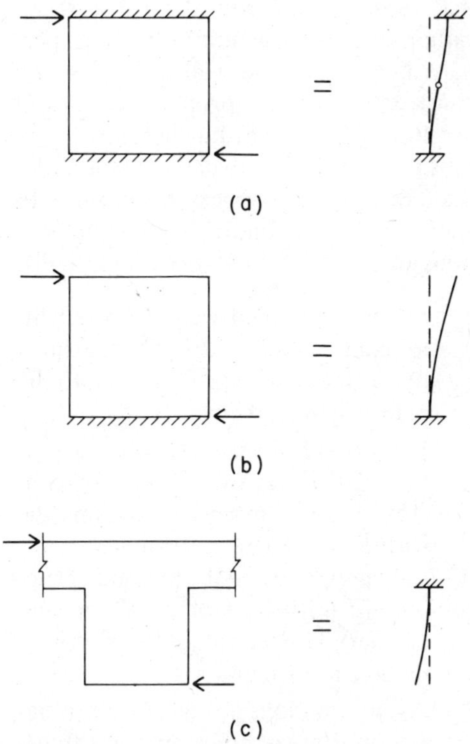

FIGURE 16.3. Shear wall support conditions: (*a*) fixed top and bottom, (*b*) and (*c*) cantilevered.

deflection of a shear wall for two assumed conditions. In *a* the wall is considered to be fixed at its top and bottom, flexing in a double curve with an inflection point at midheight. This is the case usually assumed for a continuous wall of concrete or masonry in which a series of individual wall portions (called *piers*) are connected by a continuous upper wall or other structure of considerable stiffness. In *b* the wall is considered to be fixed at its bottom only, functioning as a vertical cantilever. This is the case for independent, freestanding walls or for walls in which the continuous upper structure is relatively flexible. A third possibility is shown in *c*, in which relatively short piers are assumed to be fixed at their tops only, which produces the same deflection condition as in *b*.

In some cases the deflection of the wall

may result largely from shear distortion, rather than from flexural distortion, perhaps because of the wall materials and construction or the proportion of wall height to plan length. Furthermore, stiffness in resistance to dynamic loads is not quite the same as stiffness in resistance to static loads. The following recommendations are made for single-story shear walls:

1. For wood-framed walls with height-to-length ratios of 2 or less, assume the stiffness to be proportional to the plan length of the wall.
2. For wood-framed walls with height-to-length ratios over 2 and for concrete and masonry walls, assume the stiffness to be a function of the height-to-length ratio and the method of support (cantilevered or fixed top and bottom). Use the values for pier rigidity given in Chapter 31.
3. Avoid situations in which walls of significantly great differences in stiffness share loads along a single row. The short walls will tend to receive a small share of the loads, especially if the stiffness is assumed to be a function of the height-to-length ratio.
4. Avoid mixing of shear walls of different construction when they share loads on a deflection basis.

Item 4 in the preceding list can be illustrated by two situations as shown in Fig. 16.4. The first situation is that of a series of panels in a single row. If some of these panels are of concrete or masonry and others of wood frame construction, the stiffer concrete or masonry panels will tend to absorb the major portion of the load. The load sharing must be determined on the basis of actual calculated deflections. Better yet is a true dynamic analysis because if the load is truly dynamic in character, the periods of the two types of walls are of more significance than their stiffness.

In the second situation shown in Fig. 16.4, the walls share load from a rigid horizontal diaphragm. This situation also requires a deflection calculation in order to determine the distribution of force to the panels.

In addition to the various considerations mentioned for the shear walls themselves, care must be taken to assure that they are properly anchored to the horizontal diaphragms. Problems of this sort are illustrated in the examples in Part 4.

A final consideration for shear walls is that they must be made an integral part of the whole building construction. In long building walls with large door or window openings or other gaps in the wall, shear walls are often considered as entities (isolated, independent piers) for their design. However, the behavior of the entire wall under lateral load should be studied to be sure that elements not considered to be parts of the lateral resistive system do not suffer damage because of the wall distortions. An example of this situation is shown in Fig. 16.5. The two relatively long solid portions are assumed to perform the bracing function for the entire wall and would be designed as isolated piers. However, when the wall deflects, the effect of the movement on the shorter piers, on the headers over openings, and on the door and window framing must be considered. The headers must not be cracked loose from the solid wall portions or pulled off their supports.

16.2 TYPICAL CONSTRUCTION

The various types of common construction for shear walls are mentioned in the preceding section: wood frames with various surfacing, reinforced masonry and concrete. The only wood frame wall used extensively in the past was the plywood-covered one. Experience and testing has established acceptable ratings for other

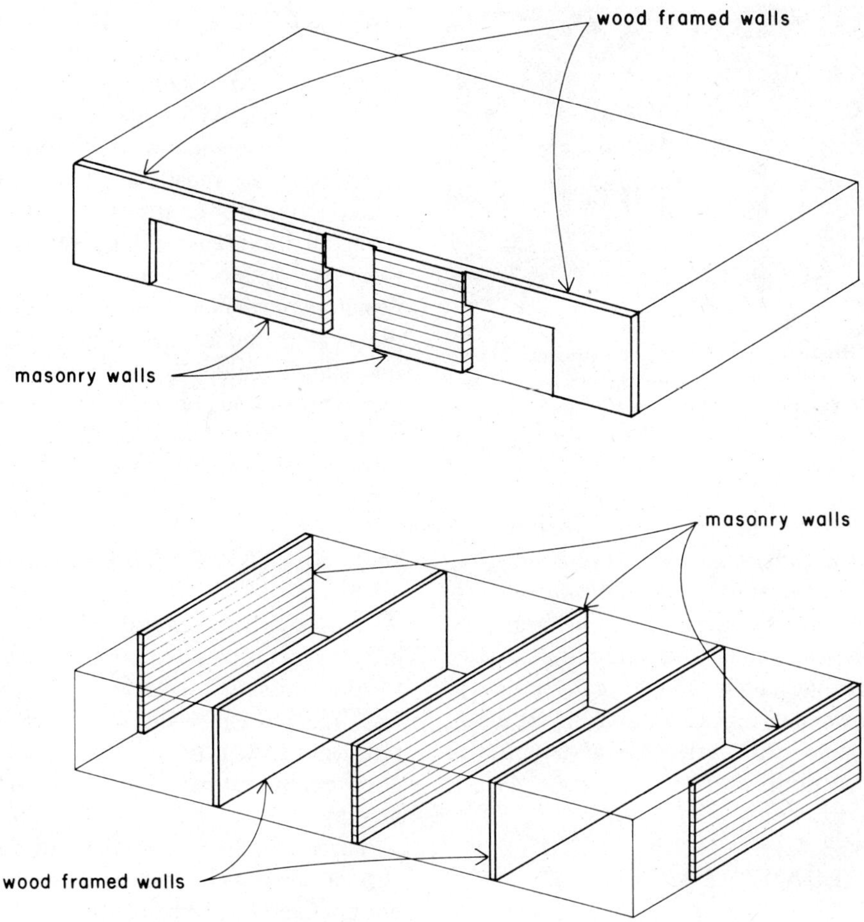

FIGURE 16.4. Interacting walls of mixed construction.

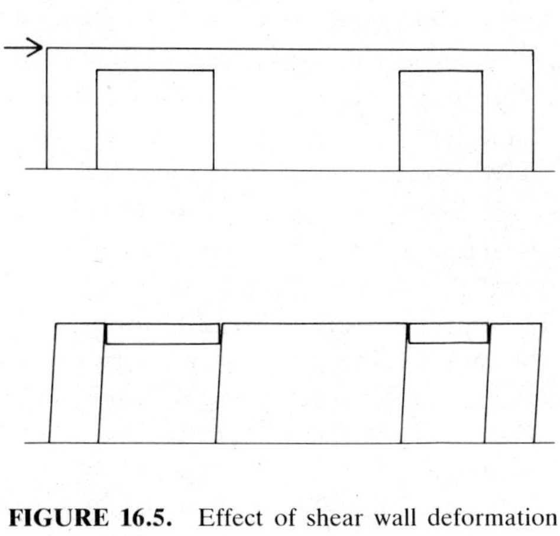

FIGURE 16.5. Effect of shear wall deformation on headers.

surfacing, so that plywood is used somewhat less when shear loads are low (see Fig. 16.6).

For all types of walls there are various considerations (good carpentry, fire resistance, available products, etc.) that establish a certain minimum construction. In many situations this "minimum" is really adequate for low levels of shear loading, and the only additions are in the area of attachments and joint load transfers. Increasing wall strength beyond the minimum usually requires increasing the size or quality of units, adding or strengthening

FIGURE 16.6. Wood frame diaphragms. Garage with side wall of stucco fastened directly to studs, end wall of plywood, let-in bracing for temporary use to brace structure during construction.

attachments, developing supporting elements to function as chords or collectors, and so on. It well behooves the designer to find out the standards for basic construction to know what the minimum consists of, so that added strength can be developed when necessary—but using methods consistent with the ordinary types of construction.

16.3 LOAD CAPACITY

Load capacities for ordinary wood-framed shear walls are given in Chapter 28; the load tables presented being reprints from the *UBC*. For seismic actions the only masonry construction ordinarily acceptable is that of reinforced masonry, of which the most common form is that using hollow precast concrete units (concrete blocks). The design of masonry construction should be done with the references that present material acceptable to the building code of jurisdiction. The reprints from the 1982 *UBC* in Chapter 33 give some criteria for masonry, but the complete design guide is Chapter 24 of the code.

Most concrete design is based on the current edition of the ACI code (*Building Code Requirements for Reinforced Concrete*, Ref. 14), although local codes are sometimes slow in accepting new changes in the ACI code editions. The current edition of the ACI code provides some criteria for seismic design, but recent developments—mostly in the form of suggested details for construction—are more stringent for areas of high seismic risk. Concrete design in general, and seismic design in particular, has become quite sophisticated and complex, and there are few simple guides. The shear wall is a relatively simple concrete element, but its design should be done in conformance with the latest codes and practices.

16.4 STIFFNESS AND DEFLECTION

As with the horizontal diaphragm, there are several potential factors to consider in the deflection of a shear wall. As shown in Fig. 16.7, shear walls also tend to be relatively stiff in most cases, falling into deep beam instead of ordinary flexural members.

There are two general cases for the vertical shear wall: the cantilever and the doubly fixed pier. The cantilever, fixed at its base, is the most commonly used. Fixity at both the top and bottom of the wall usually affects deflection only when the wall is relatively short in length with respect to its height. Walls with long lengths in proportion to their height fall into the deep beam category, in which the predominant shear strain is not affected by the fixity of the support.

As shown in Fig. 16.8, if the doubly fixed pier is assumed to have an inflection point at its mid-height, its deflection can be approximated by considering it to be the sum of the deflections of two half-height cantilevered piers. Yielding of the supports and flexure in the horizontal structure will produce some rotation of the assumedly fixed ends, which will result in some additional deflection.

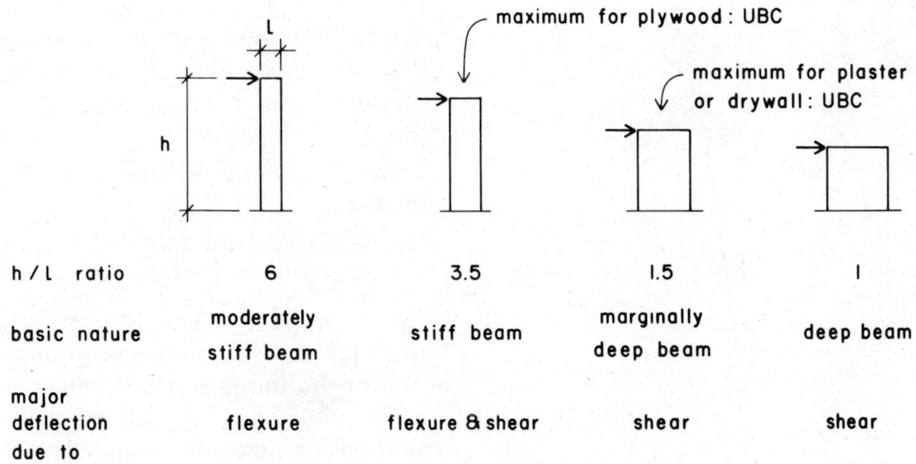

FIGURE 16.7. Behavior of vertically cantilevered elements related to height-to-length ratios.

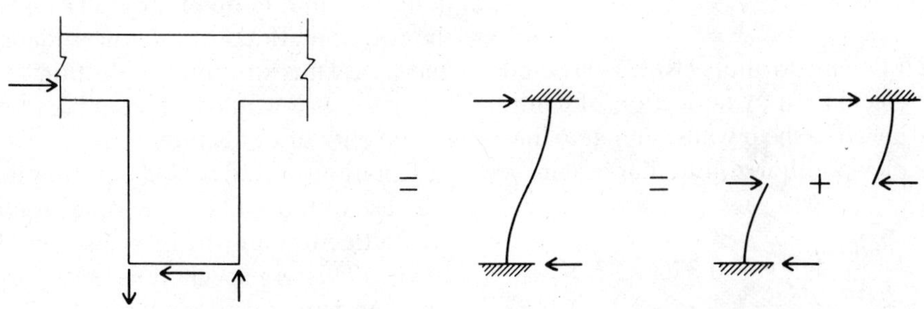

FIGURE 16.8. Deflection assumption for a fully fixed pier.

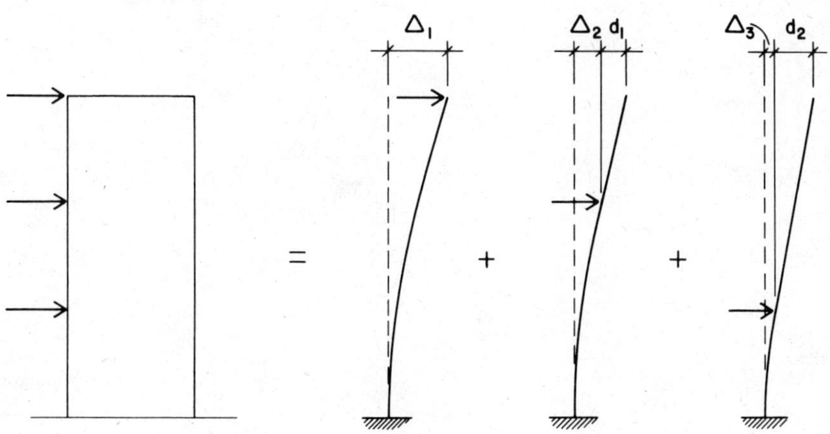

FIGURE 16.9. Deflection of a multistory shear wall.

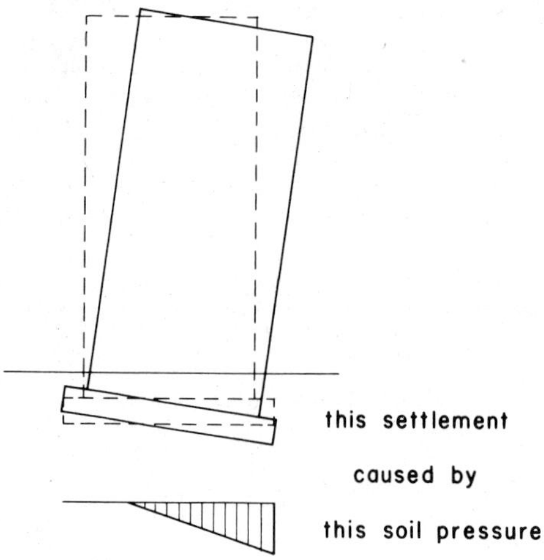

FIGURE 16.10. Shear wall tilt caused by settlement.

The following formula (Ref. 19) is used for the calculation of deflection of cantilevered plywood shear walls, similar to that for the plywood horizontal diaphragm:

$$\Delta = \frac{8vh^3}{EAb} + \frac{vh}{Gt} + 0.376he_n + d_a$$

in which the four terms account for the following:

Term 1 accounts for the change in length of the chords (wall end framing).

Term 2 accounts for the shear strain in the plywood panels.

Term 3 accounts for the nail deformation.

Term 4 is a general term for including the effects of yield of the anchorage.

As before, the reader is referred to Ref. 19 for the explanation and examples of use of the formula. The formula can also be used for the calculation of the deflection of a multistory wall, as shown in Fig. 16.9. For the loading as shown in the illustration, a separate calculation would be made for each of the three loads (Δ_1, Δ_2, and Δ_3). To these would be added the deflection at the top of the wall caused by the rotation effects of the lower loads (d_1 and d_2). Thus the total deflection at the top of the wall would be the sum of the five increments of deflection.

Rotation caused by soil deformation at the base of the wall can also contribute to the deflection of shear walls (see Fig. 16.10). This is especially critical for tall walls on isolated foundations placed on relatively compressible soils, such as loose sand and soft clay; a situation to be avoided if at all possible.

CHAPTER SEVENTEEN

Braced Frames

Although there are actually several ways to brace a frame against lateral loads, the term *braced frame* is used to refer to frames that utilize trussing as the primary bracing technique. In buildings, trussing is mostly used for the vertical bracing system, in combination with the usual horizontal diaphragms. It is also possible, however, to use a trussed frame for a horizontal system, or to combine vertical and horizontal trussing in a truly three-dimensional trussed framework. The latter is more common for open tower structures, such as those used for large electrical transmission lines and radio and TV transmitters.

17.1 USE OF TRUSSING FOR BRACING

Post and beam systems, consisting of separate vertical and horizontal members, may be inherently stable for gravity loading, but they must be braced in some manner for lateral loads. The three basic ways of achieving this are through shear panels, moment-resistive joints between the members, or by trussing. The trussing, or triangulation, is usually formed by the insertion of diagonal members in the rectangular bays of the frame.

If single diagonals are used, they must serve a dual function: acting in tension for the lateral loads in one direction and in compression when the load direction is reversed (see Fig. 17.1). Because long tension members are more efficient than long compression members, frames are often braced with a crisscrossed set of diagonals (called X-bracing) to eliminate the need for the compression members. In any event the trussing causes the lateral loads to induce only axial forces in the members of the frame, as compared to the behavior of the rigid frame. It also generally results in a frame that is stiffer for both static and dynamic loading, having less deformation than the rigid frame.

Single-story, single-bayed buildings may be braced as shown in Fig. 17.1. Single-story, multibayed buildings may be braced by bracing less than all of the bays in a single plane of framing, as shown in Fig. 17.2*a*. The continuity of the horizontal framing is used in the latter situation to permit the rest of the bays to tag along. Similarly, a single-bayed, multistoried, towerlike structure, as shown in Fig. 17.2*b*, may have its frame fully braced, while the

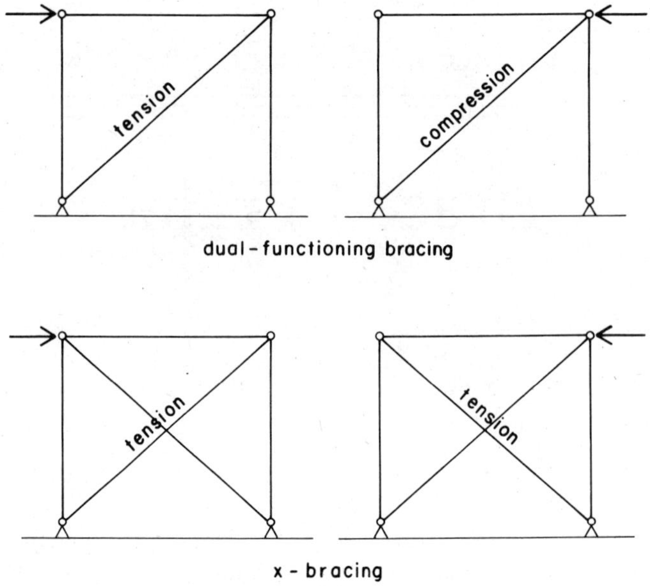

FIGURE 17.1. Dual-functioning trussing versus X-bracing.

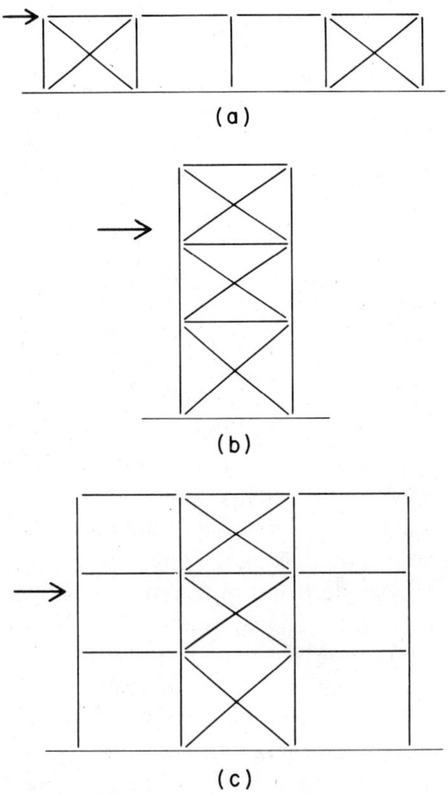

FIGURE 17.2. X-bracing of frames.

more common type of frame for the multistoried building, as shown in Fig. 17.2*c*, is usually only partly braced. Since either the single diagonal or the crisscrossed X-bracing causes obvious problems for interior circulation and for openings for doors and windows, building planning often makes the limited bracing a necessity.

Just about any type of floor construction used for multistoried buildings usually has sufficient capacity for diaphragm action in the lateral bracing system. Roofs, however, often utilize light construction or are extensively perforated by openings, so that the basic construction is not capable of the usual horizontal, planar diaphragm action. For such roofs or for floors with many openings, it may be necessary to use a trussed frame for the horizontal part of the lateral bracing system. Figure 17.3 shows a roof for a single-story building in which trussing has been placed in all the edge bays of the roof framing in order to achieve the horizontal structure necessary. As with vertical trussed frames, the horizontal trussed frame may be partly

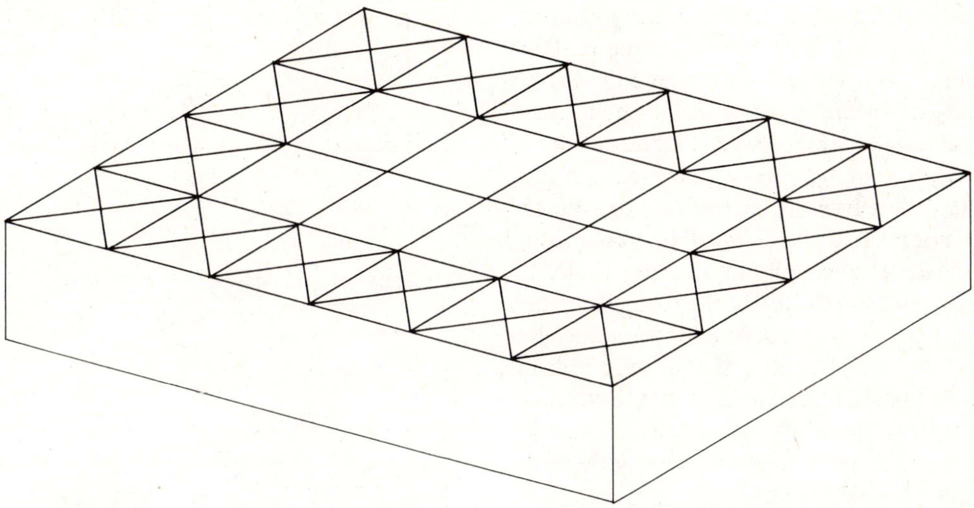

FIGURE 17.3. Horizontally braced frame—edge trussed.

trussed, as shown in Fig. 17.3, rather than fully trussed.

For single-span structures, trussing may be utilized in a variety of ways for the combined gravity and lateral load resistive system. Figure 17.4*a* shows a typical gable roof with the rafters tied at their bottom ends by a horizontal member. The tie, in this case, serves the dual functions of resisting the outward thrust due to gravity loads, and of one of the members of the single triangle, trussed structure that is rigidly resistive to lateral loads. Thus the wind force on the sloping roof surface, or the horizontal seismic force caused by the weight of the roof structure, is resisted by the triangular form of the rafter–tie combination.

The horizontal tie shown in Fig. 17.4*a* may not be architecturally desirable in all cases. Some other possibilities for the single-span structure—all producing more openness beneath the structure—are shown in Figs. 17.4*b, c,* and *d*. Figure 17.4*b* shows the so-called scissors truss, which can be used to permit more openness on the inside, or to permit a ceiling that has a form reflecting that of the gable

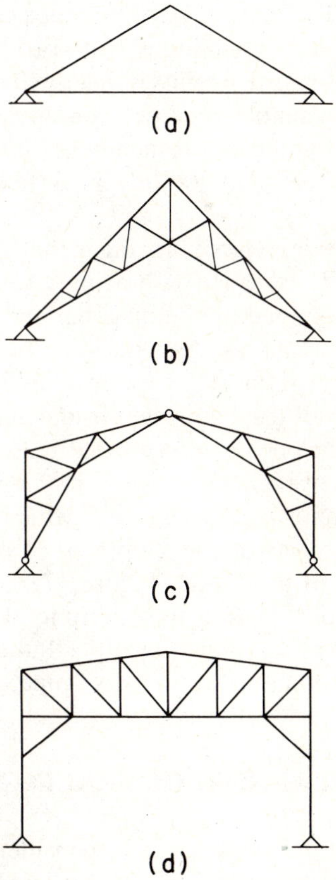

FIGURE 17.4. Forms of single-span trussed frames.

roof. Figure 17.4*c* shows a trussed bent that is a variation on the three-hinged arch. The structure shown in Fig. 17.4*d* consists primarily of a single-span truss that rests on end columns. If the columns are pin-jointed at the bottom chord of the truss, the structure lacks basic resistance to lateral loads and must be separately braced. If the column in Fig. 17.4*d* is continuous to the top of the truss, it can be used in rigid frame action for resistance to lateral loads. Finally, if the knee-braces shown in the figure are added, the column is further stiffened, and the structure has more load resistance and less deflection under lateral loading.

The knee-braces shown in Fig. 17.4*d* can also be used with a simple post and beam structure, as shown in Fig. 17.5, transforming it essentially into a rigid frame. The advantage of the knee-braces over other techniques for producing moment-resistive joints is that the jointing can be of simple pin type. This advantage is most significant for members for which moment-resistive jointing is difficult to develop.

A special type of bracing is that shown in Fig. 17.6, called K-bracing (from the form described by the diagonal and horizontal members). As with knee-bracing, the action of this structure involves a combination of truss and rigid frame actions. The principal advantage—and reason for its current popularity—of K-bracing is in its ultimate energy capacity, which typically involves the development of plastic hinging in the frame. Figure 17.7 shows the use of K-bracing for a peripheral bracing system for a low-rise office buildings, with diagonals of round steel pipe.

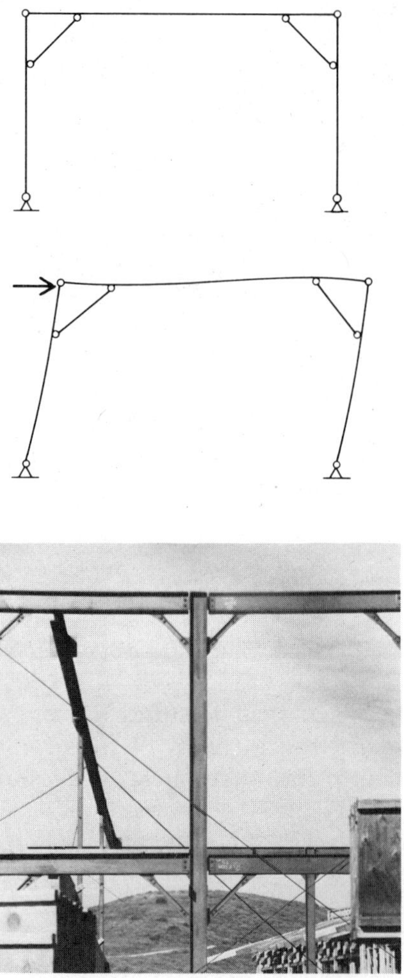

FIGURE 17.5. The knee-braced bent.

17.2 PLANNING OF BRACING

Some of the problems to be considered in using braced frames are the following:

1. Diagonal members must be placed so as not to interfere with the action of the gravity-resistive structure or with other building functions. If the bracing members are designed essentially as axial stress members, they must be located and attached so as to avoid loadings other than those required for their bracing functions. They must also be located so as not to interfere with door, window, or roof openings or with ducts, wiring, piping, light fixtures, and so on.

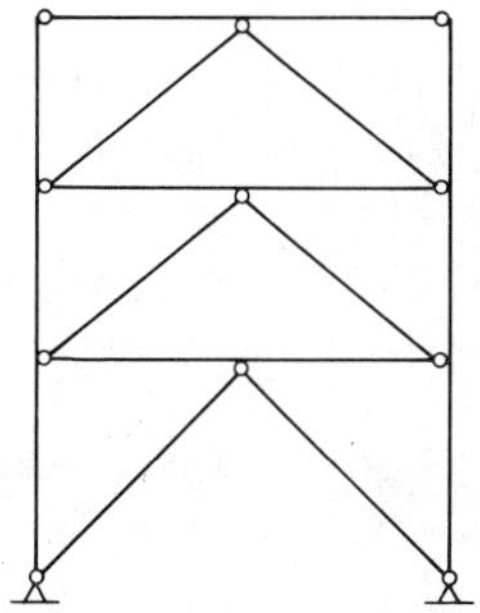

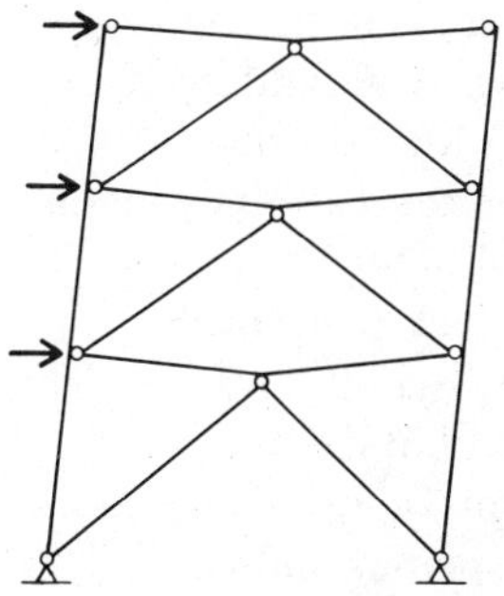

FIGURE 17.6. K-bracing.

FIGURE 17.7. Use of K-bracing for a peripheral bracing system for a low-rise office building.

2. As mentioned previously, the reversibility of the lateral loads must be considered. As shown in Fig. 17.1, such consideration requires that diagonal members be dual functioning (as single diagonals) or redundant (as X-bracing) with one set of diagonals working for load from one direction and the other set working for the reversal loading.
3. Although the diagonal bracing elements usually function only for lateral loading, the vertical and horizontal elements must be considered for the various possible combinations of gravity and lateral load. Thus the total frame must be analyzed for all the possible loading conditions, and each member must be designed for the particular critical combinations that represent its peak response conditions.
4. Long, slender bracing members, especially in X-braced systems, may have considerable sag due to their own dead weight, which requires that they be supported by sag rods or other parts of the structure.
5. The trussed structure should be "tight." Connections should be made in a manner to assure that they will be initially free of slack and will not loosen under the load reversals or repeated loadings. This means generally avoiding connections that tend to loosen or progressively deform, such as those that use nails, loose pins, unfinished bolts, and so on.
6. In order to avoid loading on the diagonals, the connections of the diagonals are sometimes made only after the gravity-resistive structure is fully assembled and at least partly loaded by the building dead loads.
7. The deformation of the trussed structure must be considered, and it may relate to its function as a distributing

element, as in the case of a horizontal structure, or to the establishing of its relative stiffness, as in the case of a series of vertical elements that share loads. It may also relate to some effects on nonstructural parts of the building, as was discussed for shear walls.

8. In most cases it is not necessary to brace every individual bay of the rectangular frame system. In fact, this is often not possible for architectural reasons. As shown in Fig. 17.2, walls consisting of several bays can be braced by trussing only a few bays, or even a single bay, with the rest of the structure tagging along like cars in a train.

The braced frame can be mixed with other bracing systems in some cases. Figure 17.8 shows the use of a braced frame for the vertical resistive structure in one direction and a set of shear walls in the other direction. In this example the two systems act independently, except for the possibility of torsion, and there is no need for a deflection analysis to determine the load sharing.

Figure 17.9 shows a structure in which the end bays of the roof framing are X-braced. For loading in the direction shown, these braced bays take the highest shear in the horizontal structure, allowing the deck to be designed for a lower shear stress.

Figure 17.10 shows a low-rise office building in which X-braced steel bents are used in combination with wood-framed shear walls for the lateral bracing system. The detail shown in Fig. 17.11 illustrates the typical use of steel gusset plates welded to the vertical and horizontal framing members for attachment of the diagonal braces. In this case the diagonal members consist of steel channel sections, turned back-to-back to form the X-braces.

Although buildings and their structures are often planned and constructed in two-dimensional components (horizontal floor and roof planes and vertical wall or framing bent planes), it must be noted that the building is truly three dimensional. Bracing against lateral forces is thus a three-dimensional problem, and although a single horizontal or vertical plane of the structure may be adequately stable and strong, the whole system must interact

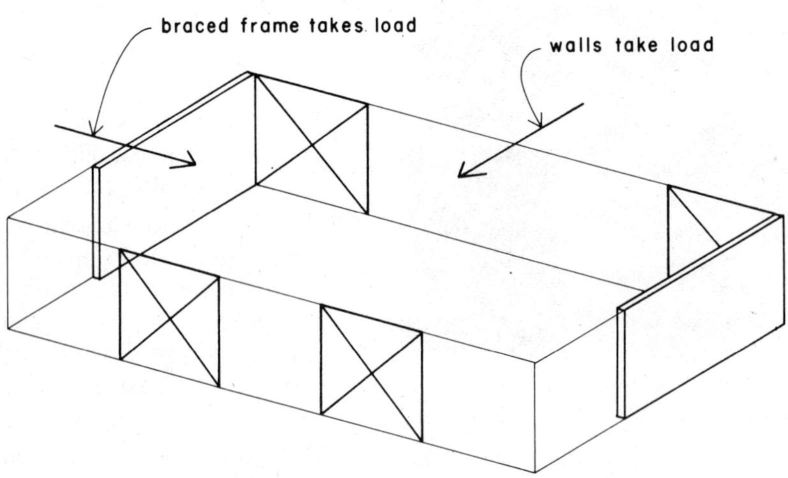

FIGURE 17.8. Mixed vertical elements for lateral resistance.

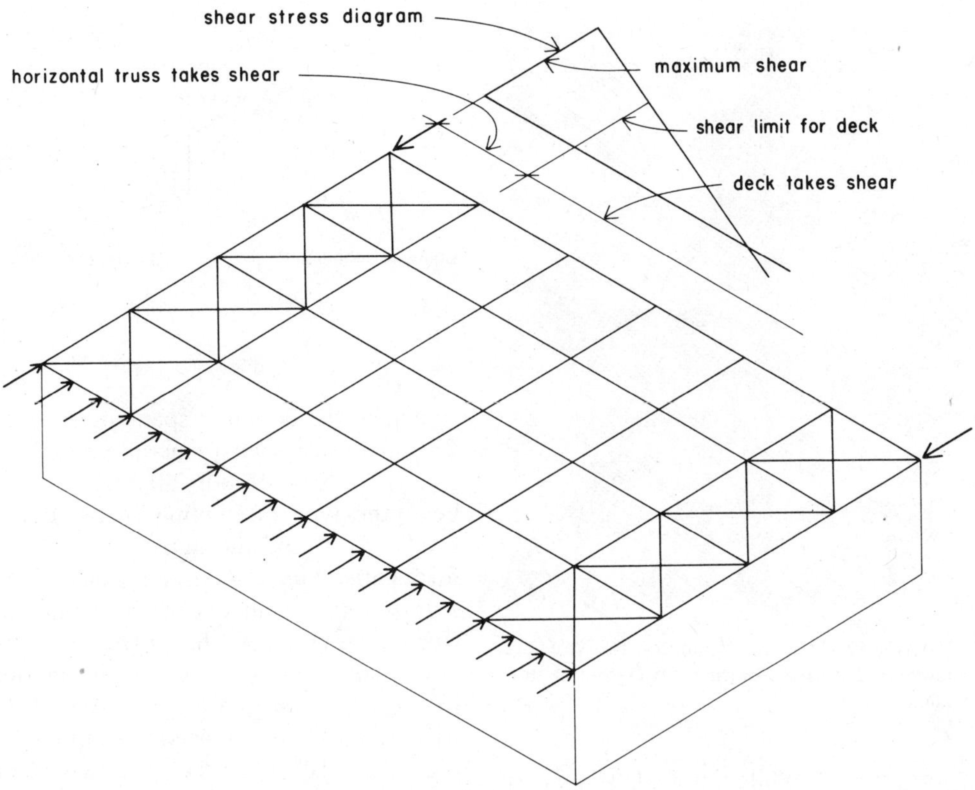

FIGURE 17.9. Mixed horizontal diaphragm and braced frame.

FIGURE 17.10. X-braced bents used for lateral resistance in a mixed steel and wood frame structure.

FIGURE 17.11. Joint detail for the structure shown in Fig. 17.10. X-bracing of back-to-back steel channels.

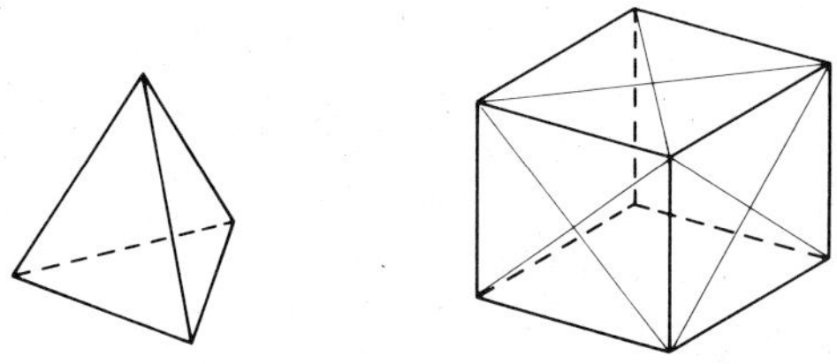

FIGURE 17.12. Three-dimensional trussing.

appropriately. While the single triangle is the basic unit for a planar truss, the three-dimensional truss may not be truly stable just because its component planes are braced.

In a purely geometric sense the basic unit for a three-dimensional truss is the four-sided figure called a tetrahedron. However, since most buildings consist of spaces that are rectangular boxes, the three-dimensional trussed building structure usually consists of rectangular units, rather than multiples of the pyramidal tetrahedral form (see Fig. 17.12). When so used, the single planar truss unit is much the same as a solid planar wall or deck unit, and general reference to the box-type system typically includes both forms of construction.

17.3 TYPICAL CONSTRUCTION

Development of the details of construction for trussed bracing is in many ways similar to the design of spanning trusses. The materials used (generally wood or steel), the form of individual truss members, the type of jointing (nails, bolts, welds, etc.), and the magnitudes of the forces are all major considerations. Since many of the members of the complete truss serve dual roles for gravity and lateral loads, member selection is seldom based on truss action alone. Quite often trussed bracing is produced by simply adding diagonals (or X-bracing) to a system already conceived for the gravity loads and for the general development of the desired architectural forms and spaces.

Figure 17.13 shows some details for wood framing with added diagonal members. Wood framing members are most often rectangular in cross section and metal connecting devices of various form are used in the assembly of frameworks (see also material in Chapter 30.) Figure 17.13*a* shows a typical beam and column assembly with diagonals consisting of pairs of wood members bolted to the frame. When X-bracing is used, and members need take only tension forces, slender steel rods may be used, a possible detail for which is shown at Fig. 17.13*b*. For the wood diagonal an alternative to the bolted connection is the type of joint shown in Fig. 17.13*c*, employing a gusset plate to attach single members all in a single plane. If architectural detailing makes the protruding members shown in Fig. 17.13*a*, or even the protruding gussets in Fig. 17.13*c*,

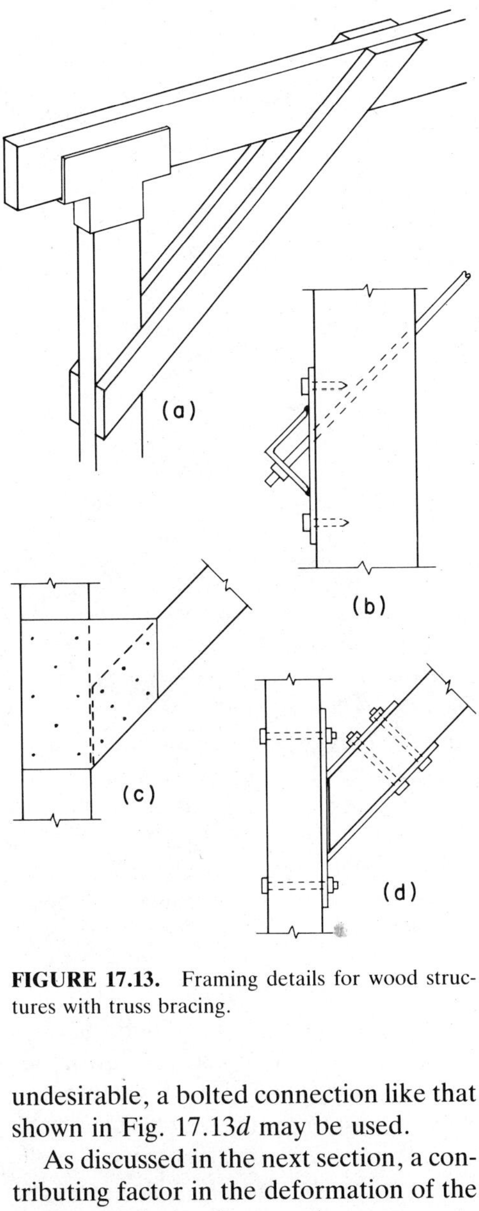

FIGURE 17.13. Framing details for wood structures with truss bracing.

undesirable, a bolted connection like that shown in Fig. 17.13*d* may be used.

As discussed in the next section, a contributing factor in the deformation of the bracing under loading may be movements within the connections. Bolted connections are especially vulnerable when used in shear resistance, since both oversizing of the holes and shrinkage of the wood contribute to a lack of tightness in the joints. In some cases it may be possible to increase the tightness of the joints by using some form of shear developer, such as steel split rings (see Fig. 17.14).

Gusset plates ordinarily consist of plywood, sheet steel, or steel plate, depending mostly on the magnitude of the loads. Plywood joints should be glued or the nails should be ring or spiral shafted to increase the joint tightness. Steel plate gussets are usually attached by either lag screws or through bolts. Thin sheet metal gussets are either nailed or screwed in place, the latter being preferred for maximum tightness.

Figure 17.15 shows some details for the incorporation of diagonal bracing in steel frames. As with wood structures, bolt loosening is a potential problem. For bolts used in tension—or for the threaded ends of round steel rods—loosening of nuts can be prevented by welding them in place or by scarring the threads. For shear-type connections, highly tensioned, high-strength bolts are preferred over ordinary, unfinished bolts. A completely welded connection will produce the stiffest joint, but on-site bolting in the field is usually preferred over welding.

Various steel elements can be used for diagonal members, depending on the magnitude of loads, the problems of incorporating or exposing the members in the construction, and the requirements for attachment to the structural frame. Figure 17.16 shows an interior view of a building in which a system of exposed truss bents is used for the roof structure as well as the lateral bracing system. Columns are round steel pipes and truss members are mostly double angles with welded gusset plate joints, as shown in the detail in Fig. 17.17.

Use of round steel pipe diagonals is shown in Fig. 17.7. Steel channels are used for X-bracing in the building shown in Fig. 17.11.

17.4 STIFFNESS AND DEFLECTION

As has been stated previously, the braced frame is typically a relatively stiff struc-

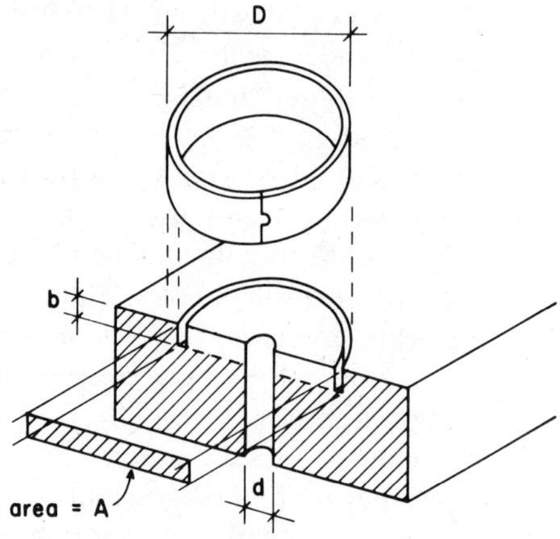

D	=	2.5"	4"
bolt size	=	$^1/_2$"	$^3/_4$"
d	=	$^9/_{16}$"	$^{13}/_{16}$"
b	=	$^3/_8$"	$^1/_2$"
A	=	1.10 in^2	2.24 in^2

FIGURE 17.14. Split-ring connector for shear development in a bolted wood joint.

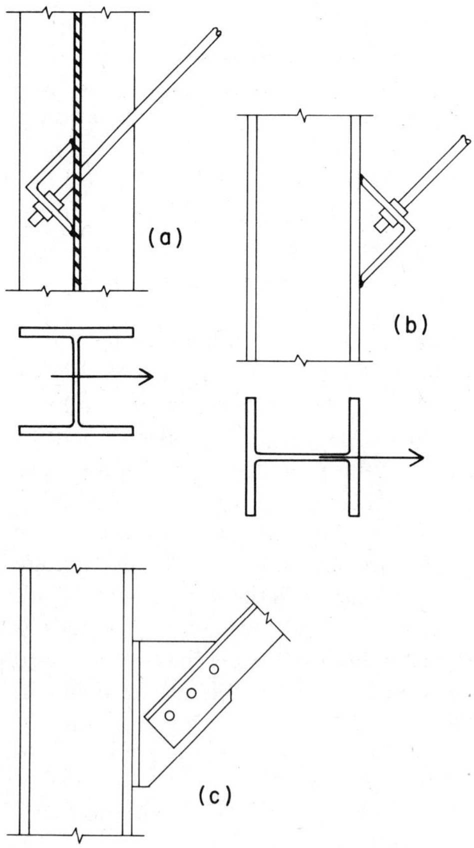

FIGURE 17.15. Framing details for steel structures with truss bracing.

ture. This is based on the assumption that the major contribution to the overall deformation of the structure is the shortening and lengthening of the members of the frame, as they experience the tension and compression forces due to the truss action. There are, however, two other potentially significant contributions to the movement of the braced frame, either or both of which may be of major concern. These are:

1. *Movement of the Supports.* This includes the possibilities of deformation of the foundations and yielding of the anchorage connections. If the foundations rest on compressible soil, there will be some movement due to soil stress, as shown in Fig. 10.4, a common problem for all lateral bracing systems. This, together with other foundation problems, is discussed in Chapter 20. Deformation of the anchorage may be due to a combination of lengthening of anchor bolts and bending of column base plates.
2. *Deformation in the Frame Connections.* This is a complex problem, having to do with the general nature

FIGURE 17.16. Building with an exposed steel structure consisting of a two-way trussed bent system. Public library for the City of Thousand Oaks, California. Architects: Albert Martin, Los Angeles.

FIGURE 17.17. Joint detail for the structure shown in Fig. 17.16.

of the connection (welds or glue versus bolts or nails, for example) as well as its form and layout and the deformation of the parts being connected.

It is good design practice in general to study the connection details for braced frames with an eye toward reduction of deformation within the connections. As has been mentioned previously, this generally favors the choice of welding, gluing, high-strength bolts, wood screws, and other fastening techniques that tend to produce stiff, tight joints. It may also favor the choice of materials or form of the frame members as these choices may affect the deformations within the joints or the choice of connecting methods.

The deflection, or drift, of single-story X-braced frames is usually caused primarily by the tension elongation of the diagonal X members. As shown in Fig. 17.18, the elongation of one diagonal moves the rectangular framed bay into a parallelogram form. The approximate value of the deflection, d in Fig. 17.18, can be derived as follows.

Assuming the change in the angle of the diagonal, $\Delta\Theta$ in the figure, to be quite small, the change in length of the diagonal may be used to approximate one side of

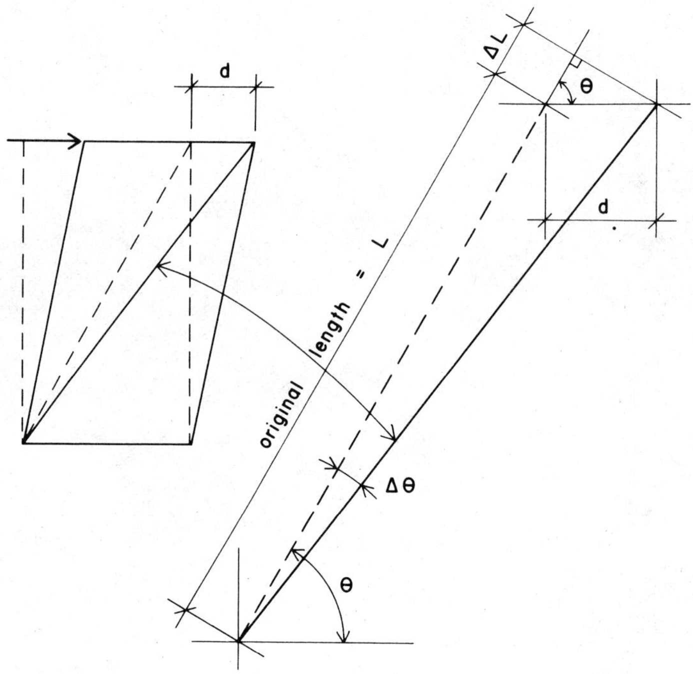

FIGURE 17.18. Horizontal deflection (drift) of a trussed structure.

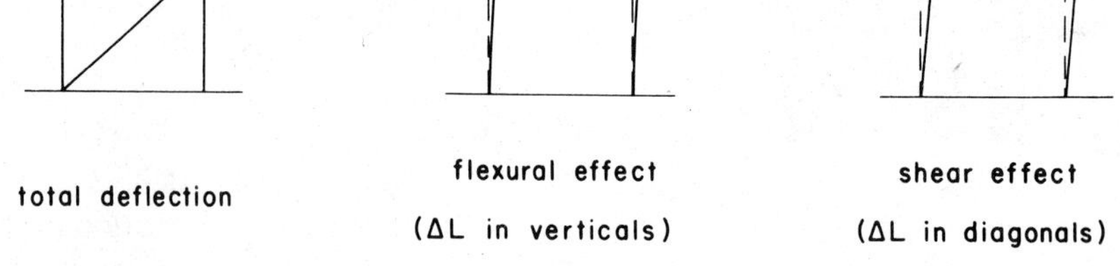

FIGURE 17.19. Horizontal deflection of a multistory trussed frame.

the triangle of which d is the hypotenuse. Thus:

$$d = \frac{\Delta L}{\cos \Theta} = \frac{T/AE}{\cos \Theta} = \frac{T}{AE \cos \Theta}$$

in which: T = the tension in the X caused by the lateral load
A = the cross-sectional area of of the X
E = the elastic modulus of the X

Θ = the angle of the X from the horizontal

The deflection of multistory X-braced frames has two components, both of which may be significant. As shown in Fig. 17.19, the first effect is caused by the change in length of the vertical members of the frame as a result of the overturning moment. The second effect is caused by the elongation of the diagonal X, as discussed for the single-story frame. These deflections occur in each level of the frame and can be calculated individually and summed up for the whole frame. Although this effect is also present in the single-story frame, it becomes more pronounced as the frame gets taller with respect to its width. These deflections of the cantilever beam can be calculated using standard formulas, such as those given in the Beam Diagrams and Formulas in Section 2 of the *Manual of Steel Construction* (Ref. 16).

CHAPTER EIGHTEEN

Moment-Resistive Frames

There is some confusion over the name to be used in referring to frames in which interactions between members of the frame include the transfer of moments through the connections. In years past the term most frequently used was *rigid frame*. This term came primarily from the classification of the connections or joints of the frame as *fixed* (or rigid) versus *pinned*, the latter term implying a lack of capability to transfer moment through the joint. As a general descriptive term, however, the name was badly conceived, since the frames of this type were generally the most deformable under lateral loading, when compared to trussed frames or those braced by vertical diaphragms. The *Uniform Building Code* (Ref. 2) uses the specific term *ductile moment-resisting space frame* and gives various qualifications for such a frame when it is used for seismic resistance. With apologies to the *UBC*, although we will assume the type of frames they thus define, we prefer not to use this rather cumbersome mouthful of a term, so will use the simpler term of rigid frame in our discussions.

18.1 GENERAL BEHAVIOR

In rigid frames with moment-resistive connections, both gravity and lateral loads produce interactive moments between the members. The *UBC* requires that a rigid frame designed for seismic loading be classified as a "ductile moment-resistive space frame." Generally, frames of steel possess this character, but frames of reinforced concrete require special consideration of the reinforcing in order to meet this qualification.

In most cases rigid frames are actually the most flexible of the basic types of lateral resistive systems. This deformation character, together with the required ductility, makes the rigid frame a structure that absorbs energy loading through deformation, as well as through its sheer brute strength. The net effect is that the structure actually works less hard in force resistance because its deformation tends to soften the loading. This is somewhat like rolling with a punch instead of bracing yourself to take it head on.

Most moment-resistive frames consist of either steel or concrete. Steel frames have either welded or bolted connections between the linear members to develop the necessary moment transfers. Frames of concrete achieve moment connections through the monolithic concrete and the continuity and anchorage of the steel reinforcing. Because concrete is basically brittle and not ductile, the ductile character is essentially produced by the ductility of the reinforcing. The type and amount of reinforcing and the details of its placing become critical to the proper behavior of rigid frames of reinforced concrete.

A complete presentation of the design of moment-resistive ductile frames for seismic loads is beyond the scope of this book. Such design can be done only by using plastic design for steel and ultimate strength design for reinforced concrete. We present only a brief discussion of this type of structure. For wind loading the analysis and design may be somewhat more simplified. However, if the structure is considerably indeterminate, an accurate analysis requires a complex and laborious calculation. We show some examples in Part Four but limit the analysis to approximate methods.

For lateral loads in general, the rigid frame offers the advantage of a high degree of freedom in architectural terms. Walls and interior spaces are freed of the necessity for solid diaphragms or diagonal members. For building planning as a whole, this is a principal asset. Walls, even where otherwise required to be solid, need not be of a construction qualifying them as shear walls.

When seismic load governs as the critical lateral load, the moment-resistive frame has the advantage of being required to carry less load than other types of lateral resistive systems. Table 23-I in the *UBC* gives the value of *K* for a rigid frame as 0.67 versus 1.33 for a box system with shear walls or braced frames. Thus, the rigid frame is required to carry only half the load specified for the box system. Even when the rigid frame is interactive with a box system, and the rigid frame is capable of carrying at least 25% of the lateral load, the *K* factor is reduced to 0.80, or 40% less than that for a box system alone.

Deformation analysis is a critical part of the design of rigid frames because such frames tend to be relatively deformable when compared to other lateral resistive systems. The deformations have the potential of causing problems in terms of movements of a disturbing nature that can be sensed by the building occupants or of damage to nonstructural parts of the building, as previously discussed. The need to limit deformations often results in the size of vertical elements of the frame being determined by stiffness requirements, rather than by stress limits.

18.2 LOADING CONDITIONS

Unlike shear walls or X-bracing, rigid frames are not generally able to be used for lateral bracing alone. Thus their structural actions induced by the lateral loads must always be combined with the effects of gravity loads. These combined loading conditions may be studied separately in order to simplify the work of visualizing and quantifying the structural behavior, but it should be borne in mind that they do not occur independently.

Figure 18.1*a* shows the form of deformation and the distribution of internal bending moments in a single-span rigid frame, as induced by vertical gravity loading. If the frame is not required to resist lateral loads, the singular forms of these responses may be assumed, and the various details of the structure may be developed in this context. Thus the direction of rotation at the column base, the sign of moment at the beam-to-column joint, the sign of the bending moment and nature of the corresponding stresses at midspan of the beam, and the location of inflection points in the beam may all relate to choices for the form and details of the members and development of any connection details. If the frame is reasonably symmetrical, the only concerns for deflection are the outward bulging of the columns and the vertical sag of the beam.

Under action of lateral loading due to

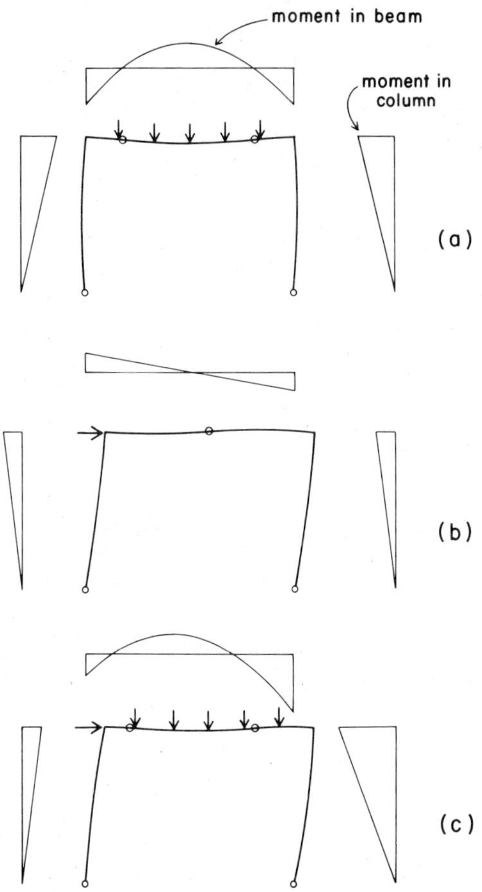

FIGURE 18.1. Gravity and lateral load effects on a single-span rigid frame.

wind or seismic force, the form of deformation and distribution of internal bending moment will be as shown in Fig. 18.1*b*. If the gravity and lateral loadings are combined, the net effect will be as shown in Fig. 18.1*c*. Observing the effects of the combined loading, we note the following:

1. Horizontal deflection at the top of the frame (called drift) must now be considered, in addition to the deflections mentioned previously for gravity load alone.
2. The maximum value for the moment at the beam-to-column connection is increased on one side and reduced on the other side of the bent. The increased moment requires that the beam, the column, and the connection at the joint must all be stronger for the combined loading.
3. If the lateral load is sufficient, the minimum value for the moment at the beam-to-column joint may be one of opposite sign from that produced by gravity loading alone. The form of the connection, and possibly the design of the members, may need to reflect this reversal of the sense of the moment.
4. The direction of the lateral load shown in Fig. 18.1*b* is reversible so that two combinations of load must be considered: gravity plus lateral load to the right and gravity plus lateral load to the left.

While single-span rigid frames are often used for buildings, the multispan or multistory frame is the more usual case. Figure 18.2 shows the response of a two-bay, two-story frame to lateral loads and to a gravity-type load applied to a single beam. The response to lateral loads is essentially similar to that for the single bent in Fig. 18.1. For gravity loads the multiunit frame must be analyzed for a more complex set of potential combinations, since the live load portion of the gravity loads must be considered to be random, and thus may or may not occur in any given beam span.

Lateral loads produced by winds will generally result in the loading condition shown in Fig. 18.2*a*. Because of its relative flexibility and size, however, a multistory building frame may quite likely respond so slowly to seismic motions that upper levels of the frame experience a whiplashlike effect; thus separate levels may be moving in opposite directions at a single moment. Figure 18.3 illustrates a type of response that may occur if the two-story frame experiences this action. Only a true dynamic

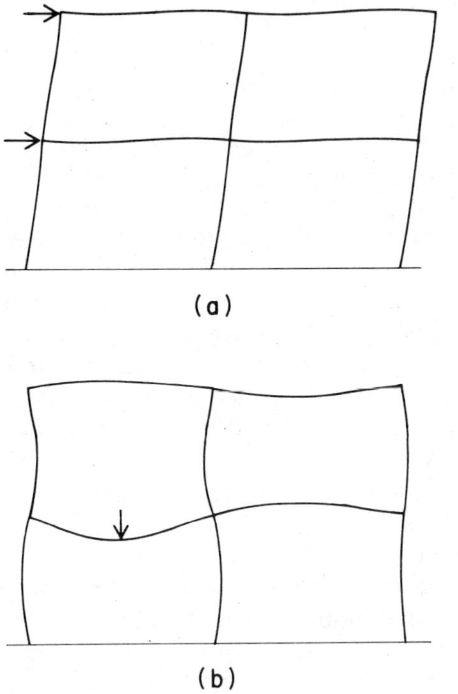

FIGURE 18.2. Frame deformations in a multiunit rigid frame.

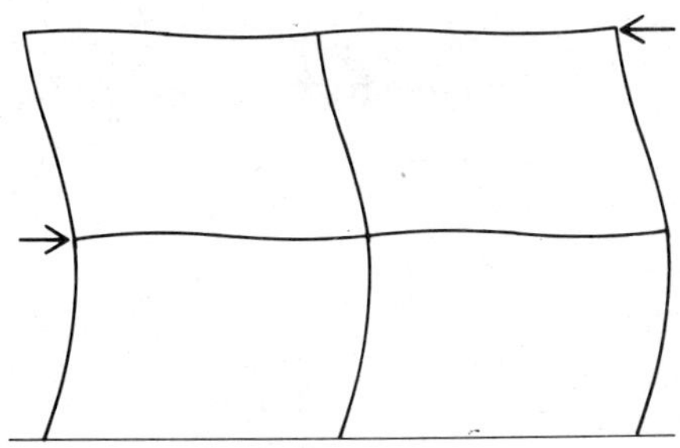

FIGURE 18.3. Effects of rapid reversal of lateral loads.

analysis can ascertain whether this action occurs and is of critical concern for a particular structure.

18.3 APPROXIMATE ANALYSIS FOR GRAVITY LOADS

Most rigid frames are statically indeterminate and require the use of some method beyond simple statics for their analysis. Simple frames of few members may be analyzed by some *hand* method, using handbook coefficients, moment distribution, and so on. If the frame is complex, consisting of several bays and stories, or having a lack of symmetry, the analysis will be quite laborious unless performed with some computer-aided method.

For preliminary design it is often useful to have some approximate analysis, which can be fairly quickly performed. Internal forces, member sizes, and deflections thus determined may be used for a quick determination of the structural actions and the feasibility of some choices of systems and components. For the simple, single-bay bent shown in Fig. 18.4, the analysis for gravity loads is quite simple, since a single-loading condition exists (as shown) and the only necessary assumption is that of the relative stiffnesses of the beam and columns. For the frame with pin-based columns the analogy is made to a three-span beam on rotation-free supports. If the column bases are fixed, the end sup-

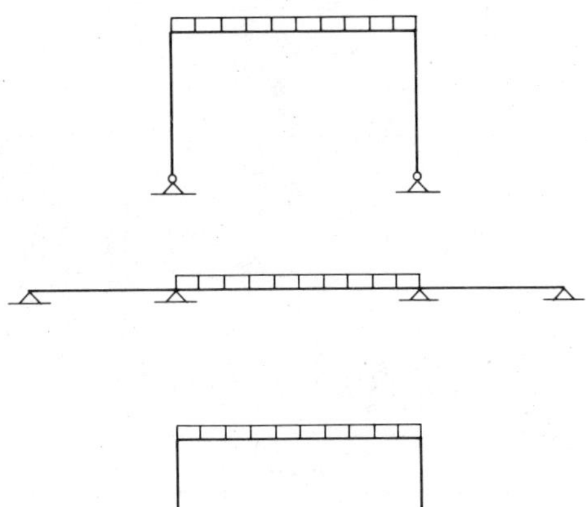

FIGURE 18.4. Continuous beam analogy for a single-span rigid frame.

ports of the analogous beam are assumed fixed.

For multibayed, multistoried frames, an approximate analysis may be performed using techniques such as that described in Chapter 8 of the ACI code (Ref. 14). This is more applicable to concrete frames, of course, but can be used for quick approximation of welded steel frames as well.

Even when using approximate methods, it is advisable to analyze separately for dead and live loads. The results can thus be combined as required for the various critical combinations of dead load, live load, wind load, and seismic load.

18.4 APPROXIMATE ANALYSIS FOR LATERAL LOADS

Various approximate methods may be used for the analysis of rigid frames under statically applied lateral loading. For ordinary frames, whether single-bay, multibay, or even multistory, approximate methods are commonly used for loading due to wind or as obtained from an equivalent static load analysis for seismic effects. More exact analyses are possible, of course, especially when performed by computer-aided methods.

For frames that are complex—due to irregularities, lack of symmetry, tapered members, and so on—analysis is hardly feasible without the computer. This is also true for analyses that attempt to deal with the true dynamic behavior of the structure under seismic load. With the increasing availability of the software, and the accumulation of experience with its use, this type of analysis is becoming more widespread in use. For quick approximations for preliminary design, however, approximation methods are likely to continue in use for some time.

For the simple bent shown in Fig. 18.5, the effects of the single lateral force may be quite simply visualized in terms of the deflected shape, the reaction forces, and the variation of moment in the members.

If the columns are assumed to be pin based and of equal stiffness, it is reasonable to assume that the horizontal reactions at the base of the columns are equal, thus permitting an analysis by statics alone. If the column bases are assumed to

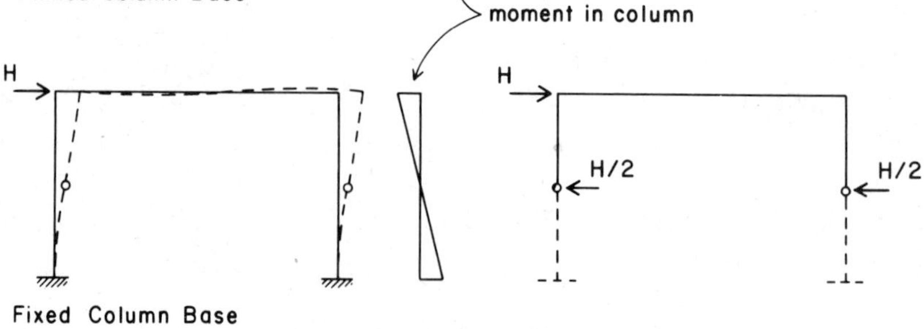

FIGURE 18.5. Effects of column base conditions on a single-span rigid frame.

be fixed, the frame is truly able to be analyzed only by indeterminate methods, although an approximate analysis can be made with an assumed location of the inflection point in the column. In truth, the column bases will most likely be somewhere between these two idealized conditions. Approximate designs are sometimes done by combining the results from both idealized conditions (pinned and fixed bases) and designing for both. Adjustments are made when the more precise nature of the condition is established by the detailed development of the base construction.

For multibayed frames, such as that shown in Fig. 18.6*a*, an approximate analysis may be done in a manner similar to that for the single-bay frame. If the columns are all of equal stiffness, the total load is simply divided by the number of columns. Assumptions about the column base condition would be the same as for the single-bay frame. If the columns are not all of the same stiffness, an approximate distribution can be made on the basis of relative stiffness.

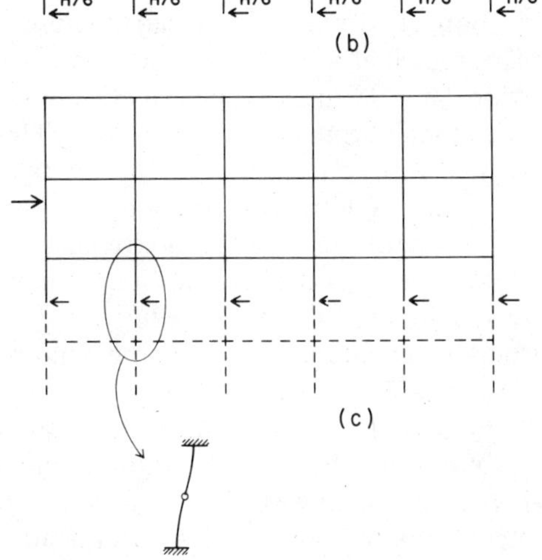

FIGURE 18.6. Distribution of lateral load in a multiunit rigid frame.

Figure 18.6*c* illustrates the basis for an approximation of the horizontal shear forces in the columns of a multistory building. As for the single-story frame, the individual column shears are distributed on the basis of assumed column stiffnesses. For the upper columns the inflection point is assumed to occur at midheight, unless column splice points are used to control its location.

18.5 GENERAL DESIGN CONSIDERATIONS

From a purely structural point of view, there are many advantages of the rigid frame as a lateral bracing system for resisting seismic effects. The current codes tend to favor it over shear walls or braced (trussed) frames by requiring less design load due to the lower assigned *K* value. For large frames the combination of slow reaction time (due to long fundamental period of vibration) and various damping effects means that forces tend to dissipate rapidly in the remote portions of the frame.

Architecturally, the rigid frame offers the least potential for interference with planning of open spaces within the building and in exterior walls. It is thus highly favored by architects whose design style includes the use of ordered rectangular grids and open spaces of rectangular or cubical form. Geometries other than rectangular ones are possible also; what remains as the primary advantage is the absence of the need for diagonal bracing or solid walls at determined locations.

Some of the disadvantages of rigid frames are the following:

1. Lateral deflection, or drift, of the structure is likely to be a problem. As a result, it is often necessary to stiffen the frame—mostly by increasing the number and/or size of the columns.
2. Connections must be stronger, especially in steel frames. Addition of heavily welded steel joints, of additional reinforcing for bar anchorage, shear, and torsion in concrete frames, and for cumbersome moment-resistive connections in wood frames, can add appreciably to the construction cost and time.
3. For large frames dynamic behavior is often quite complex, with potential for whiplash, resonance, and so on. Dynamic analysis is more often required for acceptable design when a rigid frame is used. In time, dynamic analysis may be routine and easily performed with available aids on all structures; at the present it is quite expensive and time consuming.

When rigid frames are to be used for the lateral bracing system for a building, there must be a high degree of coordination between the planner of the building form and the structural designer. As discussed in Chapter 2 and illustrated in the examples in Part 4, the building plan must accommodate some regularity and alignment of the columns that constitute the vertical members of the rigid frame bents. This may not include *all* of the building columns, but those that function in the rigid frames must be aligned to form the planes of the bents. Openings in floors and roofs must be planned so as not to interrupt beams that occur as horizontal members of the bents.

There must, of course, be a coordination between the designs for lateral and gravity loads. If a rigid frame is used for bracing, it will in most cases be used in conjunction with some type of framed floor and roof systems. Horizontal members of the floor and roof systems designed for gravity loads will also be used as horizontal elements of the rigid frames. However, in some cases, much of the horizontal structure may be used only for gravity loads. In the system that utilizes only the bents that occur in exterior walls, none of the interior portion of the framing is involved in the rigid frame actions; thus its planning is free of concerns for the bent development.

Selection of members and construction details for rigid frames depends a great deal on the materials used. The following sections will discuss separately the problems of bents of steel and reinforced concrete.

18.6 STEEL FRAMES

Steel frames with moment-resistive connections were used for early skyscrapers. Fasteners consisted of rivets, which were widely used until the development of high-strength bolts. Today, rigid frame construction in steel mostly utilizes welded joints, although bolting is sometimes used for temporary connection during erection. Figure 18.7 shows the erected frame for a low-rise office building, using a steel frame for both gravity and lateral load resistance. The principal members of the frame are wide flange (I-shape) rolled steel sections, and moment connections are welded. This is the most common form of steel rigid frame for building construction.

Another form of steel rigid frame is the trussed bent. A single-span bent is illustrated in Fig. 17.4*d*, and the use of a four-span bent is described in the building design case study in Section 27.6.

For tall buildings a currently popular system is one that uses a peripheral bracing arrangement with closely spaced, stiff steel columns and heavy spandrel beams. This type of structure offers a major advantage in its overall stiffness in resistance to lateral drift (horizontal deflection).

FIGURE 18.7. Steel frame structure with a peripheral bracing system of rigid bents.

FIGURE 18.8.

18.7 REINFORCED CONCRETE FRAMES

Poured-in-place frames with monolithic columns and beams have a natural rigid frame action. For seismic resistance both columns and beams must be specially reinforced for the shears and torsions at the member ends. Beams in the column-line bents ordinarily use continuous top and bottom reinforcing with continuous loop ties that serve the triple functions of resisting shear, torsion, and compression bar buckling. Figure 18.8 shows two buildings with exposed concrete frames that are of this type of construction. It is possible, of course, to brace such a building with poured concrete shear walls and to use the frame strictly for gravity resistance, except for collector and chord functions.

Precast concrete structures are often difficult to develop as rigid frames, unless the precast elements are developed as individual bent units, instead of the usual, single, linear members. Moment-resistive joints for these structures are usually quite difficult to develop, the more common solution being to use a shear wall system for lateral bracing.

CHAPTER NINETEEN

Interaction of Frames and Diaphragms

Most buildings consist of combinations of walls and some framing of wood, steel, or concrete. The planning and design of the seismic-resistive structure requires some judgments and decisions regarding the roles of the frame and the walls. This chapter discusses some of the issues relating to this aspect of design.

19.1 COEXISTING, INDEPENDENT ELEMENTS

Most buildings have some solid walls, that is, walls with continuous surfaces free of openings. When the gravity load resistive structure of the building consists of a frame, the relationship between the walls and the frame has several possibilities with regard to action caused by lateral loads.

The frame may be a braced frame or a moment-resistive frame designed for the total resistance of the lateral loads, in which case the attachment of walls to the frame must be done in a manner that prevents the walls from absorbing lateral loads. Because solid walls tend to be quite stiff in their own planes, such attachment often requires the use of separation joints or flexible connections that will allow the frame to deform as necessary under the lateral loads.

The frame may be essentially designed for gravity resistance only, with lateral load resistance provided by the walls acting as shear walls. This method requires that some of the elements of the frame function as collectors, stiffeners, shear wall end members, or diaphragm chords. If the walls are intended to be used strictly for lateral bracing, care must be exercised in the design of construction details to assure that beams that occur above the walls are allowed to deflect without transferring loads to the walls.

19.2 LOAD SHARING

When walls are firmly attached to vertical elements of the frame, they usually provide continuous lateral bracing in the plane of the wall, thus permitting the vertical frame elements to be designed for column action using their stiffness in the direction perpendicular to the wall. Thus 2 × 4

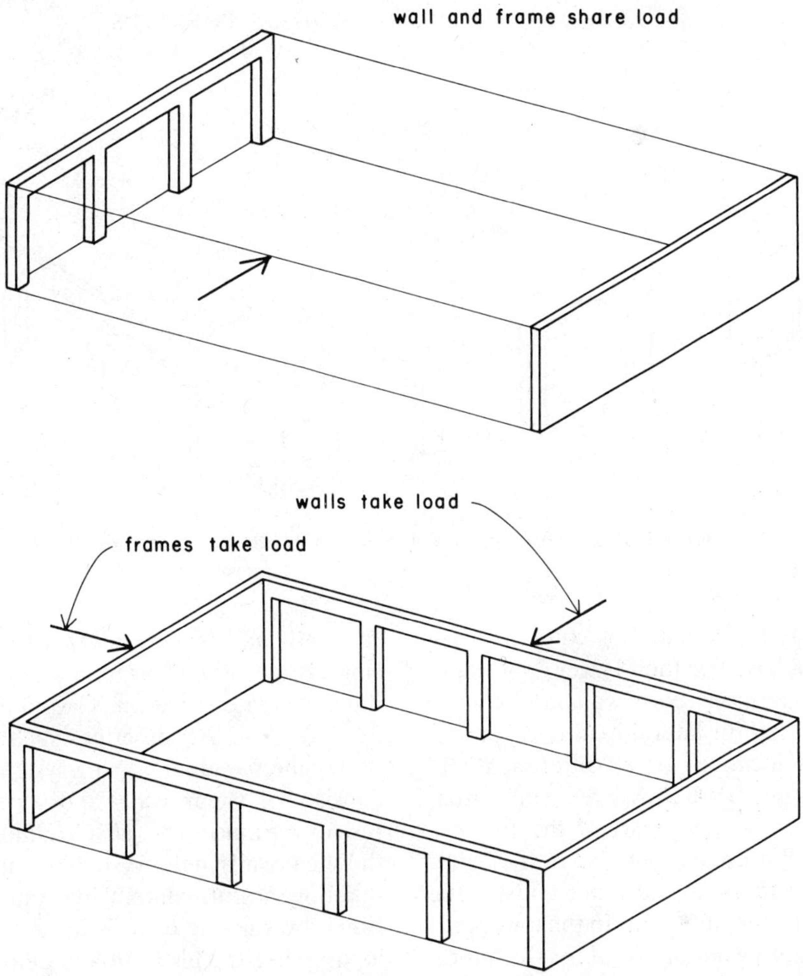

FIGURE 19.1. Mixed shear walls and rigid frames—no interaction.

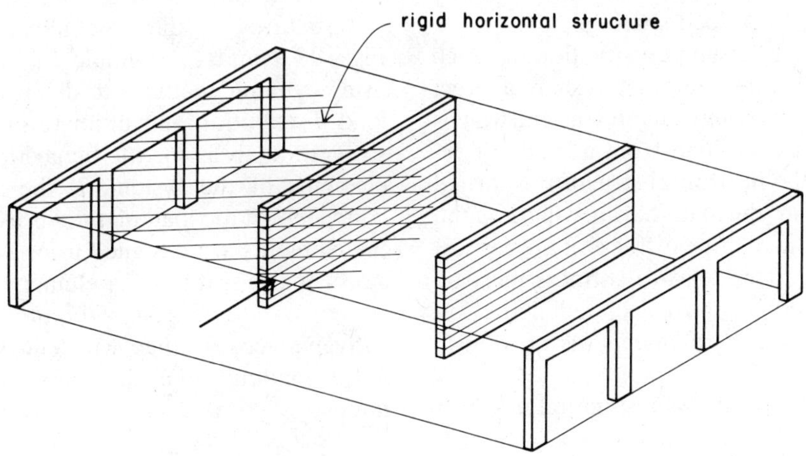

FIGURE 19.2. Load sharing by walls and frames.

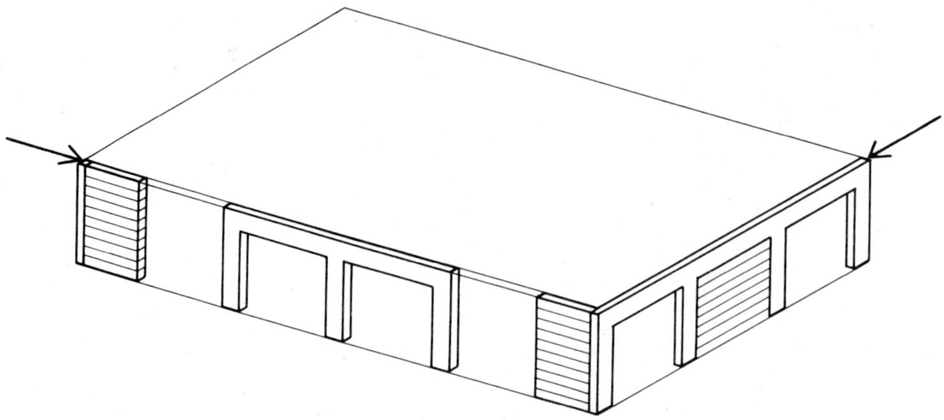

FIGURE 19.3. Mixed shear walls and rigid frames with interactions.

studs may be designed as columns using *h*/*d* ratios based on their larger dimension.

In some cases both walls and frames may be used for lateral load resistance at different locations or in different directions. Figure 19.1 shows two such situations. In the upper part of the figure a shear wall is used at one end of the building and a frame at the other end for the wind from one direction. In the lower part of the figure walls are used for the lateral loads from one direction and frames for the load from the other direction. In both cases the walls and frames do not actually interact; that is, they act independently with regard to load sharing.

Figure 19.2 shows a situation in which walls and frames interact to share a direct load. The interior walls and the end frames share the total load from a single direction. If the horizontal structure is a rigid diaphragm, the load sharing will be on the basis of the relative stiffness of the vertical elements. This relative stiffness must be established by the calculated deflection resistance of the elements, as previously discussed.

Figure 19.3 shows a situation in which the walls and frames interact to share a single direct load. This represents a highly indeterminate situation. Case 3 in Table 23−I of the *UBC* describes such a situation, called a "dual bracing system," that consists of shear walls and a moment-resistive frame. The *UBC* requires that the shear walls in this system be capable of resisting the entire lateral load but that the frame be capable of resisting 25% of the load by itself. This is an apparent redundancy in the design, but the advantage gained is that the K factor for the seismic load may be reduced from 1.33 for the walls alone to 0.80 for the dual system.

Structures of the type illustrated in Figs. 19.2 and 19.3 should be analyzed using dynamic methods to determine the load distribution and ultimate or plastic strength analysis for the behavior of the elements of the system. If the simpler equivalent static load method is used and the elements are designed using working stress analysis, the stiffer elements of the system should be designed quite conservatively, because they will tend to take more loading than the static analysis implies.

CHAPTER TWENTY

Foundations

Considerations for seismic actions must be combined with all of the other matters relating to foundation design. This chapter presents some of the issues that relate to problems of seismic action, but complete solution of foundation designs cannot be illustrated without the full context of the building design situation and the unique site conditions for a specific building.

20.1 GENERAL CONSIDERATIONS

For seismic load the foundations have a dual role. Initially, they are the origin point for load on the building, being directly attached to the ground and used by the ground to shake the building. In our analysis procedures, however, the seismic load is considered to be a result of the inertial effect of the moving building mass. From this viewpoint the loading condition becomes similar to that for the wind load. Thus we visualize the horizontal inertial force as being transmitted through the building structure, into the foundation, and finally into the ground, which is now considered to offer passive resistance, as it does actually to the wind force.

Foundation design problems have considerable variety because of the wide range of possible soil conditions, load magnitudes, building size and shape, and type of structural system. For any building it is wise to have a subsoil investigation, lab tests on representative soil samples, and a recommendation from an engineer with experience in soil behaviors and foundation problems. We illustrate some of the ordinary and simple design problems in the examples in Part 4 but do not attempt to deal with all the special problems that can occur in the design of foundations for lateral loads.

For seismic loads it is usually desirable that the entire building foundation act essentially as a single rigid unit. If elements of the foundation are isolated from one another, as in the case of individual column footings, it may be necessary to provide struts or grade walls to tie the structure into a unit. Where they exist, of course, the ordinary elements such as basement walls, grade walls, wall footings, and grade-level framing members may be used for this tying function.

For buildings with exceptional uplift or overturn effects, the foundation must usually provide an anchor through its own sheer dead weight. For a large building with shallow foundations and no basement, the necessary dead load may not exist in the foundations designed for gravity loads alone. This situation may require some additional mass in the foundation itself or the use of soil weight in the form of backfill over the foundation.

Another problem with shallow foundations is that they often bear on relatively compressible soil. Figure 10.4 illustrates the effect that can occur with a large lateral load and a strong overturning moment. With repeated applications and reversals of such a load, the soil beneath the foundation edges becomes compressed, resulting in an increasing tendency for the structure to rock, thus producing a loss of stability or a significant change in the dynamic load behavior of the structure. When the bearing strata of soil are very compressible, it is generally advisable to avoid the extreme condition of soil stress shown in the illustration.

20.2 DESIGN CRITERIA AND DATA

Chapter 29 of the *UBC* provides some data and recommendations for use in designing for sliding or lateral passive resistance of soil. The *UBC* material is quite conservative in most cases, and a complete soils investigation and report may provide more rational data and permit a less conservative design. In many cases soil conditions are general local phenomena, and local building codes often have special requirements and procedures for foundation design. Local design and construction procedures are sometimes based more on histories of successes and failures than on scientific analysis or engineering judgment.

20.3 DEEP FOUNDATIONS

Buildings on deep foundations consisting of long piles or piers offer special problems in design for lateral loads. The entire lateral force must usually be developed by the passive resistance of soil because the sliding friction present on the bottom of shallow bearing footings is not available. When the upper soil strata are considerably deformable, the fundamental period of the building and its general dynamic behavior may be significantly affected by the lateral movement of the foundation. It is highly recommended that the design of these foundations and the structures supported by them be done by an engineer with experience in these problems.

20.4 SPECIAL PROBLEMS

Many seismic load failures of building structures are precipitated or aggravated by soil movements in fill or other highly compressible soil deposits. An especially hazardous situation is that which occurs when different parts of the foundation are placed on soil of significantly different compressibility, as may occur because of a hilly site, variations in the level of the foundations, extensive regrading, or nonhorizontal soil strata. Subsidence of the fill, major differences in settlements, or lateral movement of the soil mass can produce various critical situations. Figure 20.1 illustrates some of the possibilities that can produce structural failures for both gravity and lateral load conditions.

Figure 20.2 illustrates the typical problem of tying isolated footings, which is done primarily for the purpose of ensuring that the building structure move as a single mass with respect to the ground. It is also sometimes done in order to allow for the sharing of the lateral load when the load on a single footing cannot be resisted by that footing alone.

In many cases the existence of basement walls, grade walls, wall footings, or other parts of the building substructure make the additon of ties unnecessary.

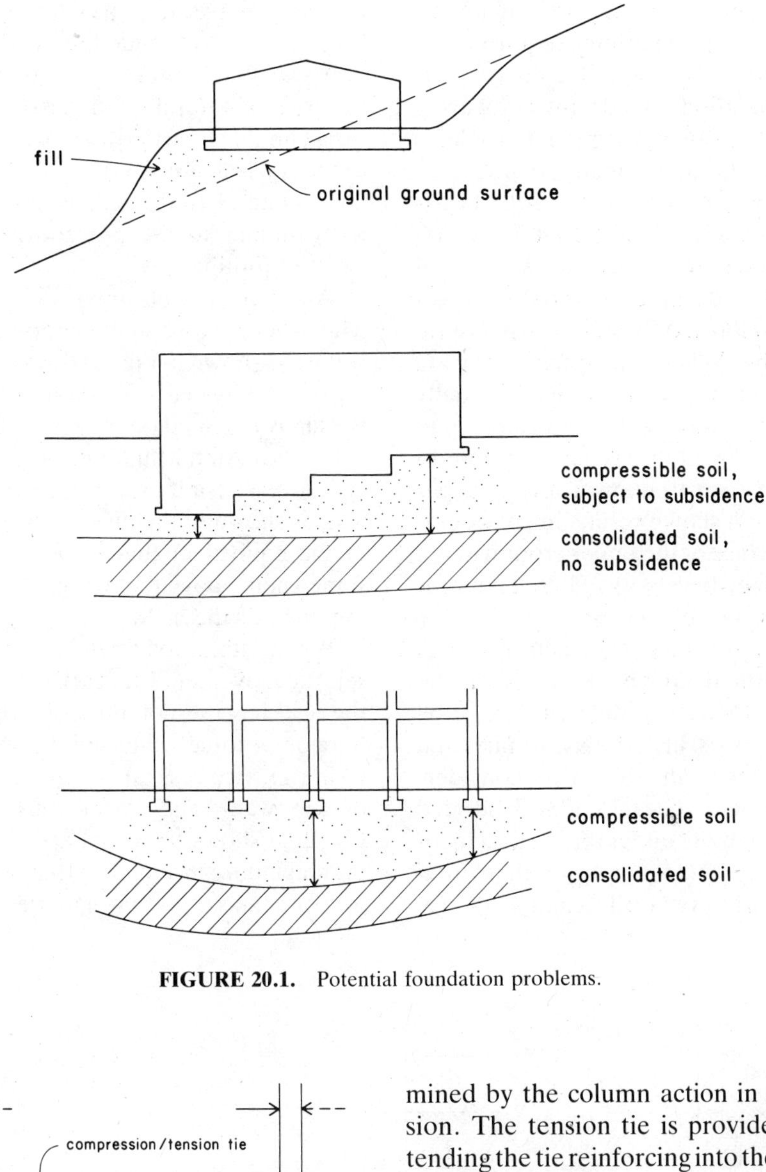

FIGURE 20.1. Potential foundation problems.

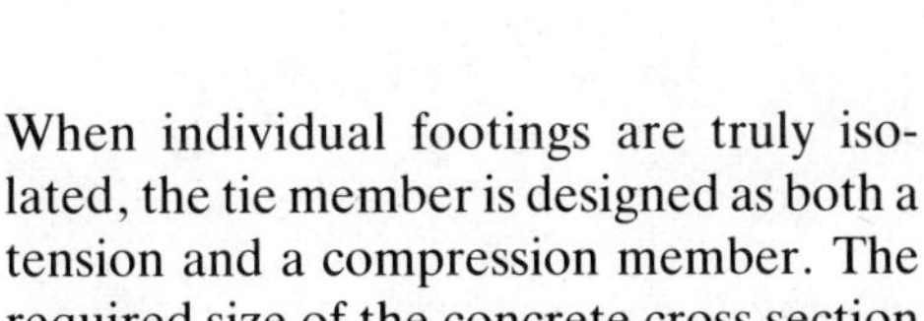

FIGURE 20.2. Tying of isolated footings.

When individual footings are truly isolated, the tie member is designed as both a tension and a compression member. The required size of the concrete cross section and the main reinforcing is usually determined by the column action in compression. The tension tie is provided by extending the tie reinforcing into the footings in dowel action.

When calculated lateral loads exist for the footings, they may be used as the design loads on the ties. Section 2312(j)3.B of the *UBC* requires ties for pile caps and caissons to be designed for a minimum load of 0.10 of the vertical load on the heaviest loaded element. For isolated footings under structural elements that are not part of the lateral resistive structural system, there is no quantifiable basis for the tie design. Such footings are usually designed using the minimum requirements for concrete compressive members.

For foundations that do not use bearing footings, there are some special problems with regard to lateral loads. Figure 20.3 shows three such situations: drilled caissons, driven piles, and pole-type structures. Direct resistance to horizontal movement by the ground in all these cases can be developed only as a passive lateral soil pressure. Where the upper soil strata are highly compressible, which is usually the case that causes piles and caissons to be used, this resistance is quite limited. In addition, the construction of all three types of foundation shown results in considerable disturbance of the upper ground mass, which further lowers its effectiveness for passive resistance.

The tying of the pole structure is usually performed above the level of the ground, by the rest of the structure or by added ties and struts. The use of piles and caissons usually involves two considerations: the actual lateral load and the need for lateral stability under the vertical load. Tying is required for both of these reasons, and the previously cited *UBC* requirement is based on the latter problem—that of stability under the vertical load. Any calculated lateral loads delivered to the tops of the pile and caisson foundations should be carried away to other parts of the construction for development of the actual lateral resistance, which is essentially similar to the procedure used for isolated footings.

Another special soil problem occurs when there is a seismic load on retaining walls. As shown in Fig. 20.4, when designing for gravity loads, a lateral soil pressure is usually assumed on the basis of an equivalent horizontal fluid pressure. If a surcharge exists, or if the ground slopes to the wall as shown in the illustration, the level of the equivalent fluid is assumed to be somewhere above the true ground level at the back of the wall.

When lateral load has been caused by seismic movement, the earth mass behind the wall has the potential of delivering a horizontal force while simultaneously offering gravity resistance to the overturn of the wall. The lower illustration in Fig. 20.4 shows the usual assumption for the soil failure mechanism that defines the potential mass whose weight develops the

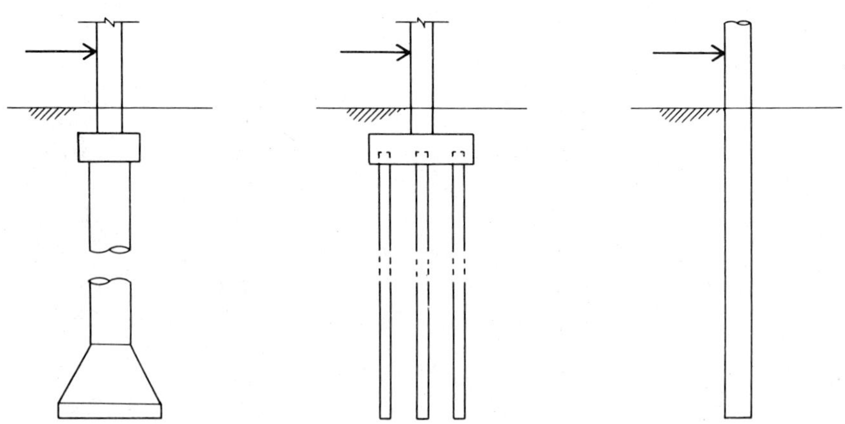

FIGURE 20.3. Deep, penetration-type foundations.

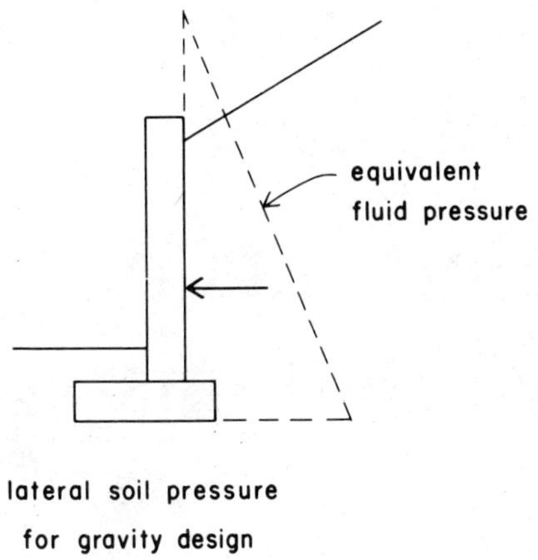

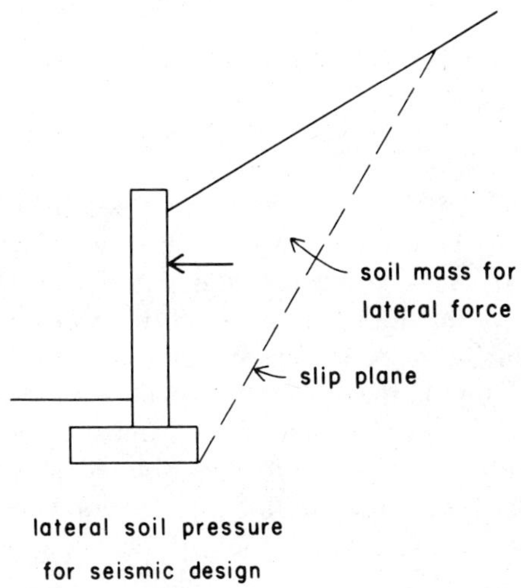

FIGURE 20.4. Lateral load conditions for tall retaining walls—gravity plus seismic loads.

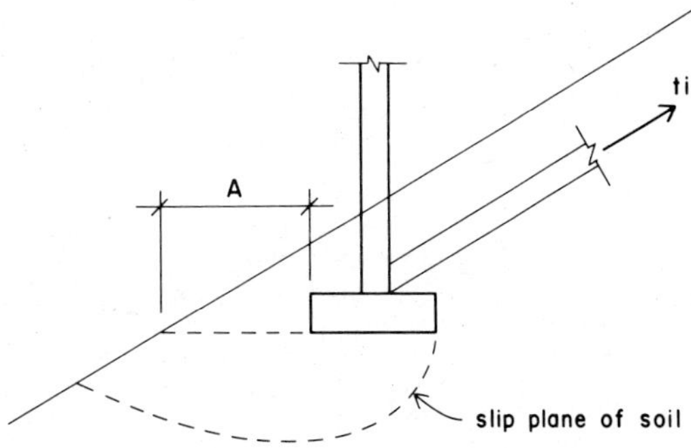

FIGURE 20.5. Bearing footings in hillside conditions.

lateral seismic force. In a conservative design, for the analysis of the wall, this force should be added to the lateral seismic forces of the wall and footing weights.

Figure 20.5 illustrates the situation of a footing in a hillside location that could be a wall footing or an isolated column footing. The dimension "A" in the figure, called the *daylight dimension*, must be sufficient to provide resistance to the failure of the soil as shown. Lateral load on the footing will further aggravate this type of failure if "A" is too small. The preferred solution, if the construction permits it, is to use a tension tie to transfer the lateral effect to some other part of the structure. Otherwise, the level of the footing should be lowered until a conservative distance is developed for the daylight dimension.

CHAPTER TWENTY-ONE

Miscellaneous Elements of Lateral Resistive Systems

In addition to the principal elements of lateral resistive systems (diaphragms, braced bents, etc.), there are typically a number of special elements or devices that are required for the complete functioning of the system. Examples of some such items are described in this chapter.

21.1 COLLECTORS

Transfer of loads from horizontal to vertical elements in laterally resistive structural systems frequently involves the use of some structural members that serve the functions of struts, drags, ties, collectors, and so on. These members often serve two functions—as parts of the gravity-resistive system or for other functions in lateral load resistance.

Figure 21.1 shows a structure consisting of a horizontal diaphragm and a number of exterior shear walls. For loading in the north–south direction the framing members labeled "A" serve as chords for the roof diaphragm. In most cases they are also parts of the roof edge or top of the wall framing. For the lateral load in the east–west direction they serve as collectors. This latter function permits us to consider the shear stress in the roof diaphragm to be a constant along the entire length of the edge. The collector "collects" this constant stress from the roof and distributes it to the isolated shear walls, thus functioning as a tension/compression member in the gaps between the walls.

In the example in Fig. 21.1 the collector "A" must be attached to the roof edge to develop the transfer of the constant shear stress. The collector "A" must be attached to the individual shear walls for the transfer of the total load in each wall. In the gaps between walls the collector gathers the roof edge load and functions partly as a compression member, pushing some of the load to the forward wall, and partly as a tension member, dragging the remainder of the collected load into the rearward wall.

Collectors "B" and "C" in Fig. 21.1 gather the edge load from the roof deck under the north–south lateral loading. Their function over the gap reverses as the

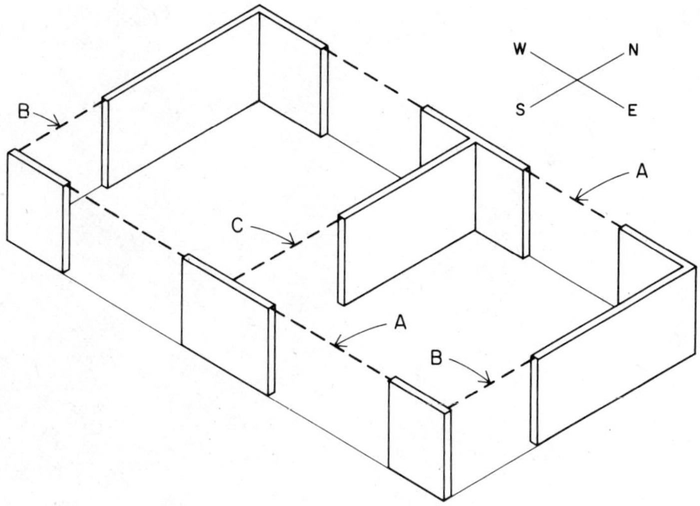

FIGURE 21.1. Collector functions in a box system.

load switches direction. They work in compression for load in the northerly direction, pushing the load into the walls. When the load changes to the southerly direction, they work in tension, dragging the load into the walls.

The complete functioning of a lateral resistive structural system must be carefully studied to determine the need for such members. As mentioned previously, ordinary members of the building construction will often serve these functions: top plates of the stud walls, edge framing of roofs and floors, headers over openings, and so on. If so used, such members should be investigated for the combined stress effects involved in their multiple roles.

The attachment of elements of the lateral resistive structure to one another, to collectors, or to supports usually involves some type of anchorage element. There is a great variety of these, encompassing the range of situations with regard to load transfer conditions, magnitude of the forces, and various materials, and details of the structural members and systems. Some of the basic types of elements and anchorage situations are described in the following sections.

21.2 TIE DOWNS

Resistance to vertical uplift is sometimes required for elements of a braced or moment-resistive frame, for the ends of shear walls, or for light roof systems subject to the force of upward wind suction. For concrete and reinforced masonry structures, such resistance is most often achieved by doweling and/or hooking of reinforcing bars. Steel columns are usually anchored by the anchor bolts at their bases. The illustrations in Fig. 21.2 show some of the devices that are used for anchoring wood structural elements. In many cases these devices have been load tested and their capacities rated by their manufacturers. When using them, it is essential to determine whether the load ratings have been accepted by the building code agency with jurisdiction for a specific building design.

The term *tie down* or *hold down* is mostly used to describe the type of anchor

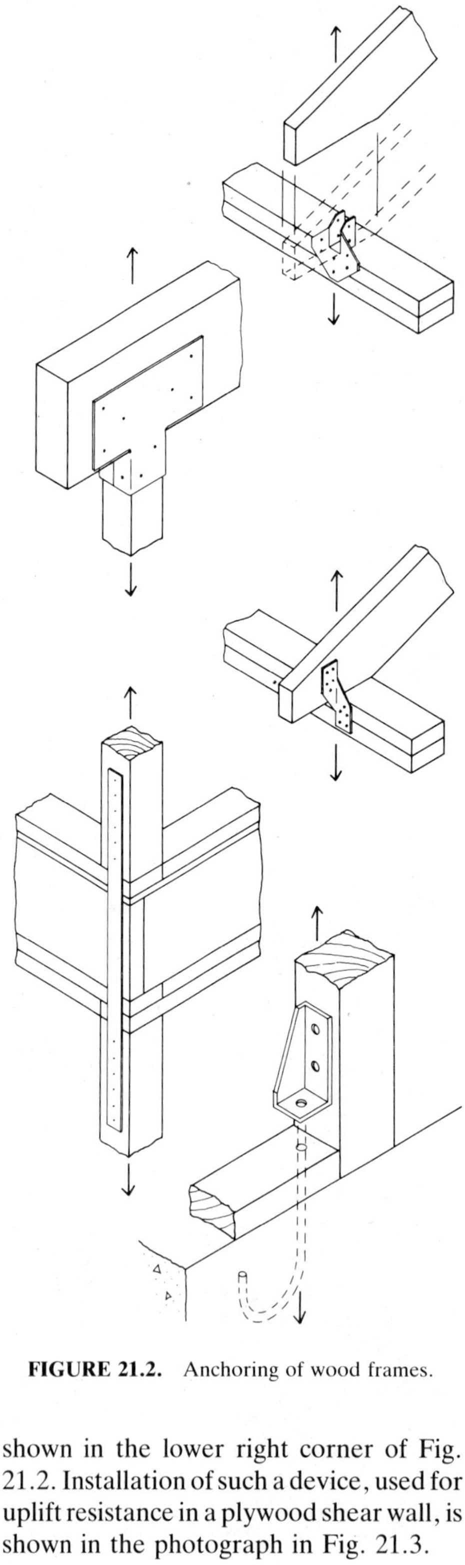

FIGURE 21.2. Anchoring of wood frames.

shown in the lower right corner of Fig. 21.2. Installation of such a device, used for uplift resistance in a plywood shear wall, is shown in the photograph in Fig. 21.3.

FIGURE 21.3. Large tie-down connector in a wood frame shear wall.

21.3 HORIZONTAL ANCHORS

In addition to the transfer of vertical gravity load and lateral shear load at the edges of horizontal diaphragms, there is usually a need for resistance to the horizontal pulling away of walls from the diaphragm edge. In many cases the connections that are provided for other functions also serve to resist this action. Codes usually require that this type of anchorage be a "positive" one, not relying on such things as the withdrawal of nails or lateral force on toe nails. Figure 21.4 shows some of the means used for achieving this type of anchorage.

21.4 SHEAR ANCHORS

The shear force at the edge of a horizontal diaphragm must be transferred from the

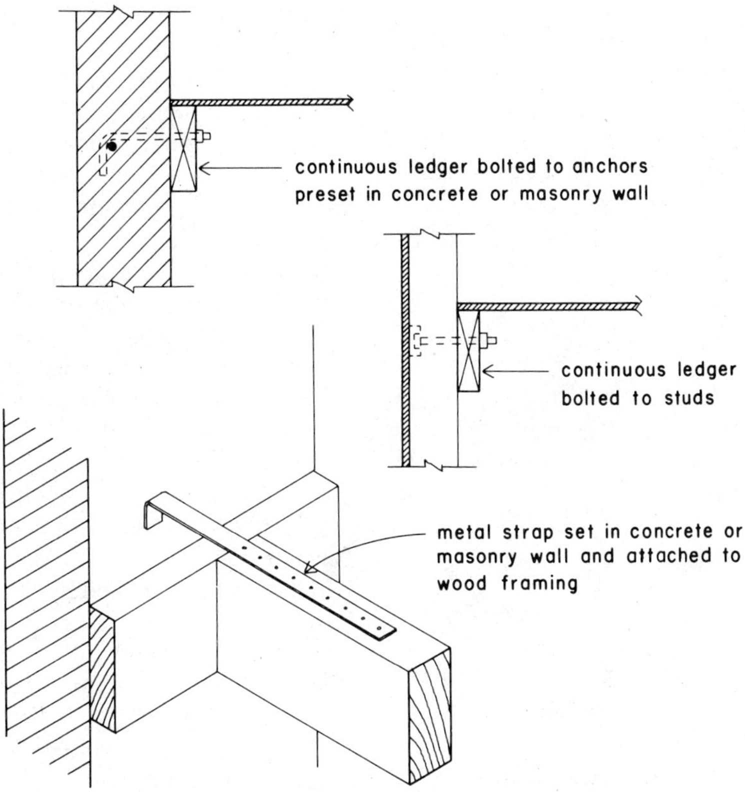

FIGURE 21.4. Anchoring of horizontal diaphragms.

diaphragm into a collector or some other intermediate member, or directly into a vertical diaphragm. Except for poured-in-place concrete structures, this process usually involves some means of attachment. For wood structures the transfer is usually achieved through the lateral loading of nails, bolts, or lag screws, for which the codes or industry specifications provide tabulated load capacities. For steel deck diaphragms the transfer is usually achieved by welding the deck to supporting steel framing. If the vertical system is a steel frame, these members are usually parts of the frame system. If the vertical structure is concrete or masonry, the edge transfer members are usually attached to the walls with anchor bolts set in the concrete or in solid-filled horizontal courses of the masonry. As in other situations, the combined stresses on these connections must be carefully investigated to determine the critical load conditions.

The complete transfer of force from the horizontal to the vertical elements of the lateral resistive system can be quite complex in some cases. Figure 21.5 shows a joint between a horizontal plywood diaphragm and a vertical plywood shear wall. For reasons other than lateral load resistance, it is desired that the studs in the wall run continuously past the level of the roof deck. This necessitates the use of a continuous edge-framing member, called a ledger, that serves as the vertical support for the deck as well as the chord and edge collector for the lateral forces. This ledger is shown to be attached to the faces of the studs with two lag screws at each stud.

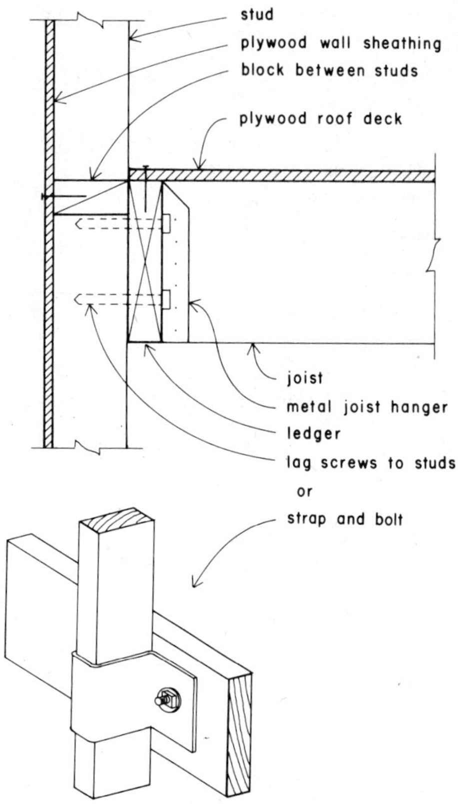

FIGURE 21.5. Load transfer from horizontal to vertical structure.

The functioning of this joint involves the following:

1. The vertical gravity load is transferred from the ledger to the studs directly through lateral load on the lag screws.
2. The lateral shear stress in the roof deck is transferred to the ledger through lateral load on the edge nails of the deck. This stress is in turn transferred from the ledger to the studs by horizontal lateral load on the lag screws. The horizontal blocking is fit between the studs to provide for the transfer of the load to the wall plywood, which is nailed to the blocking.
3. Outward loading on the wall is resisted by the lag screws in withdrawal. This is generally not considered to be a good positive connection, although the load magnitude should be considered in making this evaluation. A more positive connection is achieved by using the bolts and straps shown in the lower sketch in Fig. 21.5.

Another shear transfer problem is that which occurs at the base of a shear wall in terms of a sliding effect. For a wood-framed wall some attachment of the wall sill member to its support must be made. If the support is wood, the attachment is usually achieved by using nails or lag screws. If the support is concrete or masonry, the sill is usually attached to preset anchor bolts. The lateral load capacity of the bolts is determined by their shear capacity in the concrete or the single shear limit in the wood sill. *UBC* Section 2907(e) gives some minimum requirements for sill bolting, which should be used as a starting point in developing this type of connection.

For walls of concrete and masonry, in which there is often considerable dead load at the base of the wall, sliding resistance may be adequately developed by friction. Doweling provided for the vertical wall reinforcing also offers some lateral shear resistance. If a more positive anchor is desired, or if the calculated load requires it, shear keys may be provided by inserting wood blocks in the concrete, as shown in Fig. 21.6.

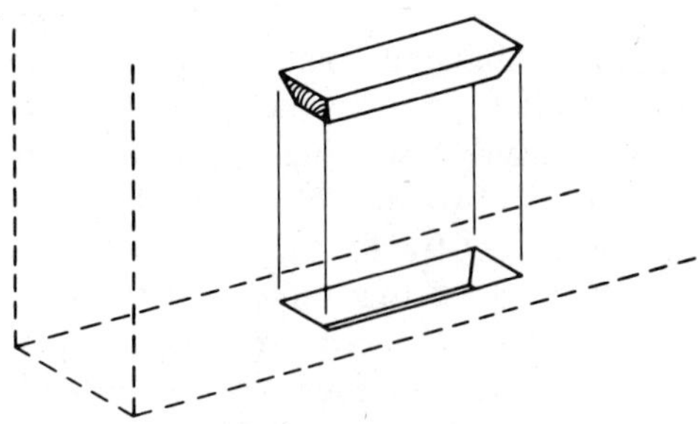

FIGURE 21.6. Shear key in concrete.

21.5 SEPARATION JOINTS

During the swaying motions induced by earthquakes, different parts of a building tend to move independently because of the differences in their masses, their fundamental periods, and variations in damping, support constraint, and so on. With regard to the building structure, it is usually desirable to tie it together so that it moves as a whole as much a possible. Sometimes, however, it is better to separate parts from one another in a manner that permits them a reasonable freedom of motion with respect to one another.

Figure 21.7 shows some building forms in which the extreme difference of period of adjacent masses of the building make it preferable to cause a separation. Designing the building connection at these intersections must be done with regard to the specific situation in each case. Some of the considerations to be made in this are discussed in the following sections.

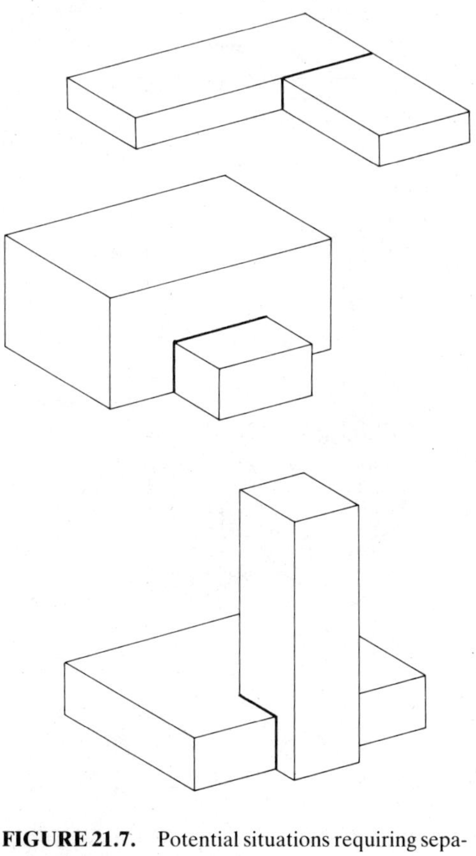

FIGURE 21.7. Potential situations requiring separation joints.

The Specific Direction of Movements

In generally rectangular building forms, such as those shown in the examples in Fig. 21.7, the primary movements are in the direction of the major axes of the adjacent masses. Thus the joint between the masses has two principal forms of motion: a shear effect parallel to the joint, and a together–apart motion perpendicular to the joint. In building forms of greater geometric complexity, the motions of the respective masses is more random, and the joint action is much more complex.

The Actual Dimensions of Movement at the Joint

If the joint is to be truly effective and if the adjacent parts are not allowed to pound each other, the actual dimension of the movements must be safely tolerated by the separation. The more complex the motions of the separate masses, the more difficult it is to predict these dimensions accurately, calling for some conservative margin in the dimension of the separation provided.

Detailing the Joint For Effective Separation

Because the idea of the joint is that structural separation is to be provided while still achieving the general connection of the adjacent parts, it is necessary to make a joint that performs both of these seemingly contradictory functions. Various techniques are possible using connections that employ sliding, rolling, rotating, swinging, or flexible elements that permit one type of connection while having a freedom of movement for certain directions or types of motion. The possibilities are endless, and the specific situation must be carefully analyzed in order to develop an effective and logical joint detail. In some cases the complexity of the motions, the extreme dimension of movement to be facilitated, or other considerations may make it necessary to have complete separation—that is, literally to build two separate buildings very close together.

Facilitating Other Functions of the Joint

It is often necessary for the separation joint to provide for functions other than those of the seismic motions. Gravity load transfer may be required through the joint. Nonstructural functions, such as weather sealing, waterproofing, and the passage of wiring, piping, or ductwork through the joint, may be required. Figure 21.8 shows two typical cases in which the joint achieves structural separation while providing for a closing of the joint. The upper

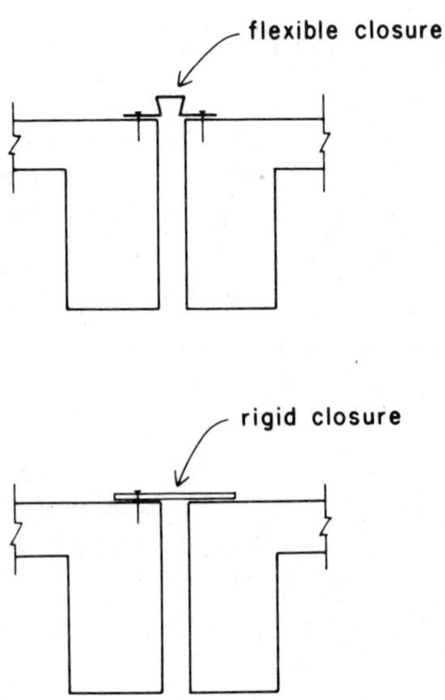

FIGURE 21.8. Closure of horizontal separation joints.

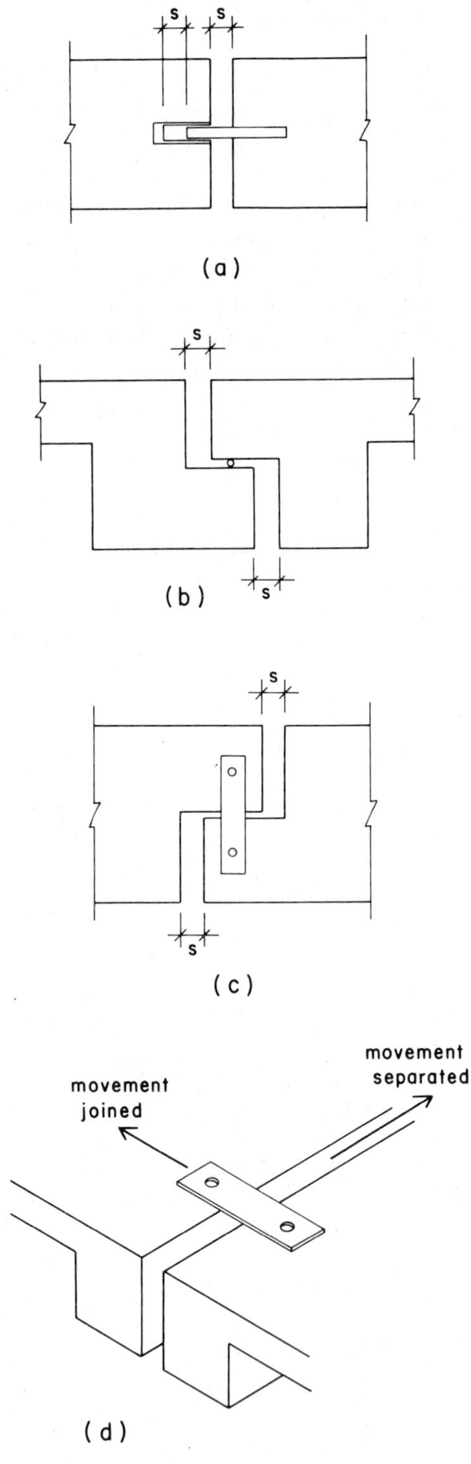

FIGURE 21.9. Means for achieving partial separation.

drawing shows a flexible flashing or sealing strip used to achieve weather or water tightness of the joint. The lower drawing symbolizes the usual solution for a floor, in which a flat element is attached to one side of the joint and is allowed to slip on the other side.

Figure 21.9 shows a number of situations in which partial structural separation is achieved. The details of such joints are often quite similar to those used for joints designed to provide separation for thermal expansion. Figure 21.9*a* shows a key slot, which is the type of connection usually used in walls where the separation is required only in a direction parallel to the wall plane. Figures 21.9*b* and *c* show means for achieving the transfer of vertical gravity forces through the joint while permitting movement in a horizontal direction. Figure 21.9*d* shows a means for achieving a connection in one horizontal direction while permitting movement in the perpendicular direction.

PART FOUR

BUILDING DESIGN EXAMPLES

The material in this part consists of examples of the design of parts of the lateral resistive system for various buildings. No attempt is made to present the complete structural design for any example building, although various problems relating to the overall design of the structure and the design of the complete building are discussed in many cases. The purpose is to illustrate a wide range of examples, including large and small buildings of various forms and employing various types of construction. For some of the example buildings, the design of more than one type of lateral bracing system is illustrated so that comparisons are possible. Most structural calculations are presented in the form of typical handwritten work as it is commonly done by the structural designer; this is done for sake of a more condensed presentation as well as to demonstrate the ordinary form of such work for the inexperienced designer.

CHAPTER TWENTY-TWO

General Design Considerations

Most buildings are still designed one at a time, with some unique situations for each. This chapter consists of some observations on the building structural design process at it occurs most often, with a team effort by an architect and a structural designer.

22.1 RESPONSIBILITIES OF THE STRUCTURAL DESIGNER

The role of the structural designer is discussed in Section 1.3, and the context within which he or she normally functions is described. We assume here that the structural designer is the person primarily responsible for the complete design—from preliminary studies to completed calculations and working drawings. However, the usual limited context within which the design of a single building is ordinarily developed typically leaves a narrow framework for decisions. The types of services that the structural designer normally must provide are the following:

1. *Preliminary Consultation.* At the earliest stages of building design the architect often needs answers to a few very broad questions, including basic uses and limits for material and systems, approximate size limits for elements of various systems, and so on.
2. *Preliminary Design.* As soon as the first schemes are drawn, it is often necessary to provide some very preliminary answers as to structural requirements. These must be carefully and conservatively done, as it is common for architects to proceed with quite serious design work from this point on.
3. *Calculations.* Whether done by hand or on a computer, some documentation must be made of the structural analysis and design work.
4. *Finished Design Requirements.* It is a somewhat difficult judgment as to how much description is enough, but essentially the finished structure must be completely and unambiguously described.
5. *Inspection.* This depends on contractural arrangements, but for accessible sites the structural designer (or

the firm) must often perform some inspections during the progress of construction.

6. *Construction Permits.* Obtaining a permit for construction usually entails passing a review of the structural work in the form of the calculations and working drawings.

22.2 PRELIMINARY DESIGN

Most designers agree that the preliminary design is the most difficult stage, in terms mostly of reliance on experience and judgment. Hard decisions must often be made at this stage, which, if reversed later, can cause much confusion and wasted effort. Resources for assistance for this work are very limited.

The structural designer must have a "quick sketch" capability to do preliminary design. This consists of the ability to quickly visualize and conceptualize the basic building scheme, to quickly analyze the situation for structural problems, and to quickly visualize and describe the range of reasonable solutions. This menu of solutions must then be bounced off of the architect and reanalyzed until the best answer is developed.

Calculations done at this stage must be simple and approximate, but produce answers that are within some range of reality. Most designers tend to round answers up at this stage; that is, use conservative estimates and otherwise play it safe.

22.3 CALCULATIONS

Calculations form a record of the structural design, at least that portion that is based on some numerical work. The larger the job, the more the designer needs this record to keep track of the work, and to provide information for other designers or for a review by someone who must check the work. For most buildings a check will be made by someone before a building permit is issued. While the designer may use brief notes for his own reference, the calculations must be readable by others and have sufficient notation to describe what they are all about, what references are used, and so on.

The format for calculations is a personal matter. Firms usually establish a standard format for work in their office. Some of the material in the other chapters in this part is presented in the form of handwritten calculations, which illustrate a particular style of documentation.

At present, it is common for computer printouts to form part of the documentation for structural design. As designers come to use the computer keyboard as a personal tool, this will undoubtedly become the new typical format for most structural calculations.

22.4 FINISHED DESIGN REQUIREMENTS

Structural design work is finished when the designer has completely described the structure and has done sufficient investigation of anticipated loading conditions so as to be able to attest to the adequate safety of the structure. Whatever it takes to accomplish all of that must be done—and in most cases documented.

In our industrial age much of what constitutes the structure comes in the form of some sort of industrial product. Steel is probably the extreme case; we use almost exclusively preformed or preassembled elements for steel structures. Designing with such elements is a limited activity that uses some amount of preestablished information regarding the properties of the products. For elements such as steel deck units or open-web joists, the preinformation is extended to the level of providing predicted capabilities, limiting the designer's work to a minimum.

Design work of any kind involves making series of decisions, often involving picking between alternatives with little difference. For complex objects like buildings this produces a situation in which rexamination can almost always produce some second thinking about choices. Knowing when to quit and go with what you have is often a difficult judgment.

22.5 CONSTRUCTION DETAILS

Part of the designer's task is to provide sufficient instruction regarding the details of the construction so that what is required is clear. For preformed products a simple catalog number may suffice. However, most buildings have a number of situations of assembly that are not necessarily standard, and some amount of both graphic and written description is required.

Through the working drawings and specifications the designer tells the builder how to make the building. Since most designers do not actually do any building, this is a delicate situation. It works best when the designer knows a considerable amount about building construction materials, products, and practices.

Since work in construction drafting is not a part of most curricula in civil engineering, young structural designers often have a severe handicap in this area.

22.6 CONSTRUCTION PERMITS

In most cases someone employed or retained as a consultant by the code-enforcing body will review the structural calculations and the construction drawings and specifications for compliance with the local code requirements and with acceptable practices of design and construction. Although it should be expected that a competent and thorough design effort will receive a good review, there is usually some room for individual judgment and personal preference so that the potential exists for some conflict between the designer and the reviewer.

Getting a building permit is somewhat like going to court. The permit-granting authorities represent the prosecution, judge, and jury. The structural designer and architect represent the counsel for the plantiff. The burden of proof is on the plantiff. As with any court case, it helps to have a good case to make (no code requirements violated; no questionable design or construction procedures) and to present it well (clear, complete structural calculations and construction drawings and specifications).

CHAPTER TWENTY-THREE

Shear Walls—One Story

The examples in this chapter are all single-story buildings employing shear walls for lateral bracing. The general problems of shear walls are discussed in Chapter 16, and the basic nature of shear walls are thoroughly illustrated there. A review of Chapter 16 is advised before any study of the material in this chapter, since the explanation of basic concepts is omitted here for sake of brevity.

23.1 SIMPLE PLYWOOD BOX

We begin with a simple example: a one-story box with a horizontal roof diaphragm and vertical, exterior bracing walls. A plan and a section of the building are shown in Fig. 23.1. The construction is light wood frame with a plywood roof deck and planar surfacing materials on wood studs for the walls.

The critical loading is that which is applied to the edges of the roof diaphragm and is transferred through the shear walls to the foundations. For this load the value of W is the sum of the dead loads of the roof and ceiling, any items sitting on or suspended from the roof (such as HVAC units and light fixtures), the upper portions of the walls (assuming them to span vertically), and any items that are attached to the upper part of the walls (such as the canopy in this example.)

If a wall is of braced construction, even though it is not used as part of the building's lateral resistive system, it is usually assumed that it transfers load to the roof only in the direction that is perpendicular to the plane of the wall. This relationship was discussed in Section 10.2 and is illustrated in Fig. 10.10. In this example each of the four exterior walls is partially braced in its own plane. They are therefore assumed to add to the W summation only for the lateral load that is perpendicular to the wall plane.

The actions of the walls as assumed in this example are shown in Fig. 23.2. In the direction perpendicular to their planes, the walls are assumed to span from floor to roof. For the interior walls the load to the roof is half of the total dead load of the wall. For the exterior walls we assume the location of the roof diaphragm to be at approximately 12 ft above the floor, so that the extension of the wall parapet is 2 ft above the roof. Assuming the function of the wall to be as described in Fig. 23.2, we thus add a total of $\frac{12}{2} + 2 = 8$ ft of wall dead load to the roof load.

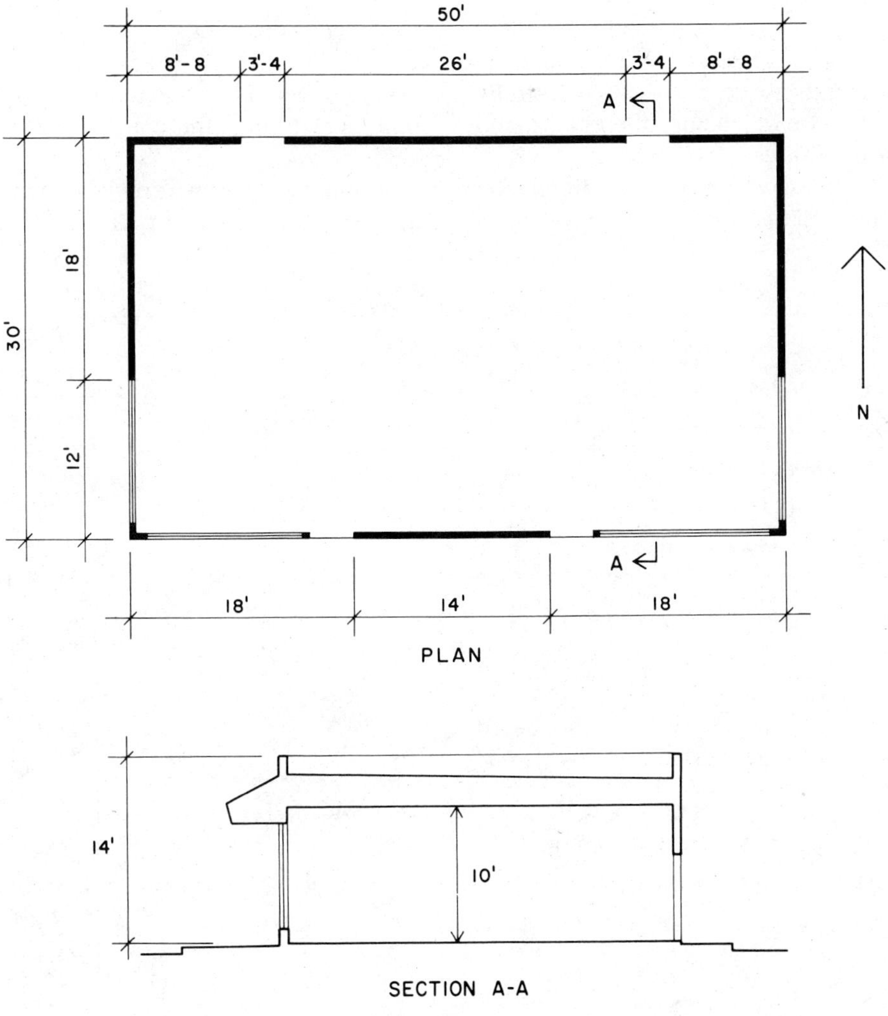

FIGURE 23.1

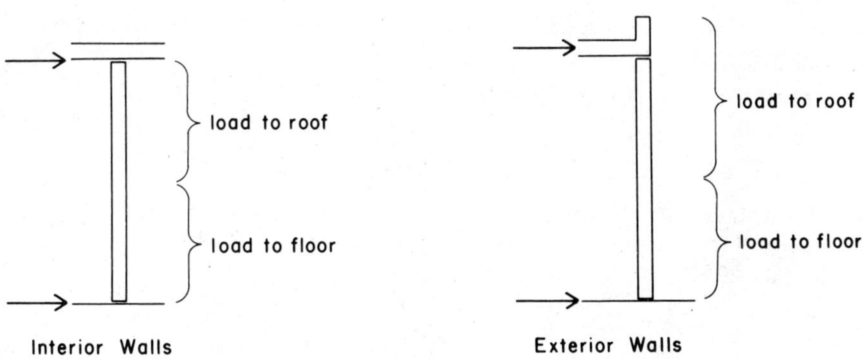

FIGURE 23.2. Distribution of wall loads.

We assume the canopy load to be added to W in both directions. We assume the HVAC units to be reasonably centrally located, so we will include their load in the W for both directions.

With these various assumptions, the tabulation of the total W for the lateral load in each direction is as shown on Calculation Sheet 23.1.1. Notice that we have assumed that all the exterior walls are solid, even though there are several door and window openings. We make this assumption because we are including the upper portions of the walls, which are

Sheet No. 23.1.1

References:

Seismic Load Analysis - Example 23.1

Use Section 2312 of the 1982 UBC.

Total Seismic Load = V = ZIKCSW.

Assume Seismic Zone 4, Z = 1.0. — UBC: 2312(c)

Assume Occupancy Importance Factor, I = 1.0. — UBC: Table 23-K

For one story building C = 0.12
but – with T_s not determined, max. CS = 0.14 — UBC: 2312(d)

For all plywood system, K = 1.0 — UBC: Table 23-I

Then: $V = (1.0)(1.0)(1.0)(0.14)W = 0.14\,W$

Assumed unit dead loads: — Chap. 32

Roof + ceiling : 12 psf
Exterior Walls : 20 psf
Interior Walls : 10 psf
Canopy : (add to wall) 100 lb/ft of wall length
Rooftop HVAC equip. : 5 k total

Dead load summation for load to roof diaphragm: — Art. 14.1 + Chap. 15; See Fig. 23-1

Source:	East-West Load:		North-South Load:	
Roof + ceiling	30 × 50 × 12 psf =	18 k	30 × 50 × 12 psf =	18 k
Ext. walls	60 × 8 × 20 psf =	9.6	100 × 8 × 20 psf =	16
Int. walls	100 × 6 × 10 psf =	6	100 × 6 × 10 psf =	6
Canopy	50 × 100 lb/ft =	5	50 × 100 lb/ft =	5
HVAC Equip.	=	5	=	5
		43.6 k		50 k

Roof diaphragm analysis: — Chap. 15

North - South

Total Load = 0.14 × 50 = 7.0 k ÷ 50 = 140 lb/ft — Fig. 23-3

Max. unit shear in plywood = $v = \frac{3500}{30} = 117$ lb/ft

Chord force = C or T = $\frac{7 \times 50}{8 \times 30} = 1.458$ k

East - West

V = 0.14 × 43.6 = 6.10 k ÷ 30 = 203 lb/ft — Fig. 23-6

$v = \frac{3050}{50} = 61$ lb/ft

Chord force = C or T = $\frac{6.1 \times 30}{8 \times 50} = 0.458$ k

Calculation Sheet 23.1.1

indeed mostly solid because the openings are generally in the lower portion of the walls.

The action of the roof deck is shown in Fig. 23.3. The roof spans from end to end of the building, transferring its end reactions to the two end shear walls. Because of the large opening, the design length of the shear wall is assumed to be 18 ft, although the structure over the opening is assumed to function in collecting load from the end of the roof. The combined seismic and gravity loads on the shear wall are shown in Fig. 23.4. The gravity load consists of the weight of the wall and some portion of the dead load of the roof and ceiling, assuming that the roof framing results in some distribution of these loads to the wall. The total dead load is assumed to be 6000 lb.

Because the wall dead load was not included in the roof load tabulation for this direction, its lateral effect should be added as shown in the illustration, although the net effect is minor in this example. We assume the wall itself to weight 5000 lb and that its lateral load is $0.14 \times 5000 = 700$ lb. The analysis of the wall is shown on Calculation Sheet 23.1.2.

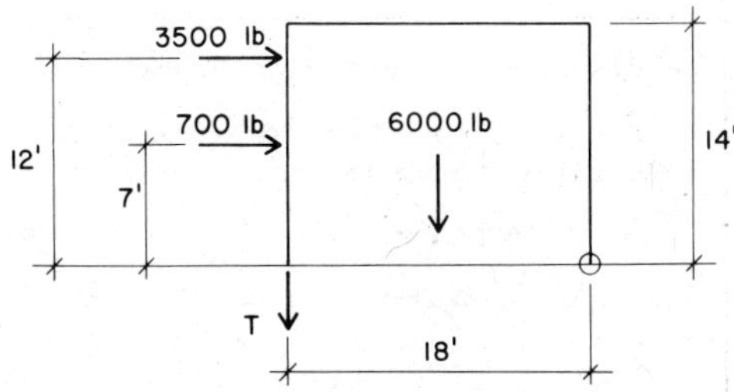

FIGURE 23.4. Loads on the end shear wall.

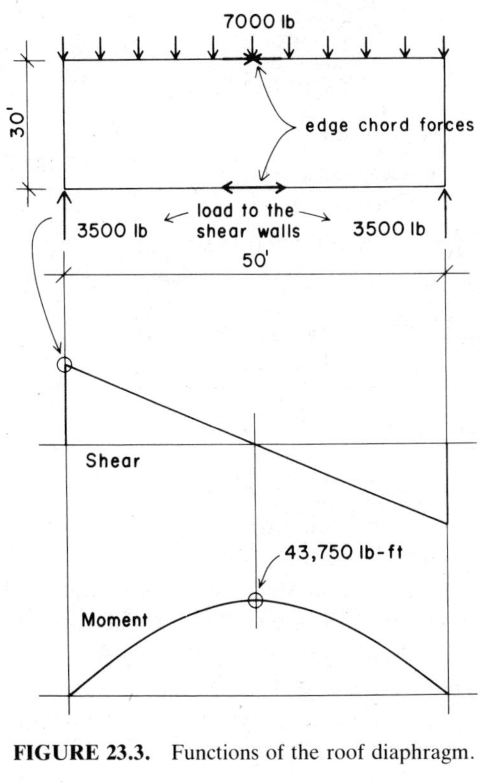

FIGURE 23.3. Functions of the roof diaphragm.

Loading in the east–west direction produces considerable torsion on the structure, due to the layout of the shear walls. Accidental torsion in the north–south direction should also be considered but will be discussed after investigation of the east–west effects.

For this example there are two approaches to the analysis for the east–west seismic force. Consider the distribution of the total force to the front and rear shear walls, as shown in Fig. 23.5. If the roof diaphragm is considered to function as a simple beam spanning between these two walls—as was assumed for the force in the north–south direction—then H_1 and H_2 will be equal, each being one-half of the total force. This assumption will result in the shear in the roof diaphragm as indicated in Fig. 23.6. This is the usual basis for design of the roof diaphragm and its chord forces.

The other approach to the east–west analysis is to consider the roof diaphragm to have enough stiffness so as to generate a torsional effect on the building, due to the eccentricity of the force (centroid of the mass) from the center of stiffness of the shear walls (see discussion in Section

Analysis of the end shear wall:	Sheet No. 23.1.2
Add DL of roof to wall load. Assume DL = 5000 lb for wall + 1000 lb for roof = 6000 lb total	References:
Add lateral load due to wall DL:	Chap. 16
$V = 0.14(5) = 0.7$ k	Fig. 23.4
Then: Total $V = 3.5 + 0.7 = 4.2$ k	
$v = \frac{4200}{18} = 233$ lb/ft	
Overturning M = 3.5 × 12 × 1.5 = 63 0.7 × 7 × 1.5 = 7.35 70.35 k-ft	
DL Moment = 6 × 9 = 54.0	
M for hold down force = 16.35 k-ft	
Required $T = \frac{16.35}{18} = 0.91$ k	
Torsional analysis:	Art. 15.1
J for the shear walls: (ft^3)	Fig. 23.7
Wall / Dist. to ℄ of stiffness / Wall length / $J = L \times d^2$ A / 25' / 2 × 18' / 22,500 B / 7.33' / 43.33' / 2,328 C / 22.67' / 14' / 7,195 32,043 ft^3	
Accidental torsion - N-S load - 5% eccen.	UBC: 2312(e)5
add to end wall: $v = \frac{Fec}{J} = \frac{7000 \times (0.05 \times 50) \times 25}{32{,}043}$ = 14 lb/ft (12% increase)	
E-W torsion on front wall:	
$v = \frac{Fec}{J} = \frac{6100 \times 7.67 \times 22.67}{32{,}043} = 33$ lb/ft	

Calculation Sheet 23.1.2

14.2.) The layout for this analysis is shown in Fig. 23.7 and the computation of the torsional resistance is given on Calculation Sheet 23.1.2, together with the additional shear stresses caused by the torsional effects. Note that, although the building has symmetry for the north–south force, an accidental eccentricity of 5% of the building dimension is required by *UBC* Section 2312(e)5.

Diaphragm shear stresses in the roof and end walls are well within the capacity of ordinary structural plywood. Assuming $\frac{1}{2}$-in. plywood for the roof deck and $\frac{3}{8}$-in. plywood for the walls, blocked diaphragms with the minimum edge nail spac-

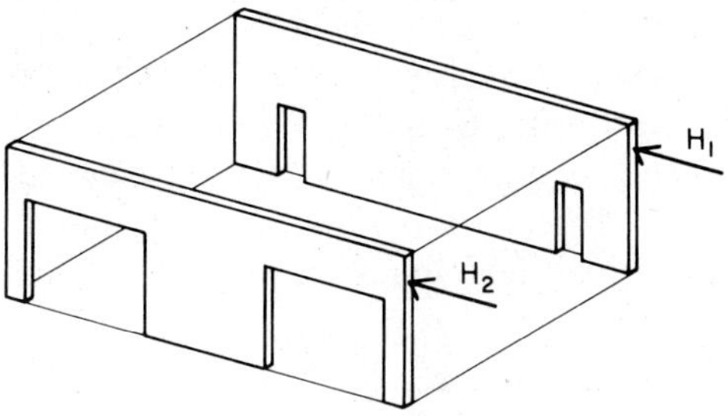

FIGURE 23.5. Lateral loading of the front and rear walls.

ing of 6 in. will safely carry the loads (see *UBC* Tables 25–J and 25–K).

It is possible to consider surfacing other than plywood for the end shear walls. If cement plaster is used on both surfaces (as plaster on the inside and stucco on the outside), its single surface value of 180 lb/ft taken from *UBC* Table 47–I can be doubled, giving a total capacity of 360 lb/ft, which is greater than the shear calculated for the walls. It should be noted that *UBC* Section 4713(a) permits this addition of the two surfaces only when they are covered with the same material.

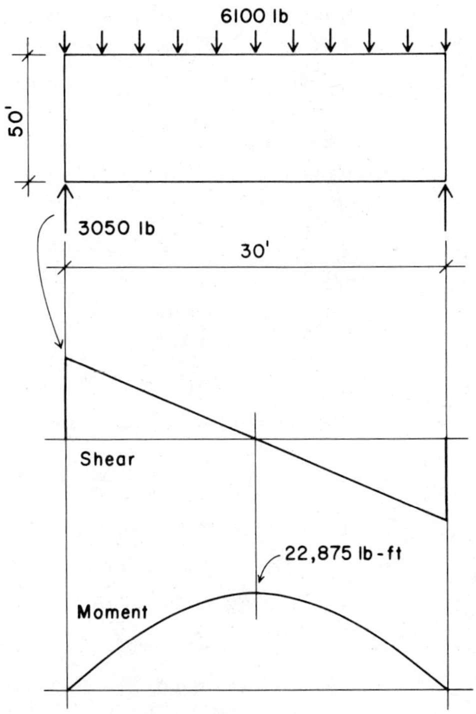

FIGURE 23.6. East–west loading for the roof.

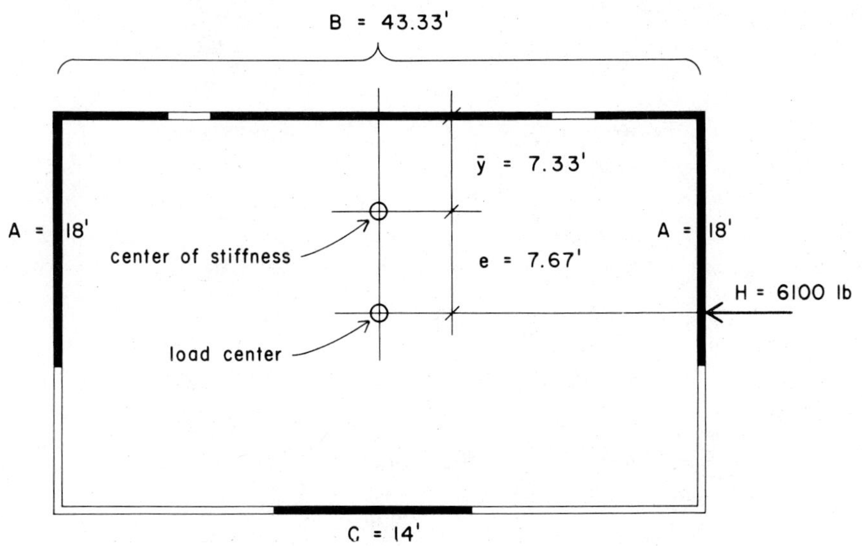

FIGURE 23.7. Torsional effect of east–west load.

The calculations show the need for tie down anchors at the ends of the shear walls, except for the single long rear wall. (See Calculation Sheet 23.1.3.) The resisting forces required are quite small and are well within the capacity of ordinary anchoring devices, such as Simpson hold downs (see Section 30.1). The real need for tie downs is questionable, however, since it is likely that the development of the wall and roof framing will result in additional dead loads on the ends of the

Calculation	Sheet No. 23.1.3
Analysis of E-W shear walls:	
Average $v = \frac{6100}{57.33} = 106$ lb/ft - use for rear	References: Chap. 16
Adding torsion to front wall: $v = 106 + 33 = 139$ lb/ft	
On peripheral basis, load on front = ½ of total	Fig. 23.5
$v = \frac{1/2 \times 6100}{14} = 218$ lb/ft - use for front	
Plywood & Nailing:	Art. 28.1
Walls - 3/8" Str. II plyw., 8d @ 6, 2x studs @ 16" o/c	
Cap. = 1.20 × 220 = 264 lb/ft	
Roof - 1/2" Str. II plyw. (min. for flat roof)	
least cap. = 180 lb/ft)	
(unblocked, panel joints alligned, min. nailing)	
Tiedowns:	
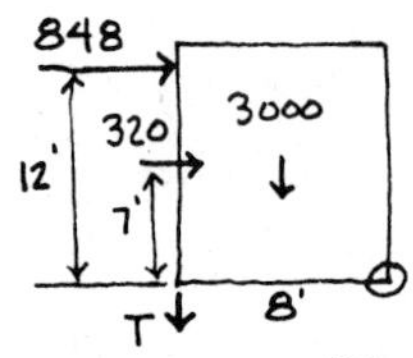	
Short rear wall: V = 106 × 8 = 848 lb	
OT Mom. = 848 × 12 × 1.5 = 15264	
320 × 7 × 1.5 = 3360	
18624 ft-lb	
DL Mom. = 3000 × 4 = 12000	
6624 ft-lb	
$T = \frac{6624}{8} = 828$ lb	
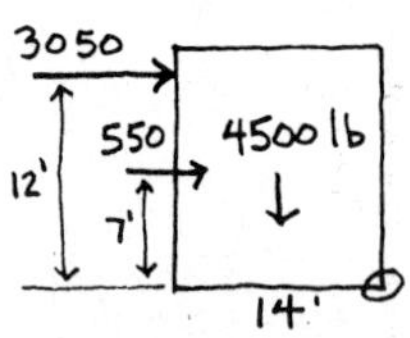	
Front wall: V = ½ total = 3050 lb	
OT Mom. = 3050 × 12 × 1.5 = 54900	
550 × 7 × 1.5 = 5775	
60675 ft-lb	
DL Mom. = 4500 × 7 = 31500	
29175 ft-lb	
$T = \frac{29175}{14} = 2084$ lb	
(Minimal tiedowns - framing loads may eliminate)	
Sill Bolts:	
End: ½" - $\frac{470}{247} = 1.90'$ o/c 5/8" - $\frac{590}{247} = 2.37'$ o/c	UBC: Table 25-F
Front: ½" - $\frac{470}{218} = 2.16'$ o/c 5/8" - $\frac{590}{218} = 2.71'$ o/c	(bolt cap.)

Calculation Sheet 23.1.3

walls. At the rear corner the connected walls will tend to hold each other down through the corner framing.

Horizontal sliding of the shear wall must be resisted by the bolts that secure the sill plate to the foundation. *UBC* Section 2907(e) requires that these bolts be a minimum of $\frac{1}{2}$ in. in diameter, a maximum of 6 ft on center, and that there be a bolt not more than 12 in. from each end of the wall. For the 18-ft-long wall these requirements will result in a minimum of four $\frac{1}{2}$-in. bolts. The lateral load capacity of these bolts is limited by the single shear capacity of the sill or the rated resistance based on the concrete strength. If the bolting is not adequate, the bolt size, the sill thickness, or the total number of bolts can be increased.

The construction details at the intersection of the roof and wall must be carefully developed to facilitate the various load transfers at this point. There are various possibilities for these details, which must be developed to satisfy the lateral load and gravity load transfers and the desired architectural features. Figure 23.8 shows one

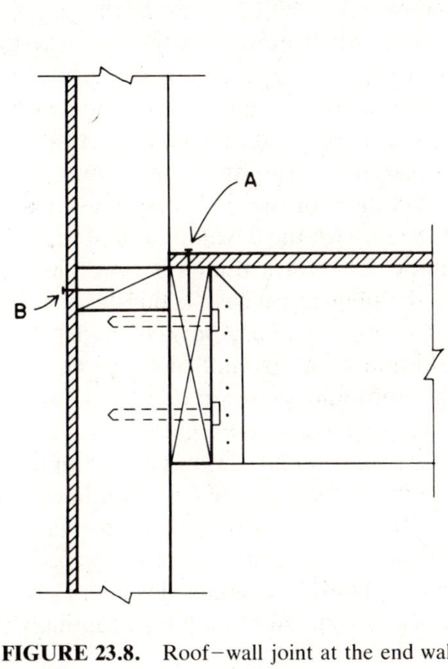

FIGURE 23.8. Roof–wall joint at the end wall.

possibility for this connection. Because the roof level varies along the wall length, a ledger is bolted to the face of the studs to provide support for the roof deck. This procedure permits the studs to be a constant length and the top plates of the wall to be horizontal. The nailing of the edge of the roof deck to the ledger, the attachment of the ledger to the studs, and the nailing of the wall plywood to the studs and blocking must all be designed for the calculated stress at the roof edge.

If the roof joists are parallel to this wall, the vertical gravity load on the ledger will be small. If the joists frame into the ledger, as shown in Fig. 23.8, the load will be somewhat higher. In either case the lateral load on the lag screws must be designed for the combination of the vertical gravity and the lateral load effects. In addition, they must resist the outward suction force on the wall by withdrawal. There are various other devices that may be used to attach the ledger to the studs for a more positive resistance to the outward force.

Figure 23.9 shows a possible detail for the connection of the roof to the wall at the front of the building. This connection is also subject to considerable variation, depending on the desired details of the parapet, the canopy structure, and the window head at this location. In the example shown the roof framing sits directly on the top plates of the stud wall, and the parapet is built as a stub wall on top of the roof deck. Nails "A" and "B" effect the transfer of the east–west load from the roof to the front shear walls.

The outward load on the wall may be carried back to the roof structure in a number of ways. The method illustrated in Fig. 23.9 is that of using the continuous ribbon to drag the load into the top plates of the wall. The roof deck is nailed to this member, and the plates are secured to it by the horizontal blocks and nails "D." An alternative would be to use metal fram-

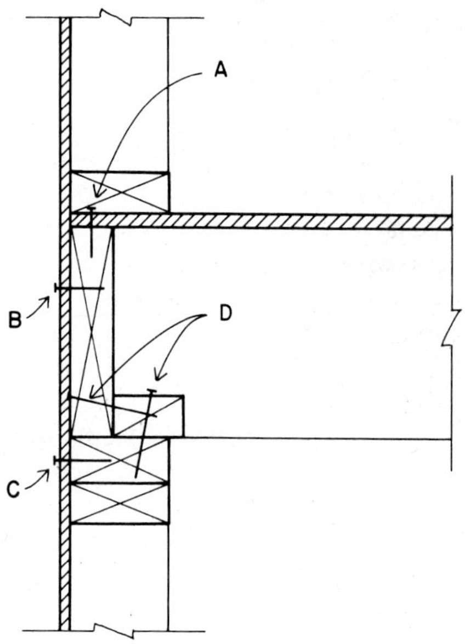

FIGURE 23.9. Alternate roof–wall joint.

ing anchors to secure the joists to the plates.

There are three elements in this detail shown in Fig. 23.9 that may contribute to the development of the chord for the roof diaphragm: the top plates of the lower wall, the continuous ribbon, or the sill of the parapet. The relatively low chord force in this example could be developed by any one of these alone; thus, their combined effect is quite redundant if all of them work. The critical consideration is the continuity of the chord force through the splices in the members, because each of them would be composed of several separate pieces in the 50-ft-long building. Assuming it to be unlikely that all of these members would be spliced at the same location, their overlapping should effectively provide the necessary continuity for the chord. If a positive splice is deemed necessary, one of the members—most likely the ribbon or the top plates—could be spliced with a bolted joint.

There are two approaches to the design of the rear wall. The first approach is to assume that the wall consists of three isolated piers with heights equal to the floor-to-roof distance, as is shown as case 1 in Fig. 23.10. For case 2 in Fig. 23.10 the continuous portion of the wall above the openings is assumed to produce a rigid-frame action, which results in the short piers being fixed at both their tops and bottoms, instead of cantilevering as in case 1. This results in a double reduction of the moment at the base of the piers. The first reduction is due to the reduced pier height: from 11 to 7 ft. The second reduction is due to the fixity at both top and bottom. If the inflection of the pier is assumed at midheight, as shown in Fig. 23.11, the moment arm for the overturn is reduced to 3.5 ft, which will considerably reduce the overturn effect at the base of the piers, while the resisting moment of the dead load remains the same.

There are some features required of the construction details for both of these cases. For case 1 the end framing members of the piers must be continuous from floor to roof for continuity of the end forces, as shown in Fig. 23.12. Because this practice is ordinarily followed in the construction of stud walls, it does not constitute any addition for the lateral load effect. For case 2 there must be both vertical and horizontal continuity of the framing at the top corners of the door openings. This is necessary for the development of the rigid frame. The vertical members would thus be the same as for case 1, but the horizontal continuity must be achieved by the addition of horizontal blocking between the studs and a tension tie across the corner, as shown in the figure.

Because both cases in this example result in no requirement for a hold down at the base, it would seem to be wise to assume case 1 and omit the blocks and straps. However, many designers prefer to use this detail at openings routinely in

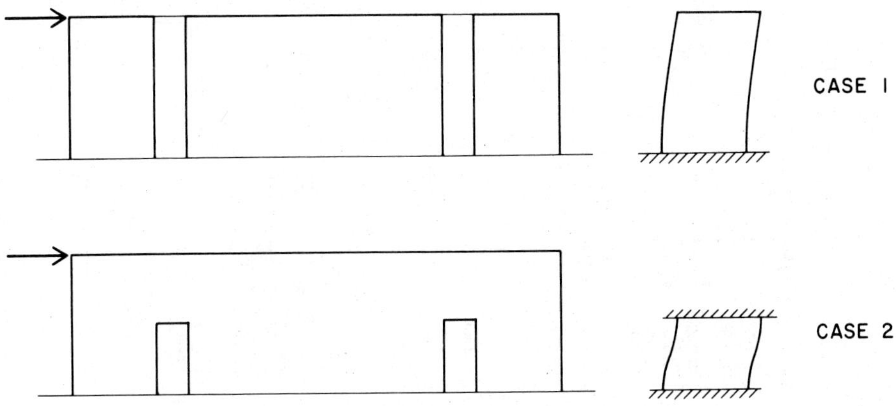

FIGURE 23.10. Alternate functioning of the rear wall.

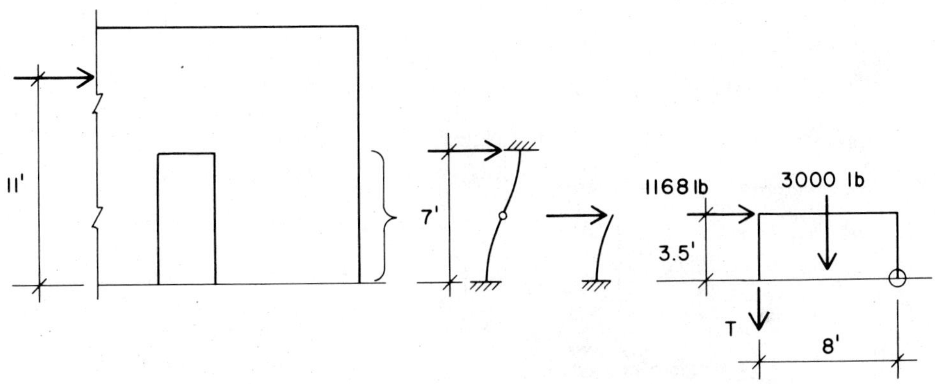

FIGURE 23.11. Functioning of a rear wall pier as a fully fixed member.

order to reduce the possibility of vertical cracking of the wall at the corner of the opening.

For all of the walls the framing at the roof edge or at the top of the wall must function as a collector, as discussed in Section 21.1. At the end wall the collector over the 12-ft-long opening accumulates a total load from the roof edge of

$$F = (131 \text{ lb/ft})(12 \text{ ft}) = 1572 \text{ lb}$$

For load in a northerly direction this force is pushed into the shear wall, and the collector functions as a compression member. For load in a southerly direction the load is dragged into the shear wall, and the collector functions as a tension member. The framing that is considered to perform these functions would be designed for both of these strucutral actions. Most likely the compression would be critical for the framing member, and the tension would be considered only if the member must be spliced.

If the end wall is constructed as shown in Fig. 23.8, the continuous ledger would be considered to perform these functions. This member is more than adequate for the compression force, and if a splice can

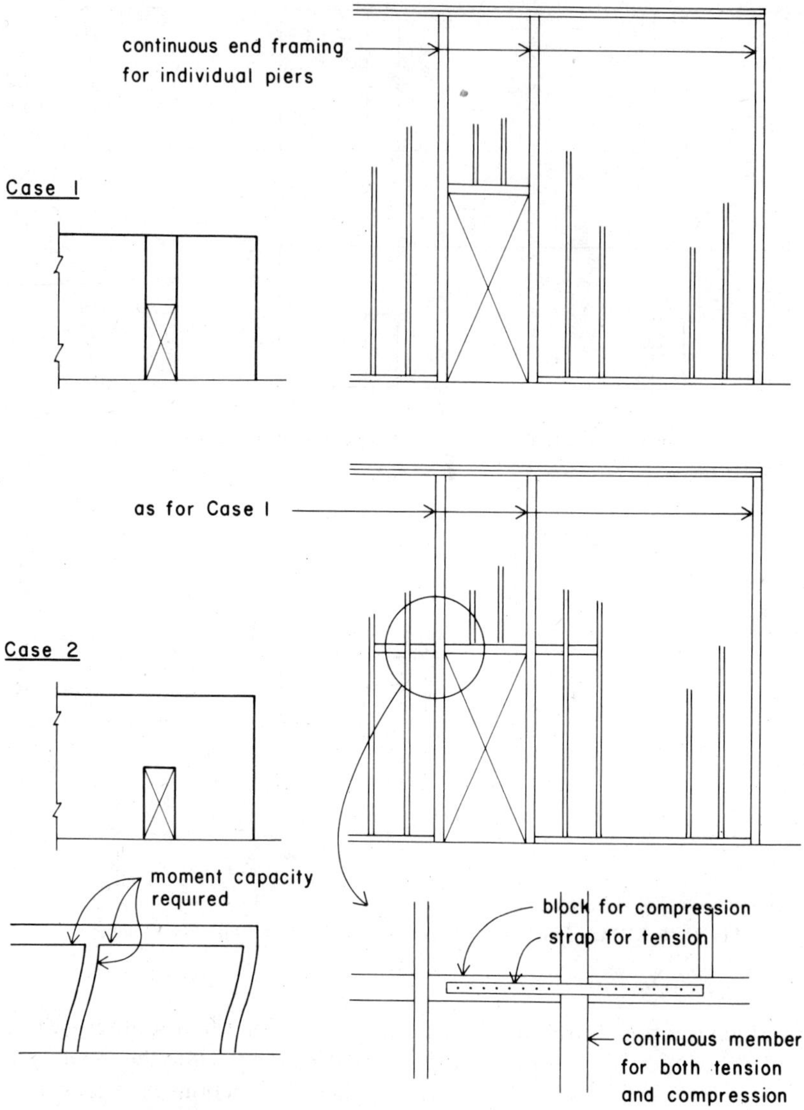

FIGURE 23.12. Framing considerations for the rear wall.

be avoided over the wall opening, no special consideration would be necessary in the construction for the collector function.

With this type of construction the wall and parapet are usually either as shown in Fig. 23.8 or as shown in Fig. 23.9. If the studs are continuous above the roof, as shown in Fig. 23.8, they become analogous to a beam with one overhanging end and their behavior will be as shown for case 1 in Fig. 23.13. The reaction R in the figure would be the edge loading for the roof

If the wall is constructed with ordinary platform framing details, as shown in Fig. 23.9, the wall action will be as shown for case 2 in Fig. 23.13. The parapet becomes a separate, freestanding wall on top of the

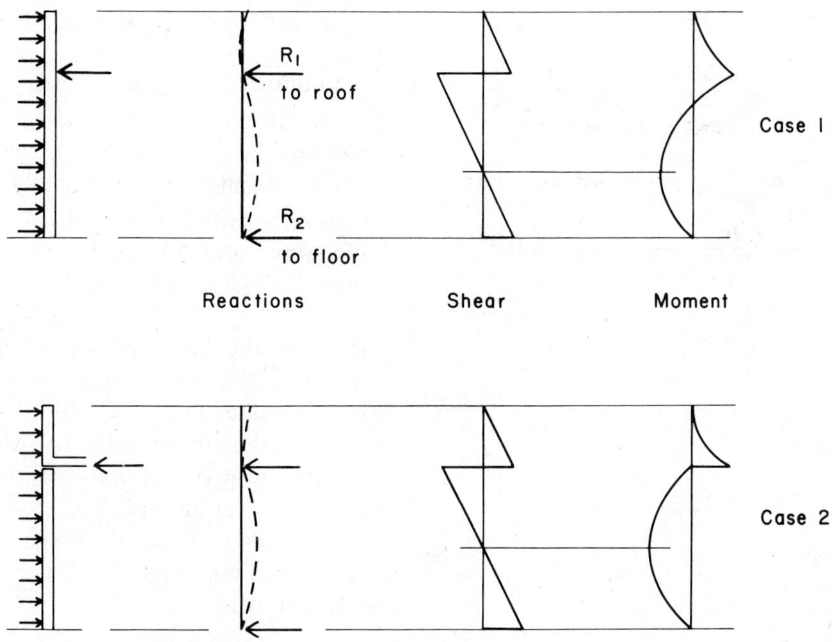

FIGURE 23.13. Alternate functioning of the parapet wall.

roof, and it must be braced independently for lateral force. The means of bracing depends on the height of the parapet as well as other considerations, such as the finishing materials used, the type of roofing, the means for roof drainage, the type of roof framing system, and so on. The illustrations in Fig. 23.14 show two possibilities for the bracing of a free-standing parapet.

In the upper figure a series of diagonal wood braces are nailed to the short studs of the parapet and to a continuous member that is attached to the roof framing through the deck. The diagonal members also function as the framing for the plywood used to achieve the cant at the roof-to-wall intersection. This technique is most useful for relatively short parapets, else the cant would tend to get quite large for a tall parapet.

In the lower illustration in Fig. 23.14, a diagonal brace is attached to the back face of the parapet and to the top of the roof structure. The brace may be of metal or wood, and the details for its attachment and the desired spacing of the braces would be critical design factors. The most undesirable feature of this technique is the attachment to the roof, which presents a problem with regard to the roofing. A second undesirable feature is the exposure of the brace and the attachments, which is likely to produce deterioration of the structure. Careful consideration should be given to both of these problems if this technique is used.

23.2 LONG PLYWOOD DIAPHRAGM

As shown in Fig. 23.15, this example is similar to that in Section 23.1, except for the longer east–west plan dimension and the higher walls. These changes produce higher stresses in the roof diaphragm and higher stresses and overturning effects in

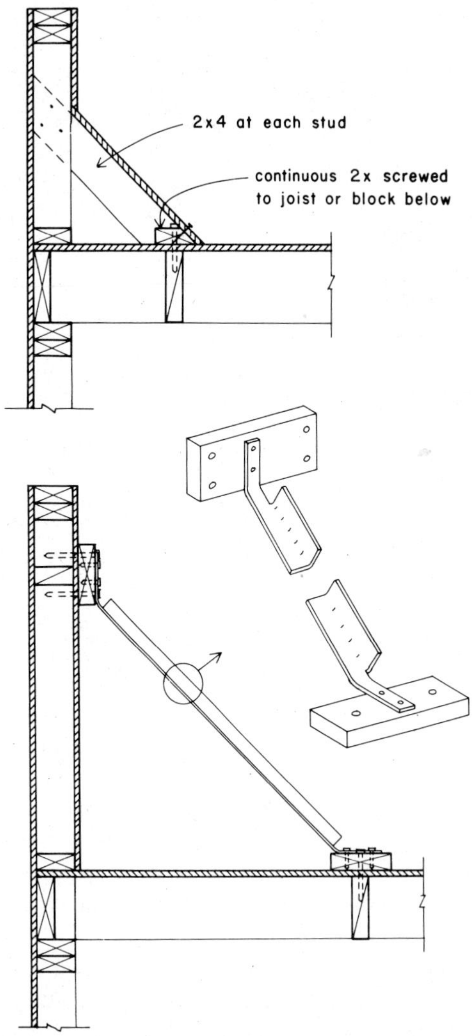

FIGURE 23.14. Bracing the parapet.

the shear walls on the ends of the building. The investigation is limited to these two problems in this example.

Calculation Sheet 23.2.1 shows the computations for the seismic load in the north–south direction and the shear stresses and chord forces in the roof diaphragm.

Using Douglas fir plywood for the roof deck (see Chapter 28 for *UBC* Table 25–J), some options are as follows:

$\frac{1}{2}$-in. Structural II with 2× framing, blocking, and 8d nails at $2\frac{1}{2}$ in. at boundaries and at 4 in. at other edges.

$\frac{5}{8}$-in. Structural II with 3× framing, blocking, and 10d nails at 4 in. at boundaries and at 6 in. at other edges.

$\frac{1}{2}$-in. Structural I with 3× framing, blocking, and 10d nails at 4 in. at boundaries and at 6 in. at other edges.

Because the high stresses occur only near the ends of the building, it is reasonable to consider the possibility of zoning the deck nailing in this case. By using the $\frac{1}{2}$-in. Structural II plywood, it is possible to use two fewer nail spacings, as given in the *UBC* table. Thus the range of nail spacings and corresponding load ratings are the following:

8d at $2\frac{1}{2}$ in. at boundaries, 4 in. at other edges: load = 530 lb/ft.

8d at 4 in. at boundaries, 6 in. at other edges: load = 360lb/ft.

8d at 6 in. at all edges: load = 270 lb/ft.

In Fig. 23.16 these allowable loads are plotted on the graph of stress variation in the deck to permit the determination of the areas in which the various nailing spacings are usable. The maximum nailing is seen to be required on only a very small area at each end of the roof. The actual dimensions of the specified nailing zones may be adjusted slightly to correspond to modules of the roof framing and plywood sheet layouts, so long as the limits of the calculated zone boundaries are not exceeded.

In the end shear wall the edge load from the roof results in a maximum stress of 700 lb/ft (see Calculation Sheet 23.2.2). Because this result is in excess of the maximum capacity of the nailing at the end of the roof diaphragm (530 lb/ft as previously shown), a direct transfer of the total load from the horizontal to the vertical

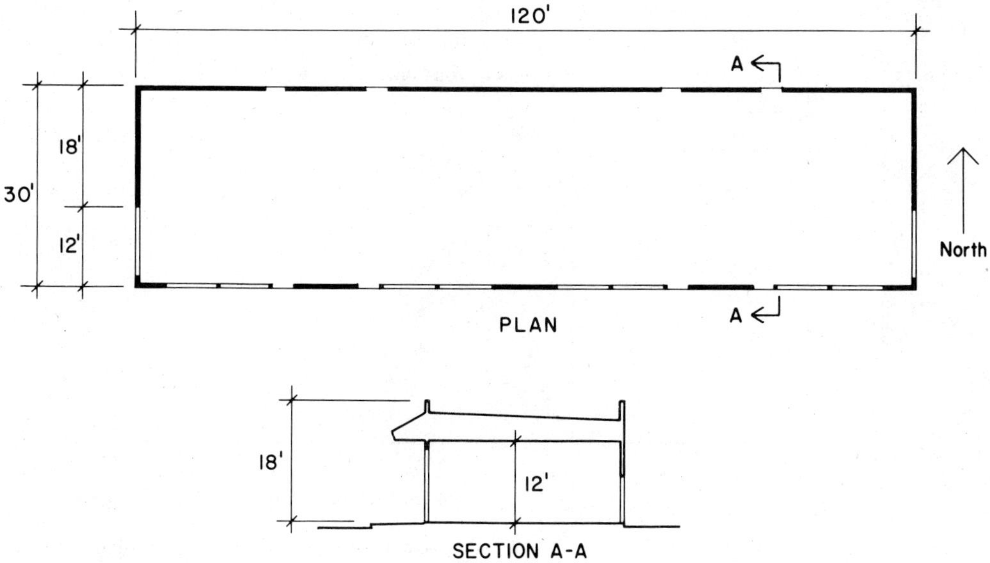

FIGURE 23.15

Seismic Load Analysis - Example 23.2	Sheet No. 23.2.1

Use Sec. 2312 of the 1982 UBC

$V = ZIKCSW = (1.0)(1.0)(1.0)(0.14)\,W = 0.14\,W$

References: Art. 14.1; Chap. 32

Unit loads:
- Roof + ceiling = 16 psf
- Ext. walls (solid) = 20 psf
- Int. walls = 10 psf
- Canopy - add to wall = 200 lb/ft of wall
- Rooftop HVAC = 15 k total

Dead load for roof diaphragm:

Source:	North - South Load:	
Roof + clg.	30' × 120' × 16 psf	= 57.6
Ext. walls	240' × 11' × 20 psf	= 52.8
Int. walls	200' × 7' × 10 psf	= 14.0
Canopy	120' × 200 lb/ft	= 24.0
HVAC		= 15.0
		163.4 k

$V = 0.14(163.4) = 22.88\text{ k}$ max. $v = 11{,}440/30 = 381$ lb/ft

Add accidental torsion: (UBC: 2312(e)5)

Approx. $J = 2 \times 18 \times (60)^2 = 129{,}600\ \text{ft}^3$ (end walls only)

at 5% e: $v = \dfrac{Fec}{J} = \dfrac{22880 \times (0.05 \times 120) \times 60}{129{,}600} = 64\ \text{lb/ft}$

Chord force: $T = C = \dfrac{22880 \times 120}{8 \times 30} = 11{,}440\ \text{lb}$

Collector load - end wall header to shear wall

$T = C = 381 \times 12 = 4572\ \text{lb}$

Calculation Sheet 23.2.1

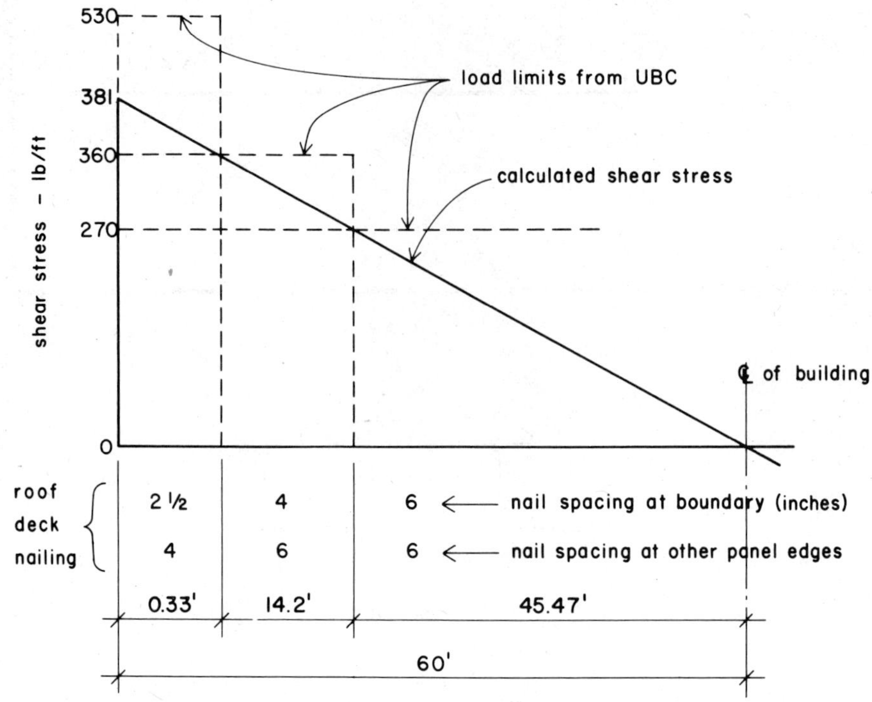

FIGURE 23.16. Zoned nailing for the roof.

Analysis of the end shear wall:	Sheet No. 23.2.2
Max. $v = \frac{11{,}440}{18} = 636 + 64 = 700$ lb / ft. (64 = accid. torsion)	References: Fig. 23.17
For overturn: $V = 700 \times 18 = 12600$ lb	
OT Mom. = $12600 \times 14 \times 1.5 = 264{,}600$; $6500 \times 9 \times 1.5 = 87{,}750$; total $352{,}350$ ft-lb	
DL Mom. = $7500 \times 9 = 67{,}500$; $284{,}850$ ft-lb	
Required $T = \frac{284{,}850}{18} = 15{,}825$ lb	
Sliding:	
Using 3x sill with ⅞" bolts @ 1370 lb each,	UBC Table 25-F
Total No. $= \frac{12600}{1370} = 9.2$ - use 10	
With end bolts 12" from end of wall:	
Spacing $= \frac{16 \text{ ft}}{9 \text{ spaces}} = 1.77' \sim 21"$ c/c	

Calculation Sheet 23.2.2

diaphragm is not possible. A collector over the wall opening must be used to push and pull the roof edge load at that location into the shear wall. Whatever parts of the roof or wall framing are used for this function, the elements used, as well as any splices in them, must be designed for this force in both tension and compression.

Using Douglas fir plywood for the wall sheathing (see Chapter 28 for *UBC* Table 25-K), possible options are:

$\frac{1}{2}$-in. Structural II with 3× boundary framing and 10d nails at 2 in. at all edges: load = 770 lb/ft.

$\frac{3}{8}$-in. Structural I with 8d nails at 2 in. at all edges: load = 610(1.2) = 732 lb/ft. (See footnote No. 3 in the *UBC* table in Chapter 28.)

$\frac{1}{2}$-in. Structural I with 3× boundary framing and 10d nails at $2\frac{1}{2}$ in. at all edges: load = 770 lb/ft.

And if it is possible to have plywood on both sides of the wall, use:

$\frac{3}{8}$-in. Structural II, both sides, with 8d nails at 4 in. at all edges: load = 2(320)(1.2) = 768 lb/ft.

Although technically permitted by the code, none of the foregoing is desirable. The closely spaced, heavy nailing is likely to cause some splitting of framing members. Placing plywood on both sides of the wall can entail some difficulty in the framing details and in the installation of wiring or other elements within the wall cavity. The heavy shear load on the wall is also likely to be a problem in terms of stress on the wall boundary framing.

Figure 23.17 shows the loading for the end shear wall. The required tie-down force is considerable and requires a very heavy anchorage device, heavy end framing in the wall, and a strong foundation. Development of sliding resistance for the total load requires heavy bolting of the wall sill to the foundation. Although the chord force can be developed with ordinary framing members, the length of the building will result in several splices, each of which needs considerable bolting to maintain the continuity of the chord.

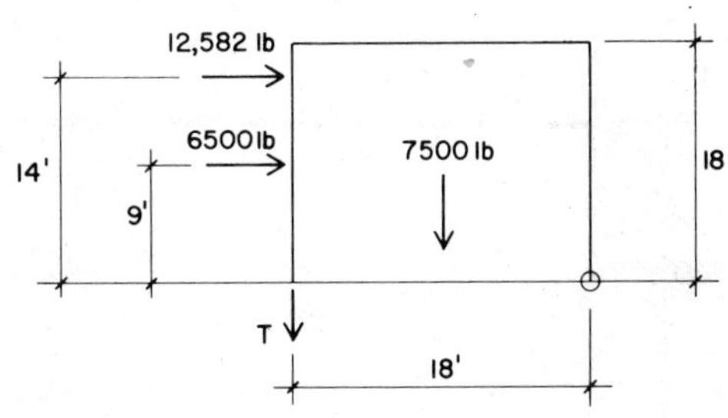

FIGURE 23.17. Loading for the end shear wall.

A final consideration is the horizontal deflection of the roof diaphragm at the center of the 120-ft span. The diaphragm depth-to-span ratio of 4 to 1 is just at the limit permitted by the *UBC* (see *UBC* Table 25−I in Chapter 33). Even though the diaphragm is within the code limit, the actual dimension of the deflection should be determined and its possible effect on interior partitions considered.

We abandon this example without attempting to solve all its problems because we actually consider it to be a poor solution. It was presented to show what bad planning can produce in the way of difficult situations and not as an illustration of proper design.

23.3 BOX WITH INTERIOR SHEAR WALL

As shown in Fig. 23.18, this example consists of the same building used in Section 23.2, with the addition of an interior shear wall. Using the same loading as in the previous example, the analysis is shown in Fig. 23.19. In this analysis the roof diaphragm is assumed to be sufficiently flexible to justify an assumption of distribution

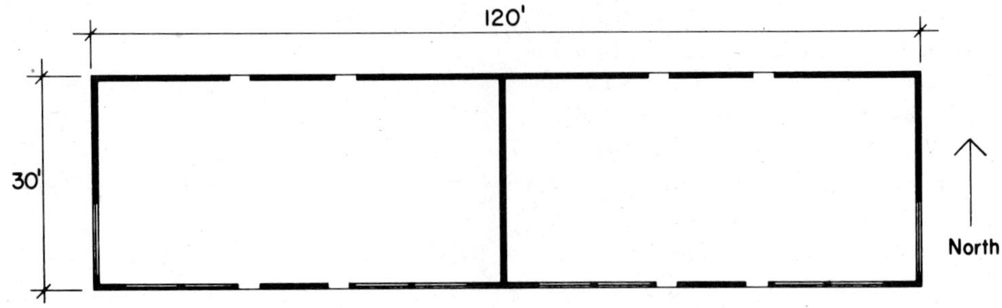

FIGURE 23.18

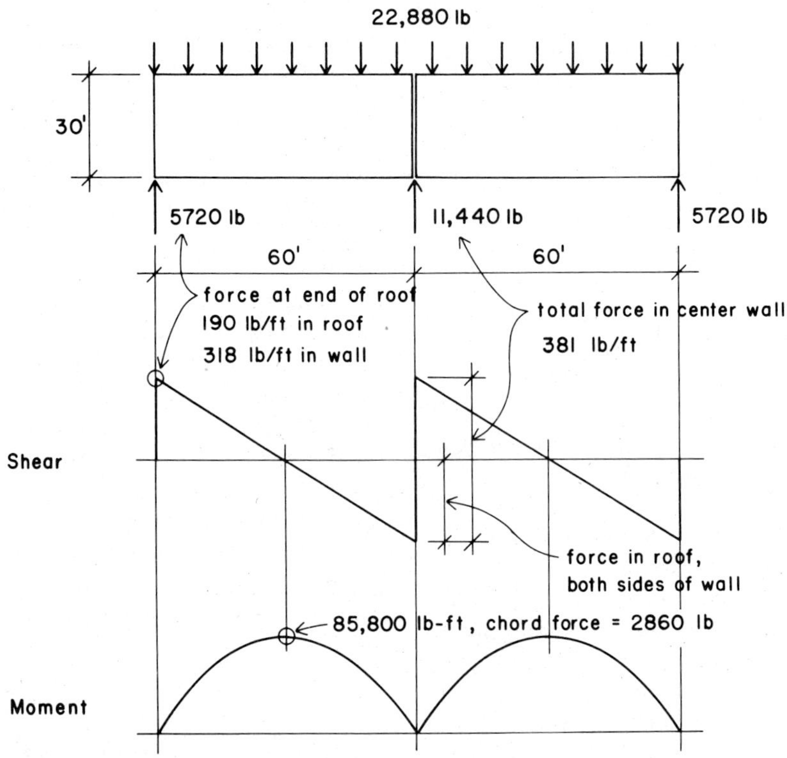

FIGURE 23.19. Roof function—peripheral distribution.

of load to the vertical elements on a peripheral basis, which is consistent with the usual practice for a wood diaphragm with the depth-to-span ratio in the example. With this assumption half of the total load is taken by the center wall, and the load on the end walls is reduced to one-half that in the previous example.

Ignoring torsional effects, the maximum stress in the roof diaphragm drops to 190 lb/ft, occurring at the ends and at both sides of the interior wall. This is less than the lowest rated capacity for $\frac{1}{2}$-in. plywood with edge blocking and minimum nailing as given in *UBC* Table 25–J (see Section 28.1). Nail zoning is therefore not a consideration, unless it is acceptable to use an unblocked diaphragm for some portions of the roof.

The shear stress in the end wall drops to 318 lb/ft. This is still a significant stress, but it can be achieved with $\frac{3}{8}$-in. Structural II plywood with 8d nails at 4 in. at all panel edges, which is quite reasonable. The overturn and tie-down requirement are also considerably reduced.

The center shear wall must carry a considerable force—the same as that on the end shear walls in the previous example. However, the wall is longer in plan, which results in a lower shear stress and less overturn effect. The design of this wall must include the full consideration of its use in the building. Possible issues are its use for fire separation, acoustic separation, and load bearing for the roof structure. For our analysis we consider it to be a single stud wall and to serve as a bearing wall. If it were not a bearing wall, the net overturning effect would be higher. It happens in this example that the same plywood and nailing can be used for the end walls and the interior wall. From the *UBC* table, the capacity is 384 lb/ft, which includes the allowable 20% increase noted in the footnote to the table.

If the interior wall is not a bearing wall, there may be some tie-down requirement. However, the connection of this wall to the walls at the front and rear of the building should be adequate for this function.

If sliding friction is ignored and the entire horizontal force is taken on the sill bolts, the choice is whether to use a lot of small bolts or a few large ones. This is a matter of individual preference by designers and builders. If the maximum code spacing of 6 ft is used, the two walls will have a minimum of 4 bolts in the end wall and 6 in the center wall. Thus, the minimum required capacity for this minimum number of bolts is as follows:

$$\text{end wall:}\quad \text{Load} = \frac{5720}{4} = 1430 \text{ lb/bolt}$$

$$\text{Center wall: Load} = \frac{11440}{6} = 1907 \text{ lb/bolt}$$

If the bolt size required for these loads is not excessive, the minimum number of bolts may be used—which is the usual preference of both the concrete and framing contractors.

A construction detail that must be developed for this example is the connection between the roof diaphragm and the center shear wall. This connection must transfer the total force from the roof to the wall. Because there are several options for both the roof and wall construction, the potential variations for this connection become quite numerous. Figure 23.20 illustrates some of the possibilities, based on the assumption that the wall is a single stud wall and that the roof framing consists of joists perpendicular to and supported by the wall. There are three basic structural functions to be considered for this situation: the vertical gravity load transfer, the hold down against wind uplift, and the transfer of diaphragm force parallel to the wall.

Detail "A" in Fig. 23.20 shows the ordinary construction used if the gravity load alone is considered. The joists are either butted end to end, or they are lapped on top of the wall and are toe nailed to the top plate. Vertical blocking is usually provided between the joists for their stability as well as to provide a nailer for the plywood edges perpendicular to the joists. This joint provides only minor resistance to uplift (relying on withdrawal of the toenails) and virtually no capacity for transfer of the shear.

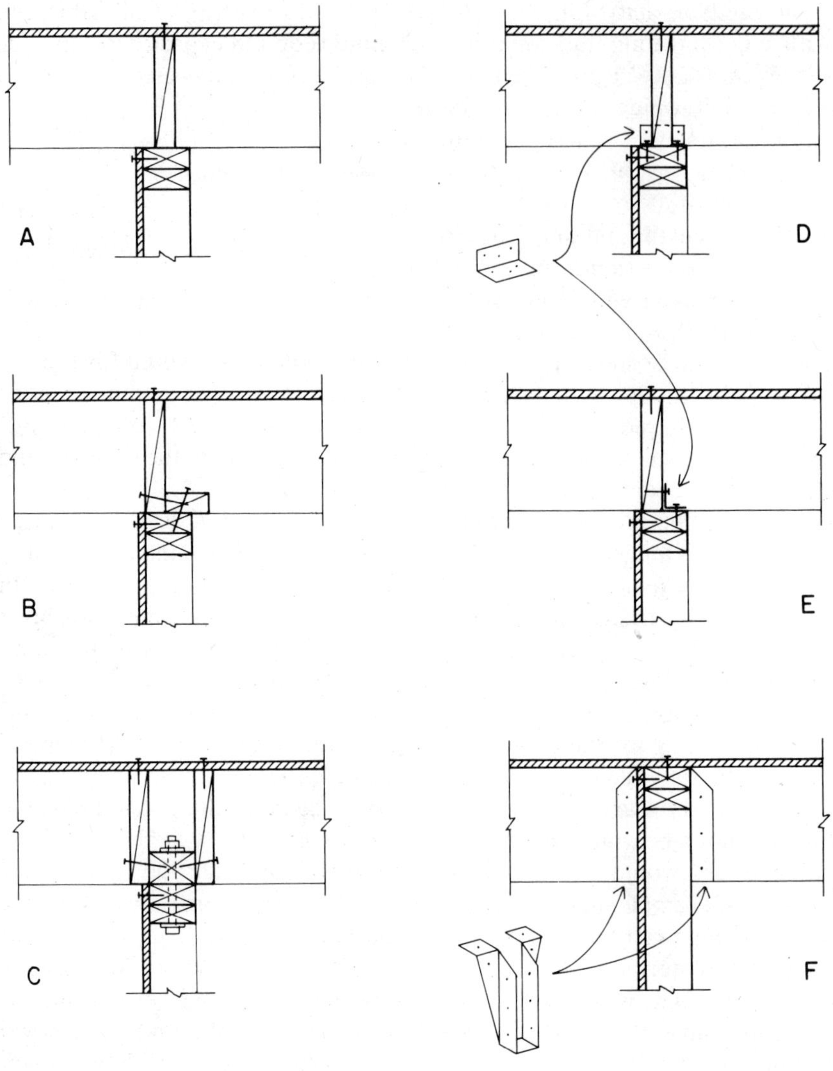

FIGURE 23.20. Connection details—roof to interior wall.

In detail "B" of Fig. 23.20, a second, horizontal block is added to facilitate the transfer of diaphragm shear. The vertical block is nailed to the horizontal block, and the horizontal block is nailed to the top plate. With the roof deck nailed to the top of the vertical block, the transfer is achieved from the roof deck to the top plate of the wall. With the wall surfacing nailed to the top plate, the transfer from horizontal to vertical diaphragms is then complete.

It should be noted that, although the stress in the roof deck is only 190 lb/ft at this location (see Fig. 23.19), the total load transfer to the wall is twice this, resulting from a delivery of load to both sides of the wall by the deck. Thus all the nailing shown in detail "B" of Fig. 23.20—the deck to the vertical block, the blocks to one another, and the block to the top plate—must be designed for the load transfer of 381 lb/ft. If a roof plywood panel edge occurs at this point, the nailing will be adequate because there will actually be two edge nailings at the joint. If this is an interior support point for a plywood panel, the usual minimum nailing with nails at 12-in. centers will not be sufficient, and a nailing must be specified that is capable of the load transfer. It is questionable whether the nailing of the blocks for this load magnitude could be achieved without splitting the blocks, so this option is probably not the best for the example.

A variation on detail "B" is shown in detail "C" of Fig. 23.20. A second vertical block is added, and the horizontal block is bolted to the top plate. This results in an extended range of load capacity because there are now two rows of nails and the bolts are much stronger than nails in lateral load resistance. However, with fairly closely spaced joists, this detail would require a considerable number of bolts through the top plates.

Another approach to this connection is shown in details "D" and "E" of Fig. 23.20, in which metal framing anchors are used for the attachment to the top plates. In detail "D" the anchors are attached to the joists, and in detail "E" they are attached to the blocks. Assuming that the anchor devices used have a rated load capacity adequate for the load transfer, either of these options is acceptable for the wind shear function. Detail "D" is slightly better for wind uplift because there is more attachment between the roof deck and the joists than there is between the deck and the blocks.

A third technique for this connection is shown in detail "F" of Fig. 23.20, in which the top plate of the wall is raised to the level of the roof deck. This procedure allows for the simplest transfer of the wind shear because both the roof deck and the wall sheathing are directly nailed to the top plate. The joists are supported by saddle-type metal hangers hung from the top plate. One problem with this detail is that the upper panels of the wall sheathing must be installed before the roof framing can be placed, which is not the usual sequence of the construction.

We do not attempt to judge which of these, or of other possible options, is the best solution for this connection. From the viewpoint of structural design, anything that "works" is all right. In real situations there are many issues to consider in addition to the necessary structural functions. Thus the influence of roof drainage, wall surface finishes, ceiling construction, ductwork installation, and so on may provide the deciding factors for choice between viable alternatives.

23.4 BUILDING WITH MASONRY WALLS

The building in this example is the same as that in Section 23.1, except that the walls are of reinforced concrete block. The differences resulting from the change in construction are as follows:

1. The K factor for determination of the seismic force is 1.33, whereas the *UBC* allows a K of 1.0 for an all wood frame and plywood building. This, combined with the weight of the heavier walls, results in considerably more load for the roof diaphragm and shear walls.
2. Because of the weight of the walls and the doweling of the vertical reinforcing, overturn and sliding are not a problem. However, the foundations should be designed for the combination of the gravity and seismic loads.
3. The rear wall should be designed as case 2 in Fig. 23.10 if the masonry is continuous over the door openings. Ordinary required reinforcing will

provide sufficient strength for the wall actions for this case.

4. Table 24–H of the *UBC* requires that the lateral load for seismic design be increased by 50% when calculating the shear stresses in the walls.

5. Section 2310 of the *UBC* requires that the roof-to-wall connection be designed for an outward load on the wall of 200 lb/ft or the calculated force, whichever is greater.

6. For determination of the distribution of loads to the piers in the east–west direction, the pier stiffnesses are determined on the basis of flexural stiffness. This is different from the procedure used for plywood walls, in which pier stiffnesses are assumed to be proportionate to the plan lengths of the walls.

The determination of pier and total wall stiffnesses for the front and rear walls is illustrated on Calculation Sheet 23.4.1. Because of the large opening width and short header height in the front wall, the pier is assumed to be cantilevered from a fixed base (case 1 in Fig. 23.10). For the rear wall, due to the narrow openings and tall headers, the piers are assumed to be fixed at both top and bottom. Stiffnesses for piers for both of these conditions are given in the tables in Chapter 31. The stiffness of individual piers (R in the calculations) is determined as a function of the ratio of height to length (h/d in the calculations). The total stiffness of the rear wall is the sum of the R values for the three piers.

The R values for the piers would also be used to determine the J for torsional resistance; just as the plan lengths of the plywood walls were used in the example in

Sheet No. 23.4.1

References: Chap. 16, Chap. 31, Fig. 23.21

Analysis of piers for stiffness distribution:

Wall	Pier	h	d	h/d	Pier R	No. of piers	Σ R
Rear	1	7'	8.67'	0.807	0.8485	2	1.697
	2	7'	26'	0.269	3.025	1	3.025
				Total stiffness of rear wall =			4.722
Front	3	12'	14'	0.857	0.4928	1	0.4928
				Total stiffness of E-W walls =			5.2148

For front wall - use 1/2 of total E-W load (peripheral)

For rear wall - use proportionate stiffness:

$$\text{Wall load} = \text{E-W load} \times \frac{4.722}{5.2148}$$

Pier loads:

$$\text{Pier 1 - load} = \text{Wall load} \times \frac{0.8485}{4.722}$$

$$\text{Pier 2 - load} = \text{Wall load} \times \frac{3.025}{4.722}$$

Calculation Sheet 23.4.1

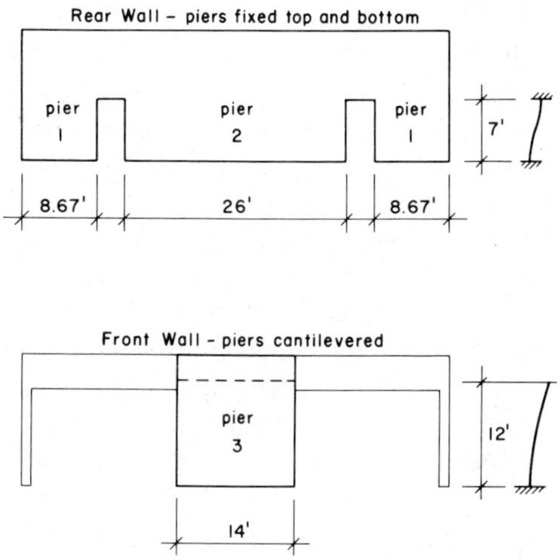

FIGURE 23.21. Pier functions in the masonry wall.

Section 23.1. It is doubtful that any type of roof diaphragm other than a poured concrete deck would be sufficiently stiff to induce any significant torsion on the building with reinforced masonry walls. However, the minimum eccentricity required by the *UBC* should be used for the wall design.

Although we have not shown the complete calculations for load determination for this building, a hypothetical load for the rear wall is assumed in order to demonstrate the procedure for finding the loads on the individual piers in the rear wall. As noted in comment 4 above, this load must be increased by 50% for the design of the wall.

The effect of the gravity load plus the overturning moment on the wall foundation should be investigated, especially if the supporting structure is a shallow grade beam or wall. Design for this condition, as well as design of the wall itself, is illustrated in other examples in Chapter 24 in which the load magnitudes are considerably higher.

23.5 STEEL DECK DIAPHRAGM

The building in this example is the same as that in Section 23.1, except that the roof structure is a light steel frame with a formed sheet steel deck and the walls are of reinforced concrete block. The maximum shear stress is quite low for the steel deck. Assuming the maximum deck span to be 6 ft, the deck units will probably be placed in a minimum unit length of 3 spans, or 18 ft. We will therefore assume that the gravity design will be satisfied with a type B deck of 22-gauge steel. For the lateral load design decisions to be

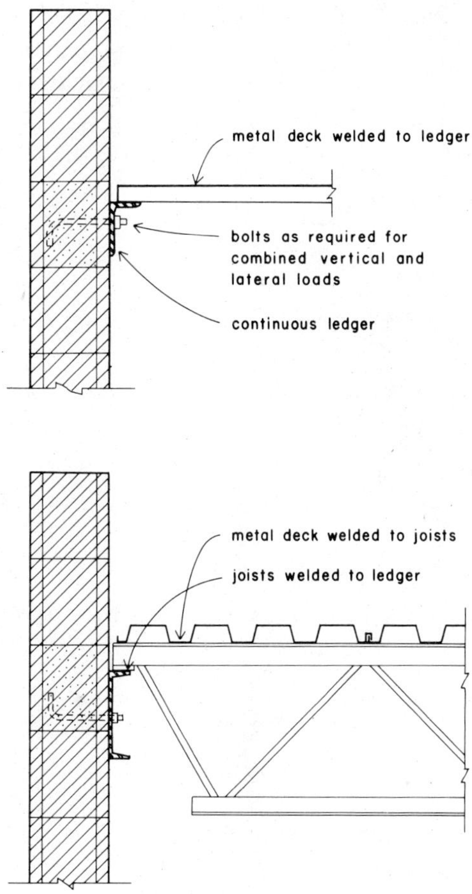

FIGURE 23.22. Roof–wall joints.

made include the method of attachment of the deck to its steel supports and the means for connecting adjacent deck units at their common edges, called the *side seams*.

Data for the lateral load design of steel decks can be obtained from various sources. In Chapter 29 we show an excerpt from the *Inryco Lateral Diaphragm Data Manual 20-2*. The table shown is for type B deck with a deck unit width of 24 in., side seams button-punched at 36-in. centers, and attachment to supports with 3 puddle welds per unit, as shown in the sketch accompanying the table. For our example with the 22-gauge deck on a 6-ft span, the allowable lateral shear is given in the table as 240 lb/ft (labeled "q" in the table.) The F values given in the table ($27.87 + 187.0R$ for our example) are for determination of the relative flexibility of the diaphragm and calculation of the diaphragm deflection, as discussed in Section 15.4.

Some possibilities for the roof-to-wall connections are shown in Fig. 23.22, where it is assumed that the roof framing consists of open web steel joists spanning the 30-ft dimension and supported at the front and rear by a ledger channel bolted to the wall. At the end walls the end of the metal deck is supported by a ledger angle that is bolted to the wall. The deck-to-joist, deck-to-ledger, joist-to-ledger, and ledgers-to-walls connections must be adequate to transfer the combinations of gravity and wind loads. The roof-to-wall connection must also be capable of resisting the minimum outward load on the wall of 200 lb/ft, as specified by Section 2310 of the *UBC*.

CHAPTER TWENTY-FOUR

Shear Walls—Multistory

The examples in this chapter illustrate the use of shear walls in multistory buildings of various size and construction type. For most of the examples illustrations are also developed of alternative bracing systems in Chapters 25 and 26. The general problems of shear walls are discussed in Chapter 16, and the basic nature of shear walls are thoroughly illustrated there. A review of Chapter 16 is advised before any study of the material in this chapter since the explanation of basic concepts is omitted here for sake of brevity.

24.1 TWO-STORY BOX

This example consists of a simple two-story box, as shown in the plan and section in Fig. 24.1. The lateral load in the north–south direction is resisted by the two end shear walls. As shown in Fig. 24.2, the loading on these walls consists of the two end reactions from the roof and second floor diaphragms.

Using *UBC* formula 12–3A from Section 2312(d) we determine an approximate value for T as

$$T = \frac{0.05h_n}{\sqrt{D}} = \frac{0.05(20)}{\sqrt{30}} = 0.183 \text{ sec}$$

This produces a calculated value for C of

$$C = \frac{1}{15\sqrt{T}} = \frac{1}{15\sqrt{0.183}} = 0.156$$

Because this is in excess of the maximum required value of 0.12 for C, we use the same factor for the total lateral force as for the one-story box system building. The total dead loads for the lateral forces applied through the roof and second-floor diaphragms are shown in the table in Calculation Sheet 24.1.1. We will assume an all wood frame structure with plywood diaphragms; thus the base shear is determined as for the one-story building, with $V = 0.14W$.

This total force must be distributed to the roof and second floor in accordance with the requirements of Section 2312(e) of the *UBC*. The force at each level, F_x, is determined from Formula 12–7 as follows:

$$F_x = \frac{(V)(w_x h_x)}{\sum_{i=1}^{n} w_i h_i}$$

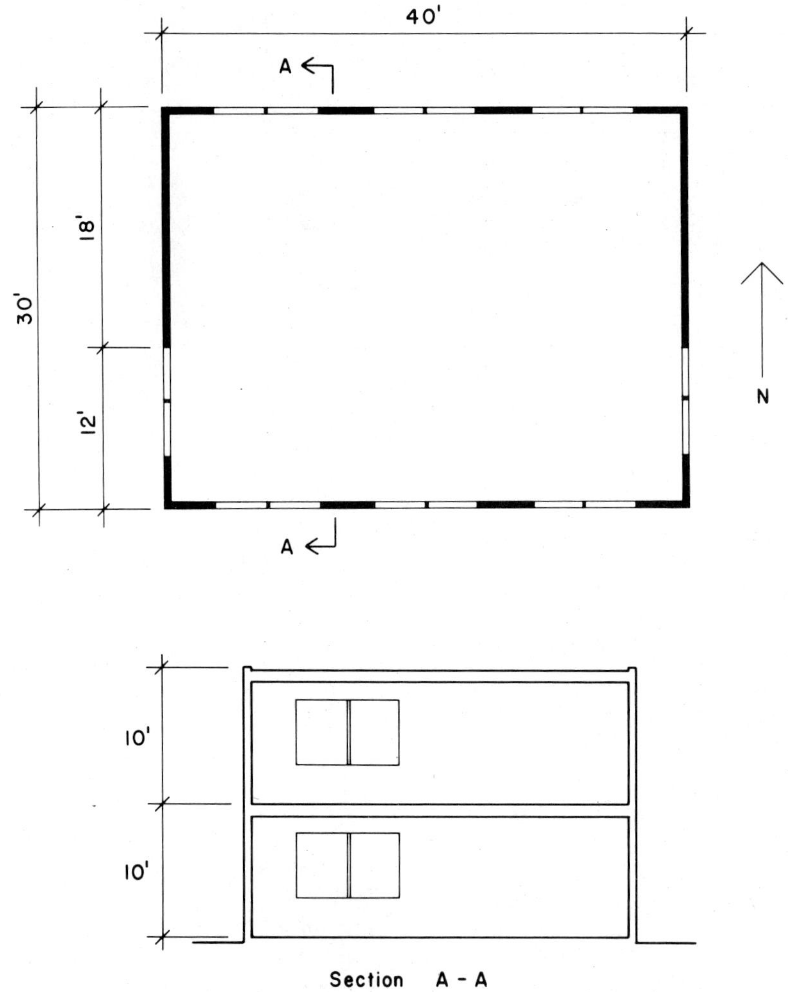

FIGURE 24.1

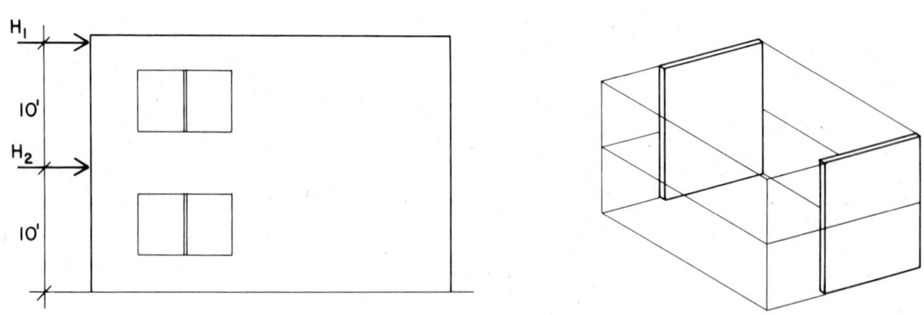

FIGURE 24.2

Sheet No. 24.1.1

References: Art. 14.1, Chap. 32

Dead load for N-S forces:

Level:	Source:	Unit load:	Load:		
Roof	Roof + clg.	12 psf	30' × 40' × 12	=	14,400
	Ext. walls (ave)	15 psf	5' × 80' × 15	=	6,000
	Int. walls	10 psf	5' × 60' × 10	=	3,000
	HVAC unit	8 k			8,000
		Total for level:			31,400 lb
2nd Fl	Floor	10 psf	30' × 40' × 10	=	12,000
	Ext. walls (ave)	15 psf	10' × 80' × 15	=	12,000
	Int. walls	10 psf	10' × 60' × 10	=	6,000
		Total for level:			30,000
		Total for base shear: W		=	61,400 lb

Base shear: $V = 0.14W = 0.14(61.4) = 8.60\text{ k}$

Distribution: (UBC 2312 (e))

$$F_x = V\left[\frac{w_x h_x}{\sum_{i=1}^{n} w_i h_i}\right]$$

Level:	w_x	h_x	$w_x h_x$	F_x
Roof	31.4	20	628	5.82 k
2nd fl.	30.0	10	300	2.78 k
			928	

Analysis of single wall:

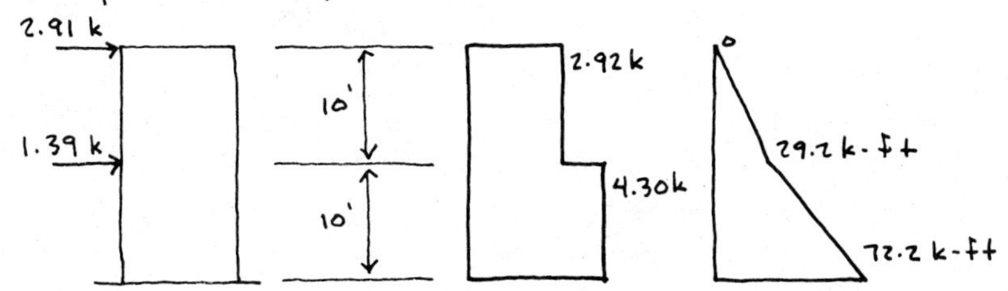

Calculation Sheet 24.1.1

in which

- F_x is the force to be applied at each level (x)
- w_x is the total dead load at level x
- h_x is the height of level x above the base of the structure

(Notice that F_t has been ommitted from the formula because T is less than 0.7 sec.)

The determination of the F_x values is shown in Calculation Sheet 24.1.1 together with the design loading, shear diagram, and moment diagram for the individual shear walls. Figure 24.3 shows the loadings for the overturn analyses of the two stories. The assumed dead loads shown reflect the use of the walls as bearing walls and thus include some of the roof and second-floor loads. The shear stress and overturn analyses for the two stories are shown on Calculation Sheet 24.1.2. Assuming the walls to be of plywood, it

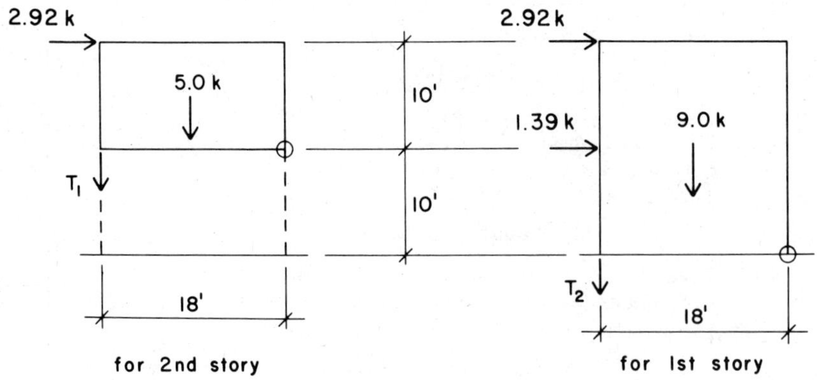

FIGURE 24.3. Loading of the end shear walls.

Unit shear: 2nd story – $v = \frac{2920}{18} = 162$ lb/ft 1st story – $v = \frac{4300}{18} = 239$ lb/ft Overturn: 2nd floor – OT mom. = 2.92 × 10 × 1.5 = 43.8 k-ft DL mom. = 5 × 9 = 45.0 Net for T: 0 1st floor – OT mom. = 2.92 × 20 × 1.5 = 87.6 k-ft 1.39 × 10 × 1.5 = 20.85 Total = 108.45 k-ft DL mom. = 9 × 9 = 81.0 Net for T: 27.45 k-ft Required $T = \frac{27450}{18} = 1525$ lb	Sheet No. 24.1.2 References: Fig. 24.3

Calculation Sheet 24.1.2

would be necessary to use the minimum code-required nailing at the second story and fairly heavy nailing at the first story. The tie-down force is significant, but it is within the capacity of a reasonably sized tie-down device.

Stresses in the horizontal diaphragms are not critical unless there are large openings that reduce their effective width. Figure 24.4 shows a plan for the second floor with a large stair opening next to one end wall. The stair opening reduces the width

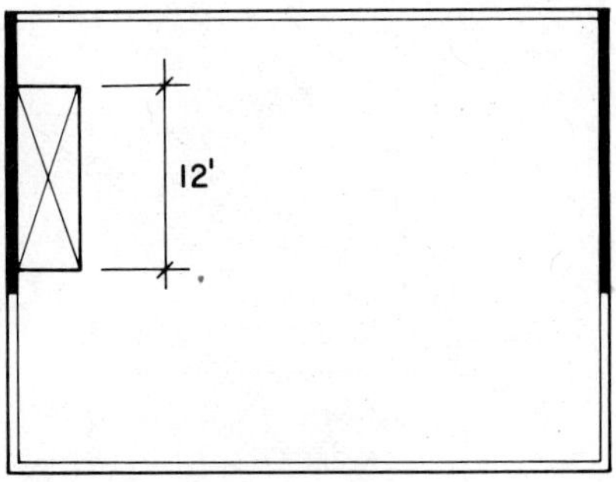

FIGURE 24.4

of the second-floor diaphragm to 18 ft at the end, which results in a maximum stress of

$$v = \frac{2920}{18} = 162 \text{ lb/ft}$$

This is still not very high for the deck, which would probably be a minimum of $\frac{1}{2}$-in plywood. As shown in Fig. 24.5, it would be desirable to reinforce the corners of the opening with straps and blocking.

The location of the opening creates a problem in achieving the transfer of force from the floor to the end wall. With the framing as shown in Fig. 24.6, the header built into the wall may be used to effect the transfer if the header is continuous or is adequately spliced. Care should be taken to assure that the plywood on the end wall is nailed to this header with nails having the capacity to achieve the transfer from the second-floor diaphragm.

Figure 24.6 shows a typical detail for the construction of the end wall at the second floor. Nails "A" in the figure are

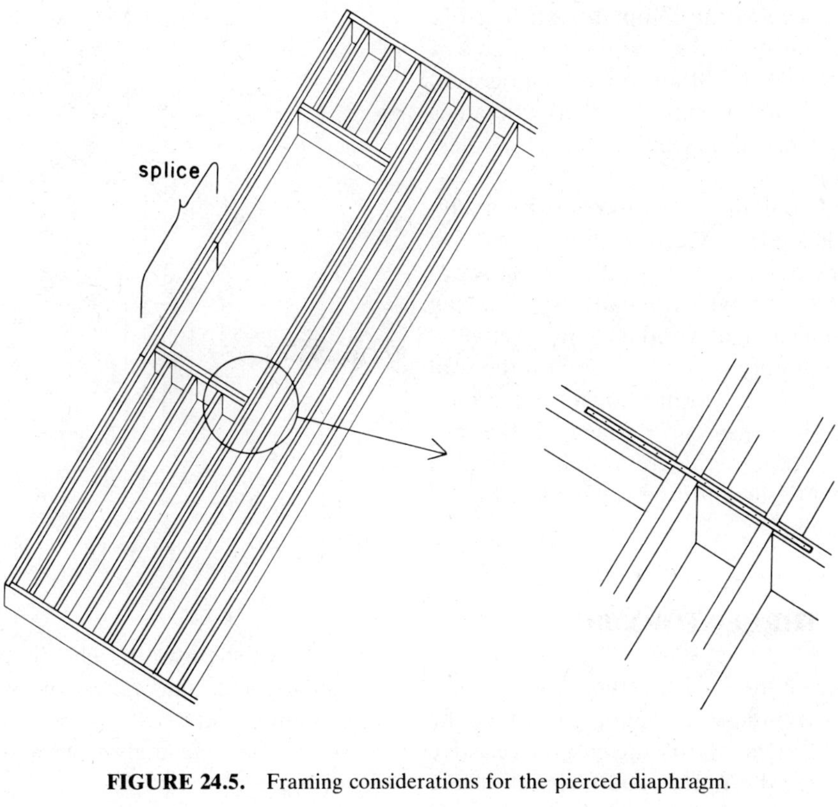

FIGURE 24.5. Framing considerations for the pierced diaphragm.

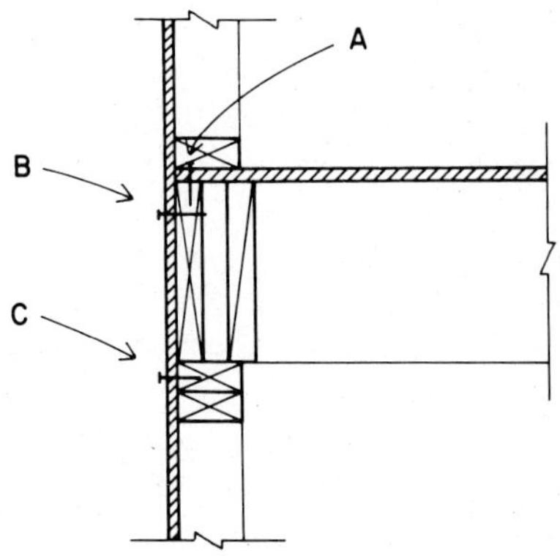

FIGURE 24.6. Wall detail for load transfers.

used to transfer the load from the floor plywood to the header. However, the load should preferably be delivered to the top plates of the first-story wall because they more fully provide continuity at the top edge of the lower shear wall.Thus both nails "B" and "C" in the figure should be used to achieve the complete transfer from the second-floor deck to the top plates of the lower wall. Other means can be used to achieve this result, but for the detail as shown this nailing seems to be the simplest solution.

Although the previous overturn analysis indicated no requirement for a tie down for the second-story wall, we show in Fig. 24.7 two typical means for achieving such a tie in light wood framing. Actually, the continuity of the plywood on the wall surfaces at the corner constitutes a considerable tie effect. However, the ties shown, or others, are usually provided when a tie-down force is calculated to be required.

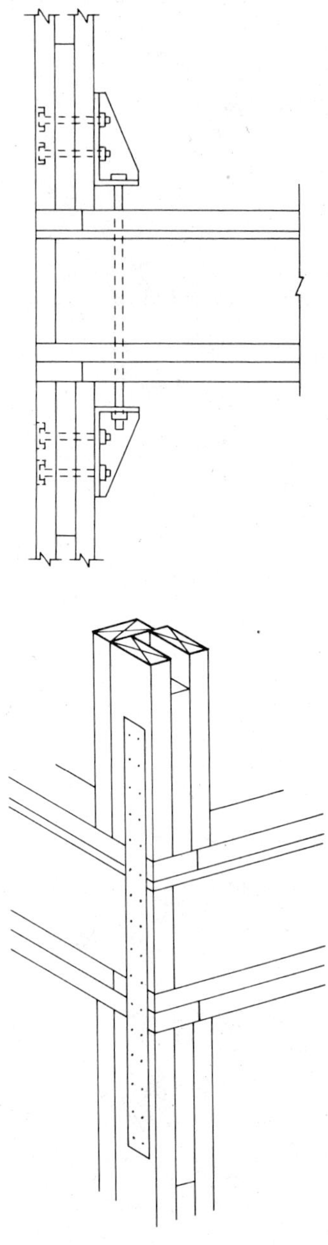

FIGURE 24.7. Anchorage for the upper shear wall.

24.2 THREE-STORY BOX

As shown in Fig. 24.8, this example consists of a three-story building in which the vertical elements of the lateral resistive system in the north–south direction are two interior shear walls. The discussion is limited to the design of these two walls and their foundations.

For the seismic analysis we assume the following:

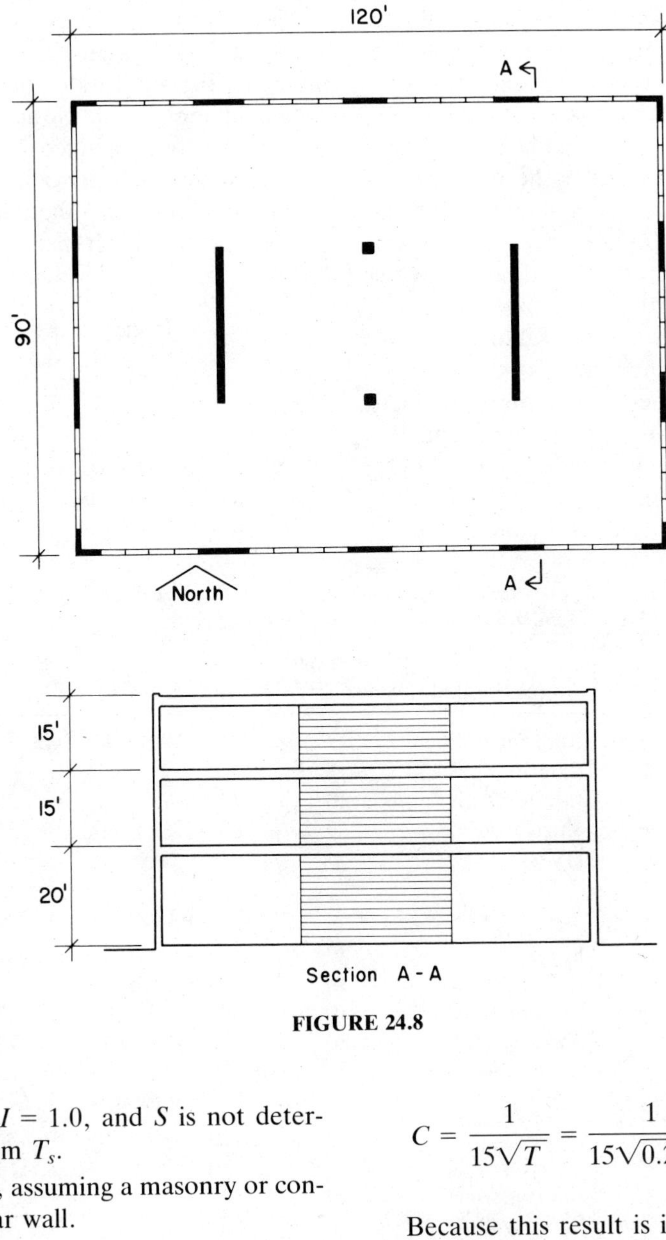

FIGURE 24.8

$Z = 1.0$, $I = 1.0$, and S is not determined from T_s.
$K = 1.33$, assuming a masonry or concrete shear wall.

Section 2312(d) of the *UBC* permits a calculation of the value of T using the simplified formula

$$T = \frac{0.05h_n}{\sqrt{D}} = \frac{0.05(50)}{\sqrt{90}} = 0.263\,\text{sec}$$

which results in a calculated value for C of

$$C = \frac{1}{15\sqrt{T}} = \frac{1}{15\sqrt{0.263}} = 0.13$$

Because this result is in excess of the code maximum value of 0.12, we will use the maximum CS value of 0.14. Thus for the base shear, $V = 0.1862W$. As for the example in Section 24.1, the forces at the various levels must be determined from the F_x values using *UBC* formula 12–7.

The tabulation of the building dead loads for the determination of the values of w_x and the total lateral force, V, are shown in Table 24.1. Using these values, the determination of the values for F_x at each level is as shown in Table 24.2. For the building without a basement, the shear wall foundation shown in Fig. 24.9, with a large grade beam under the wall is proposed. The grade beam adds to the total dead weight for resisting overturn, lowers the bearing level in the soil, and slightly spreads the width at the base. The distances used for h_x in Table 24.2 are measured from the bottom of the footing. The weight of the foundation is included in the determination of the total value for V, but it has not otherwise been included in the determination of the F_x values because its moment effect is quite small.

Using the F_x values determined in Table 24.2, the loading and the shear and moment diagrams for the shear wall are as shown in Fig. 24.10. Notice that the F_x values have been divided by two for the load on a single shear wall.

At the third story the shear force of 37.7 k produces a unit shear on the wall of

$$v = \frac{37{,}700}{30} = 1257 \text{ lb/ft}$$

Assuming an 8-in. block wall with 45% solid block, the stress on the net wall section is

TABLE 24.1. Dead Load for the North–South Seismic Force

Level	Source of Load	Unit Load (psf)	Load	
Roof	Roof and ceiling	12	120 × 90 × 12	= 129.6 k
	Exterior walls	15	420 × 7.5 × 15	= 47.2
	Interior walls	10	300 × 7.5 × 10	= 22.5
	Shear walls	80	60 × 7.5 × 80	= 36.0
	HVAC units			= 15.0
			Subtotal	= 250.3 k
Third floor	Floor	40	120 × 90 × 40	= 432.0
	Exterior walls	15	420 × 15 × 15	= 94.5
	Interior walls	10	300 × 15 × 10	= 45.0
	Shear walls	80	60 × 15 × 80	= 72.0
			Subtotal	= 643.5 k
Second floor	Floor	40	120 × 90 × 40	= 432.0
	Exterior walls	15	420 × 17.5 × 15	= 110.2
	Interior walls	10	300 × 17.5 × 10	= 52.5
	Shear walls	80	60 × 17.5 × 80	= 84.0
			Subtotal	= 678.7 k
First floor	Shear walls	80	60 × 10 × 80	= 48.0
	(Assume remainder of first story load to first floor)			
Total for V at first floor level:				1620.5 k
Foundation				100.0
Total for V at bottom of footing:				1720.5 k
For F_x: V = 0.1862(1720.5) = 320 k				

TABLE 24.2. Seismic Loads

Level	w_x (kips)	h_x (ft)	$w_x h_x$	F_x (kips)
Roof	250.3	58	14,517	75.4
Third floor	643.5	43	27,671	143.8
Second floor	678.7	28	19,004	98.8
First floor	48	8	384	2.0
$\sum_{i=1}^{n} w_i h_i =$			61,576	

$$F_x = 320 \left(\frac{w_x h_x}{61{,}576} \right)$$

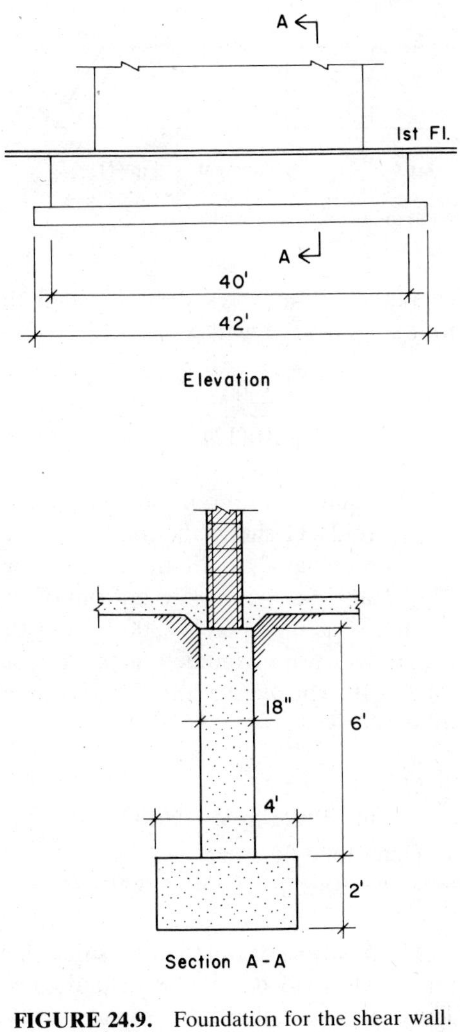

FIGURE 24.9. Foundation for the shear wall.

$$v = \frac{1257}{(0.45)(12)(7.625)} = 30.5 \text{ psi}$$

This is not a critical stress for the reinforced concrete block wall (see *UBC* Table 24–H in Chapter 33.) A footnote to *UBC* Table 24–H requires that the shear load be increased by 50% for shear stress calculation. Applying this increase, together with the allowable one-third increase due to seismic load, the actual design shear stress for the wall is

$$v = (1.5)\frac{30.5}{1.33} = 34.4 \text{ psi}$$

which is still below the lowest allowable stress in the table.

The ends of the shear wall would most likely be designed as columns to carry the roof framing. If this is done, the end column forces due to the seismic overturn moment can probably be easily accommodated. If a tie-down force is required, the doweling of the reinforcing will provide it.

At the second story the shear force of 109.6 k produces a unit shear on the wall of

$$v = \frac{109{,}600}{30} = 3653 \text{ lb/ft}$$

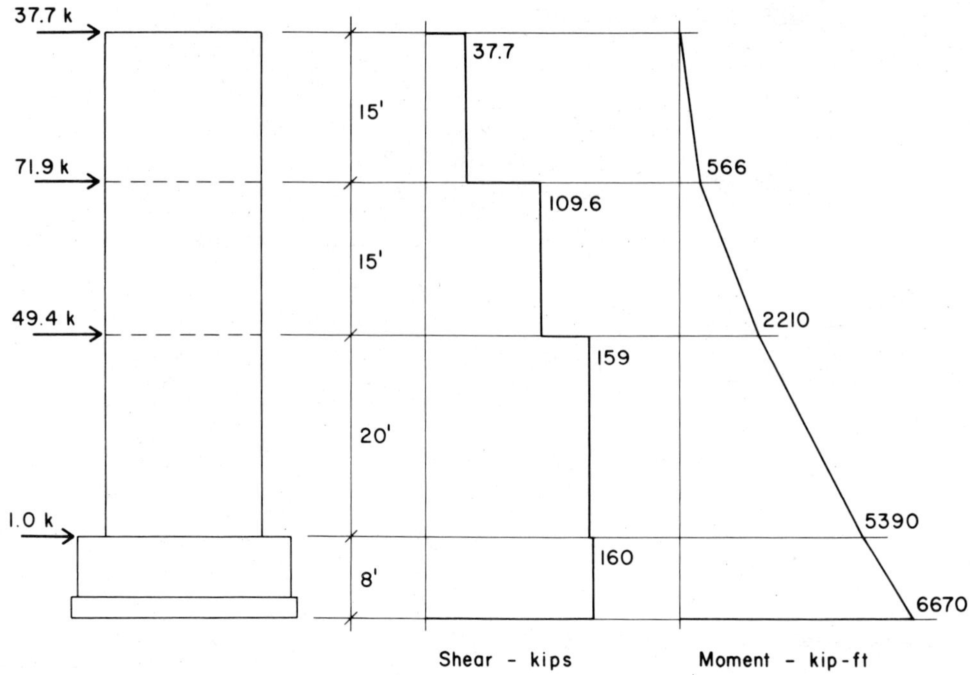

FIGURE 24.10. Functioning of the multistory shear wall.

which is high for an 8-in. block wall.

Assuming a 12-in. block wall with 45% solid blocks, the stress on the net section is

$$v = \frac{3653}{(0.45)(12)(11.625)} = 58 \text{ psi}$$

With the adjustments discussed previously, the actual design stress will be slightly higher, but it is still within the conceivable range of the masonry wall. Overturn moments can also probably be handled with the use of reinforced masonry columns at the ends of the walls.

At the first story the seismic shear and moment are probably beyond the range of a feasible masonry wall. An alternative would be to use a reinforced concrete wall with end columns. The unit shear on the wall is

$$v = \frac{159{,}000}{30} = 5300 \text{ lb/ft}$$

With a 10-in.-thick concrete wall the unit stress in the wall is

$$v = \frac{5300}{(10)(12)} = 44 \text{ psi}$$

which is quite low for reinforced concrete.

Figure 24.11 shows the loading condition for the analysis of the foundation. The total dead load at the bottom of the footing is assumed to be 300 k, and the actual overturning moment is taken from Fig. 24.10. The overturn analysis is thus as follows:

Overturning M:
(6670)(1.5 SF) = 10,005 k-ft

Dead load M:
(300)(21) = 6,300 k-ft

This analysis shows that the safety factor is not provided by the dead load alone. The options are, therefore, to increase the

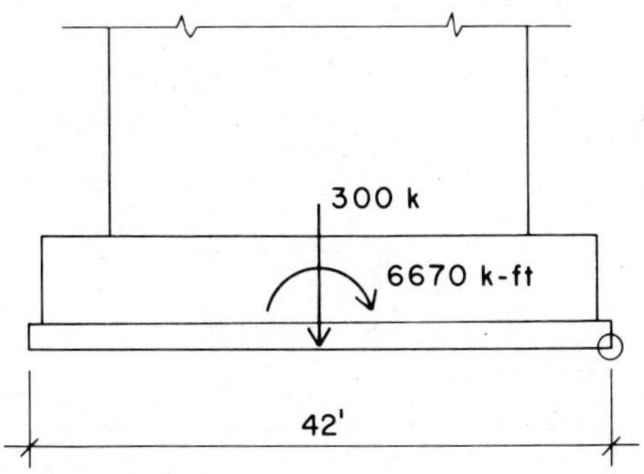

FIGURE 24.11. Base condition for the shear wall foundation.

size of the wall foundation or to develop ties to other parts of the building foundation to provide additional resistance to the overturn.

Although the structure may be adequate for wind, if the seismic load must be accommodated, it is advisable to reconsider the general design of the structure. Some possible options are the following:

1. Reduce the building dead load as much as possible to lower the lateral shear force.
2. Increase the length of the shear walls or add more shear walls.
3. Use an interactive shear wall and ductile rigid frame. The major gain that this will achieve is in the reduction of the K factor: from 1.33 to 0.80, a reduction of 40%.

The use of the F_x forces obtained from formula 12–7 of the *UBC* as the lateral loads at the various levels is intended to bring the static analysis into closer agreement with true dynamic behavior in the multilevel structure. These forces are appropriate for use in the design of the vertical elements of the lateral resistive system but not for the design of the horizontal structures at the various levels. Formula 12–9 of the *UBC* is used to obtain the forces to be used for the design of the diaphragms and collectors at each level. The implementation of this formula is illustrated in Table 24.3. The entries in the table, labeled "A" through "E" at the top of the table, are determined as follows:

A is the same as w_x in Table 24.2.

B is the summation of the F_x values from Table 24.2.

C is the summation of the w_x values from Table 24.2.

TABLE 24.3. Horizontal Diaphragm and Collector Loads: Example 5

	A	B	C	D	E
Level	w_{px} (kips)	$\sum_{i=x}^{n} F_I$ (kips)	$\sum_{i=x}^{n} w_I$ (kips)	$\frac{B}{C}$	F_{px} (kips)
Roof	250.3	75.4	250.3	0.301	75.4
Third floor	643.5	219.2	893.8	0.245	157.7
Second floor	678.7	318.0	1572.5	0.202	137.1

$$F_{px} = \frac{\sum_{I=x}^{n} F_I}{\sum_{I=x}^{n} w_I}(w_{px}) \quad \text{or} \quad \frac{B}{C}(A)$$

D, as indicated, is the ratio of entry B to entry C.

E is obtained from *UBC* Formula 12−9, or with reference to the entries in Table 24.3.

$$E = \frac{B}{C}(A)$$

24.3 STEPPED SHEAR WALLS

Figure 24.12 shows plans and one elevation for a building similar to that discussed in the preceding section. In this example the lateral bracing system consists of exterior shear walls that are placed at the corners of the plan. This is an optimal location for the development of torsional resistance, as compared to the use of interior shear walls. For architectural planning the use of exterior walls offers the advantage of freer use of the interior space; however, the corners may be desirable locations for windows, making this scheme less useful for some situations.

Inspection of the elevation shown in Fig. 24.12 will reveal that the corner shear walls are stepped; that is, their plan length is graduated from top to bottom. This is also evident from the plans for the third floor and ground-level floor. The profile of a single shear wall is thus as shown in Fig. 24.13. The shear force resistance and overturn resistance at each level is thus somewhat in proportion to the loading. There is no particular advantage to this scheme from a structural design point of view, but it may well present advantages in terms of the ability to utilize more of the wall for windows in the upper floors.

A particular problem that must be considered for the stepped wall is that illustrated in Fig. 24.14. At the building corner the vertical ends of the walls are continuous from story to story. However, at the interior ends the corner of an upper wall

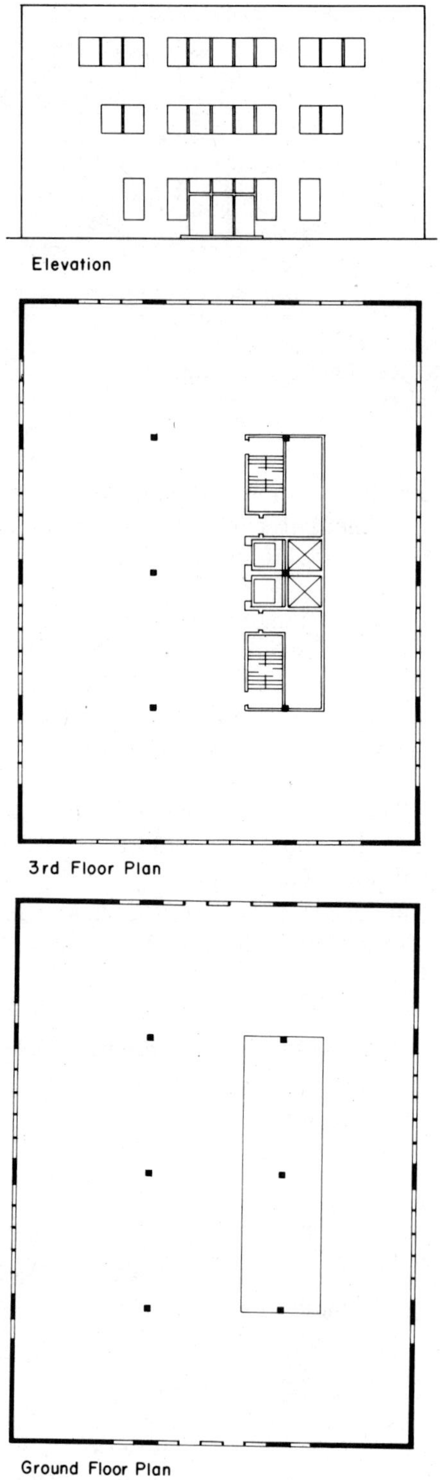

FIGURE 24.12. Building with stepped corner shear walls.

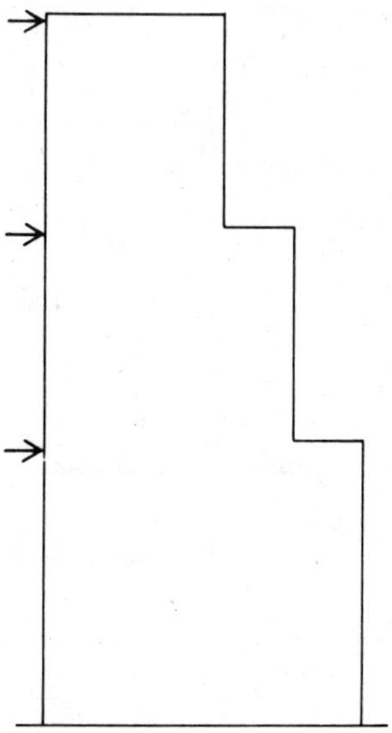

FIGURE 24.13. Loading of the stepped wall.

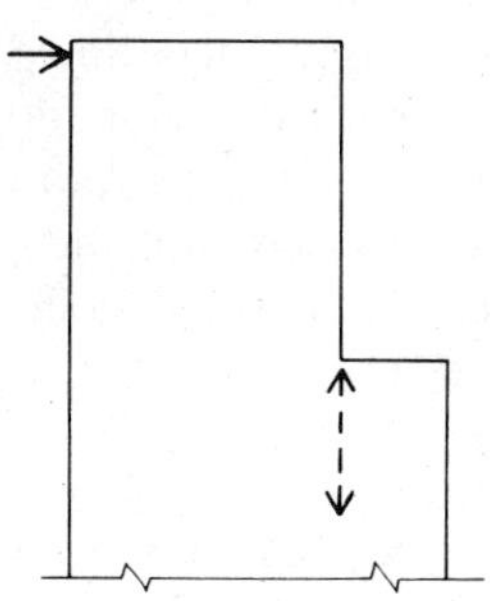

FIGURE 24.14. Load transfer from the upper wall.

falls within the interior portion of the wall below. This requires that the end forces in the upper wall (both compression and net tension) be resolved within the lower wall. For masonry or concrete walls this is primarily only a matter of the placement of the necessary reinforcing. For a wood frame wall with plywood facing it is probably necessary to incorporate a post within the lower wall under the end of the upper wall, and if uplift resistance is required, to use a detail similar to those shown in Fig. 24.7.

Stepped walls may also be used at locations other than at the building corners. A somewhat similar technique is that of simply incorporating *more* shear walls at lower levels. Thus another solution for this building may be to use the corner shear walls as shown plus the interior shear walls, but to have the interior shear walls only at the ground-floor level where the forces are the highest.

24.4 THREE-STORY BUILDING WITH PERIPHERAL BRACING

Figure 24.15 shows an upper-floor plan for an office building in which the lateral bracing system consists primarily of a continuous peripheral system. In this system the exterior wall structure constitutes a continuous system of elements completely circling the outside of the plan. For the example shown the vertical elements of the system are individual short piers of reinforced masonry. Although an individual pier is not highly resistive—especially to overturn effects—the piers make up in number what they lack in individual resistance.

As with the example in the preceding section, the chief advantages of this scheme are the high torsional resistance of the system and the freeing of the interior space for planning. Because of the long narrow plan, the systems in the two directions are somewhat different. In the east–west direction the vertical bracing consists entirely of the short exterior masonry piers. In the north–south direction there is a combination of exterior and interior shear walls, with some of the interior walls (see Fig. 24.16) extending up only to the level of the third floor.

As shown in Fig. 24.17, the roof structure consists of a series of trussed joists spanning the entire width of the building—approximately 60 ft. This is a reasonable structure for the roof and permits a full utilization of the open plan in conjunction

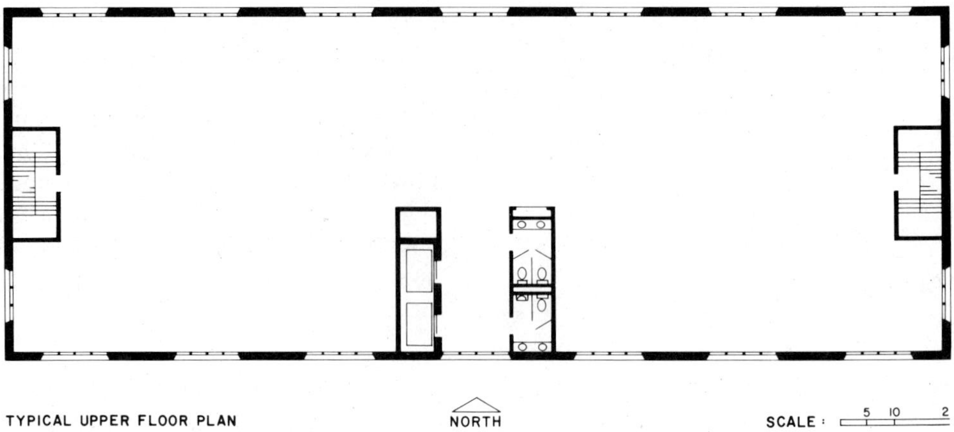

FIGURE 24.15. Upper-floor plan—three-story office building.

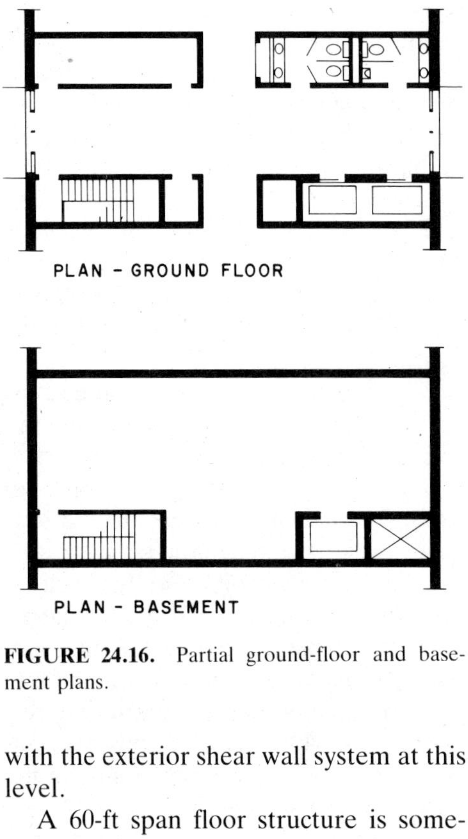

FIGURE 24.16. Partial ground-floor and basement plans.

with the exterior shear wall system at this level.

A 60-ft span floor structure is somewhat less practical, so the floor system shown in Fig. 24.17 utilizes a series of interior columns, placed to allow for a central corridor. The special framing required at the building core is shown in Fig. 24.18. The framing plans generally indicate the use of some type of deck–joist–beam system, with a concrete slab and beam system for the core area over the small central basement. Choice of materials and details for the roof and floor systems would need to incorporate considerations for diaphragm shear and the development of chords, collectors, and drag struts.

24.5 TWELVE-STORY BUILDING

Figure 24.19 shows a typical upper-floor plan for a multistory apartment building that utilizes shear walls as the bracing system in both directions. In the north-south direction the interior walls between the apartment units are used, together with the two end walls at the stairs. There are thus a total of 14 walls, all approximately 20 ft long in plan.

In the east–west direction the vertical bracing consists of the interior corridor walls. While there is usually a desire to have a minimum of permanent interior wall construction for some occupancies (such as the office building in the preceding section), it is reasonable to consider this structure for buildings such as apart-

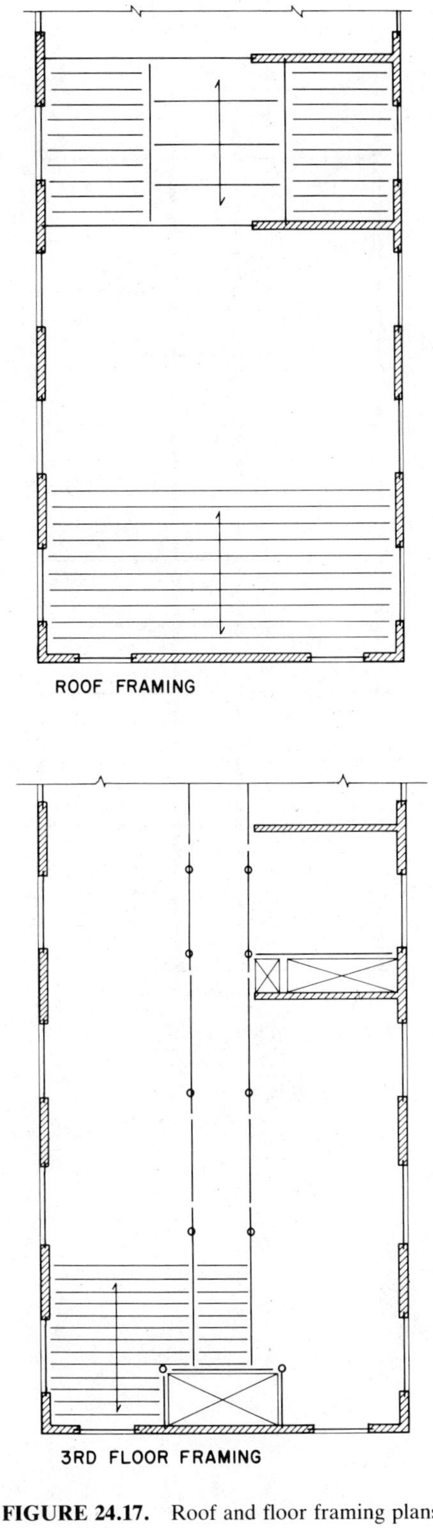

FIGURE 24.17. Roof and floor framing plans.

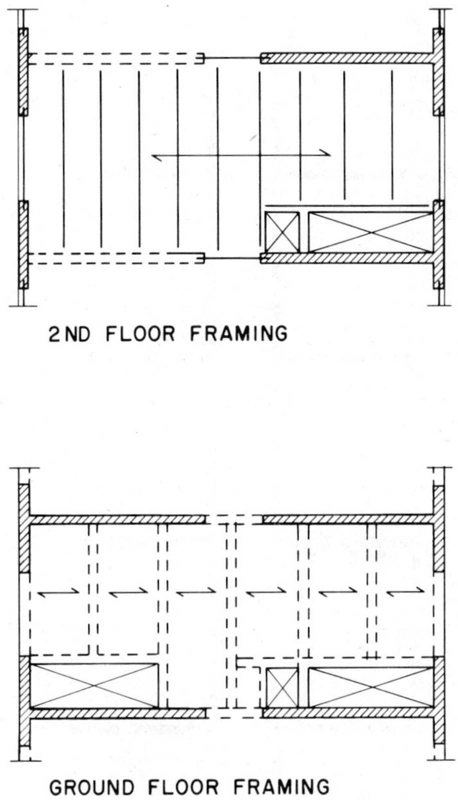

FIGURE 24.18. Floor framing plans at building core.

ments, hotels, dormitories, jails, and hospitals.

The exterior wall structure consists of a column and beam system. The ends of the shear walls are used as columns in this system. The typical floor structure could be a concrete slab or a deck and joist system. In either of these the shear walls would also be used as bearing walls. The solid concrete slab is a popular system for this situation, primarily because it permits a minimum floor-to-floor distance.

The building section is shown in Fig. 24. 20. There are 12 floors of apartments plus a basement and a rooftop structure housing the elevator and ventilating equipment. For analysis we consider the shear wall base to be at the basement-floor

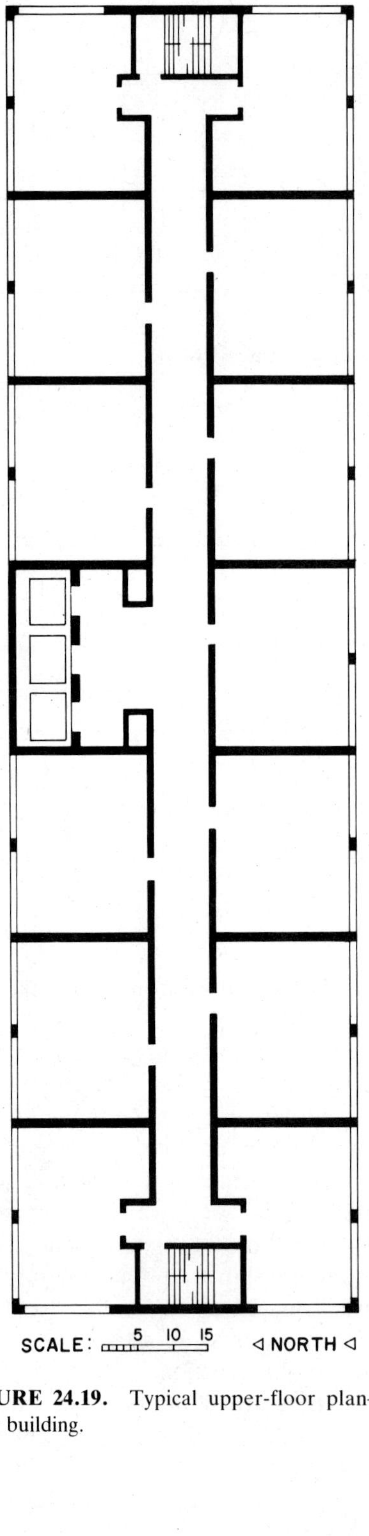

FIGURE 24.19. Typical upper-floor plan—12-story building.

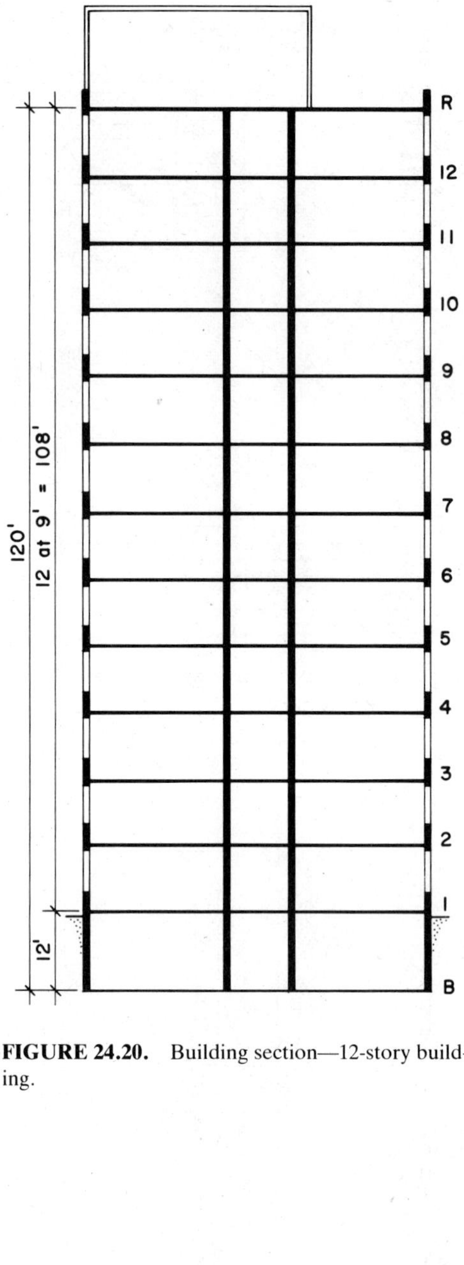

FIGURE 24.20. Building section—12-story building.

Factors for base shear:	Sheet No. 24.5.1

References: UBC 2312 (d)

E-W direction:

$$T = \frac{0.05\,h_n}{\sqrt{D}} = \frac{0.05\,(120)}{\sqrt{175}} = 0.454 \text{ sec.}$$

$$C = \frac{1}{15\sqrt{T}} = \frac{1}{15\sqrt{0.454}} = 0.090$$

Use max. $CS = 0.14$

$V = 1.33 \times 0.14\,W = 0.1862\,W$

N-S direction:

$$\frac{0.05\,(120)}{\sqrt{48}} = 0.866 \text{ sec.}$$

$$\frac{1}{15\sqrt{0.866}} = 0.0716$$

$CS = 1.5 \times 0.0716 = 0.1074$

$1.33 \times 0.1074\,W = 0.1428\,W$

Loads per level:

Floor DL	175' × 48' × 75 psf	=	630 k
Shear walls	20' × 8½' × 125 psf × 24 walls	=	510
Other walls	600' × 8½' × 25 psf	=	128
	600' × 8½' × 15 psf	=	77
	Assume as w_x at each level:		1345 k

With constant w_x at each level, use:

$$F_x = \frac{(V - F_t)\,h_x}{\sum h_i}$$

UBC 2312 (e)

Base V per wall:

$W = 1345 \times 13 \text{ levels} \div 14 \text{ levels}$

$= 1249$ k

$V = 0.1428\,(1249) = 178$ k

$F_t = 0.07\,TV = 0.07 \times 0.866 \times 178$

$= 10.8 \sim$ say 11 k

Then:

$$F_x = \frac{(V - F_t)\,h_x}{\sum h_i} = \frac{(178 - 11)\,h_x}{858}$$

$= 0.1946\,h_x$

Distribution: (N-S)

Level	h_x - ft	F_x - kips
R	120	23.35
12	111	21.60
11	102	19.85
10	93	18.10
9	84	16.35
8	75	14.60
7	66	12.84
6	57	11.09
5	48	9.34
4	39	7.59
3	30	5.84
2	21	4.09
1	12	2.34
B	$\sum h_i = 858$	

Calculation Sheet 24.5.1

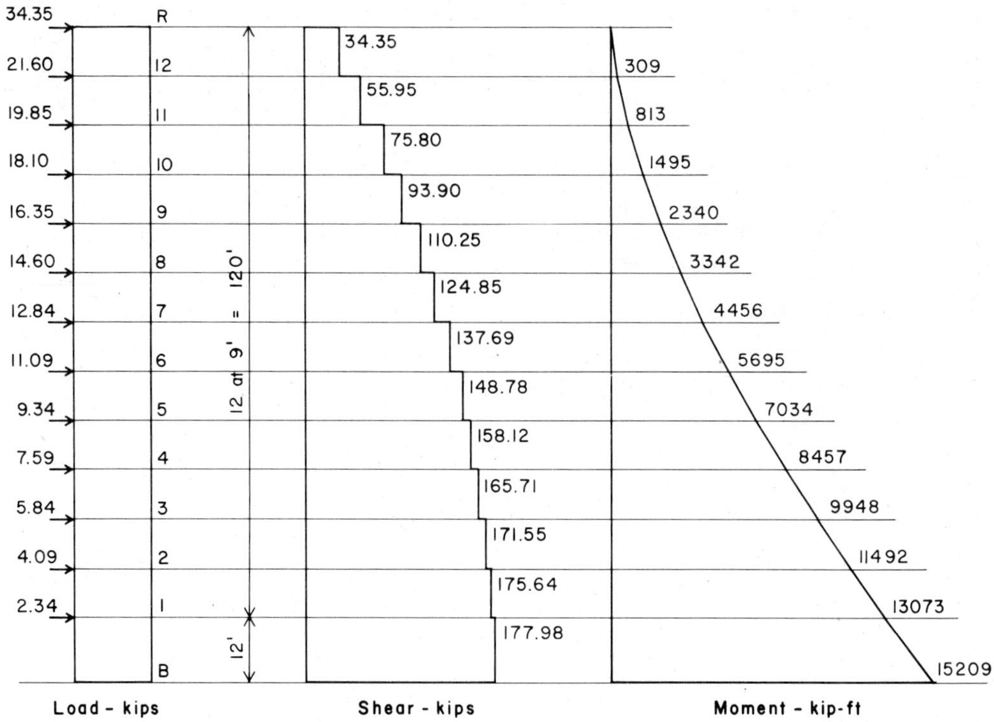

FIGURE 24.21. Functioning of the 12-story wall.

level and the top of the building to be at the roof level. Although it must also be braced, and its weight added to the roof load, the penthouse structure is not really part of the general lateral system for the building.

Because of the long narrow plan, this building has considerably different responses in the two directions. Using the *UBC* criteria for determination of the required *C* values, it may be observed that the lateral forces will be different in the two directions. Calculation Sheet 24.5.1 shows an approximate analysis for the north–south shear walls. The loading for a typical shear wall, together with the shear and moment diagrams, is shown in Fig. 24.21.

CHAPTER TWENTY-FIVE

Braced Frames

Trussing has long been a favored means for lateral bracing of framed structures, especially for wind loads. Only in relatively recent times have the shear wall and the rigid frame seen major use for the bracing of framed structures. In various ways current code criteria tend to favor the shear wall and the rigid frame over trussing. There are, nevertheless, situations in which trussing—or X-bracing—offers a viable alternative, if not some possibilities for unique solutions.

25.1 ONE-STORY BRACED FRAME

The plan and framing elevation in Fig. 25.1 show a one-story building in which the lateral resistive system consists of a braced frame. For east–west load the frame is braced by the X-braces in the end bays of the front and rear walls. The computations for the seismic analysis for east–west load are shown on Calculation Sheet 25.1.1.

Although the X-brace essentially produces an indeterminate structure, the usual practice for static load is to assume that the relative slenderness of the diagonals will result in the buckling of the compression member at a very low load, after which the total load is assumed to be taken by the tension member. For the reversible direction loading, both members are thus designed as tension members, with each designed to take the full load in a single direction. For dynamic loading, however, the buckling of the compression member represents a major energy-absorbing effect and a major source of damping for vibratory motions.

The calculations show that the brace may consist of a fairly small-diameter round steel rod. However, for reasons of the wall construction details or the development of the connection of the brace to the frame, it may be desirable to use something other than the round rod.

If the X-braces are fully stressed to the maximum tension stress permitted by the codes, the resulting lateral deflection at the roof level may be more than can be reasonably tolerated for the wall construction. For this reason X-braces are sometimes oversized, resulting in lower stresses and less lateral drift of the structure. If the oversizing of the members is significant, their stiffness may make the assumption of compression buckling at low load questionable.

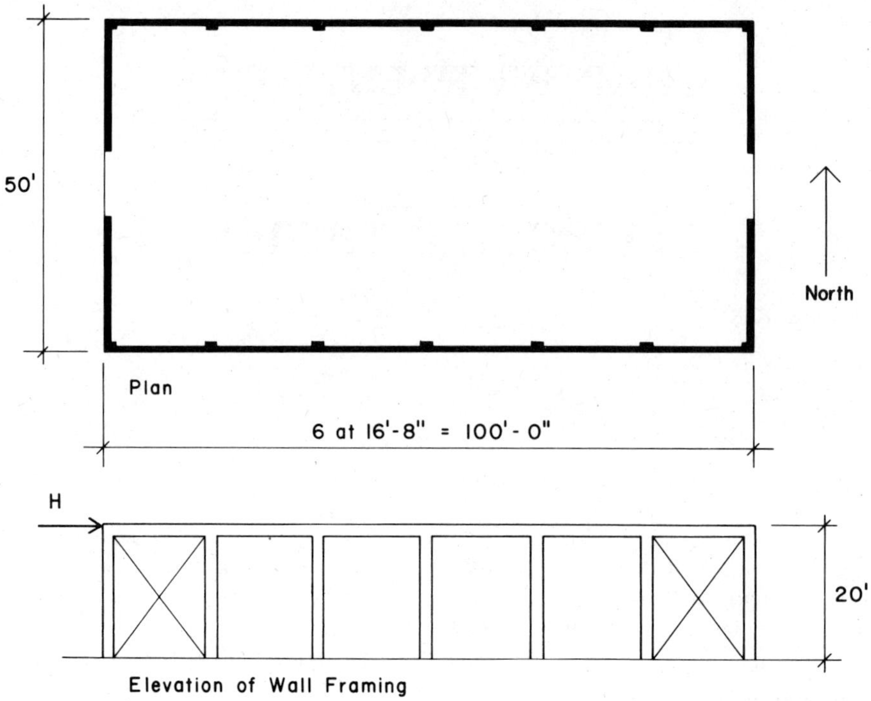

FIGURE 25.1

Loads:	Sheet No. 25.1.1
	References: Chap. 17

Loads:

Roof DL $\quad 50' \times 100' \times 20$ psf $= 100$ k

Ext. walls $\quad 10' \times 300' \times 20$ psf $= 60$ k

$\qquad\qquad\qquad\qquad 160$ k

$V = 0.1862\ W = 0.1862 \times 160 = 29.8 \sim 30$k

Load / wall $= 15$k

Load / bent $= 7.5$k

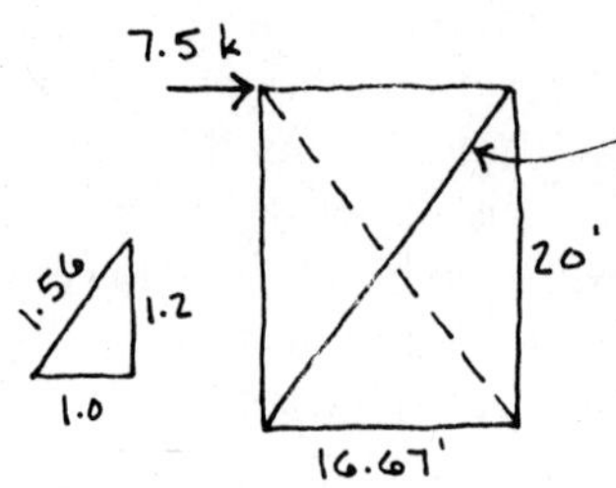

$T = 1.56\ (7.5) = 11.7$ k

Calculation Sheet 25.1.1

25.2 THREE-STORY BUILDING WITH INTERIOR BENTS

This example consists of the same building as shown in Section 24.2. In this case the three-story high shear walls are replaced by X-braced bays, as shown in Fig. 25.2. The loading for these braced frames and the geometry of the braces are shown in Fig. 25.3.

Because the required *K* factor is the same for shear walls and braced frames, the only difference in the lateral loads between this example and that in Section 24.2 is in the weight reduction caused by the elimination of the heavy shear walls. Thus the total lateral forces for the various levels are approximately 90% of those determined in Table 24.2.

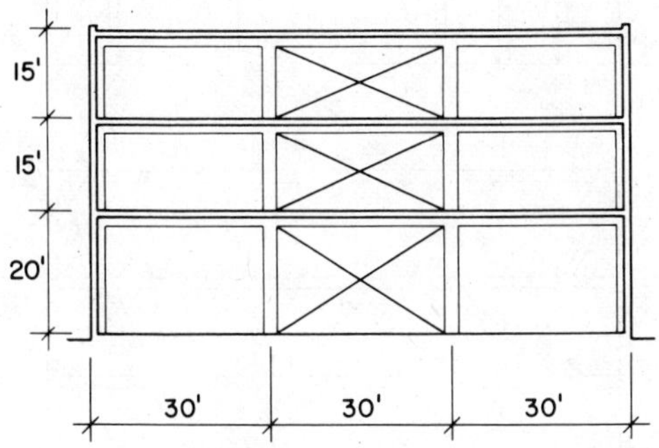

FIGURE 25.2

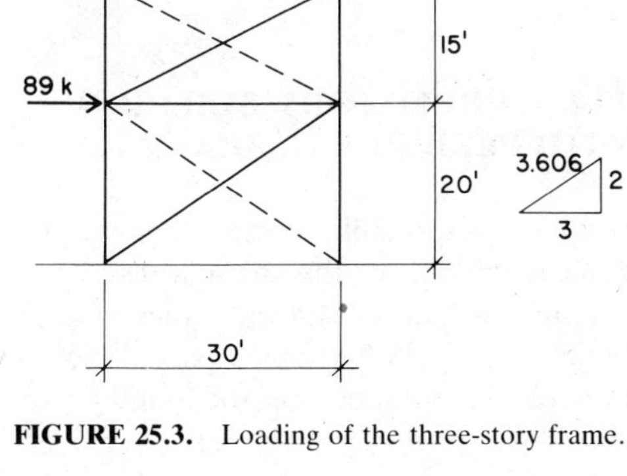

FIGURE 25.3. Loading of the three-story frame.

It should be noted that Section 2312(j)1G of the *UBC* requires that the members of a braced frame be designed for 1.25 times the load as determined from *UBC* formula 12−1 ($V = ZIKCSW$) and that the connections for the members be designed for this increased load or the full strength of the members, whichever is higher. In addition, the usual one-third increase in allowable stress is not allowed for the connections.

Assuming the braces to be of A36 steel, the forces and required areas for the braces are as follows (see Fig. 25.3.):

Third story:

Story shear: 68 k (total for the two bents)

$$\text{Brace tension: } \frac{2.236}{2}(68)\left(\frac{1}{2}\right) = 38.0 \text{ k}$$

$$\text{Required area: } \frac{38.0 \times 1.25}{1.33 \times 21.6} = 1.65 \text{ in.}^2$$

Second story:

Story shear: 198 k

$$\text{Brace tension: } \frac{2.236}{2}(198)\left(\frac{1}{2}\right) = 110.7 \text{ k}$$

$$\text{Required area: } \frac{110.7 \times 1.25}{1.33 \times 21.6} = 4.82 \text{ in.}^2$$

First story:

Story shear: 287 k

$$\text{Brace tension: } \frac{3.606}{3}(287)\left(\frac{1}{2}\right) = 172.5 \text{ k}$$

$$\text{Required area: } \frac{172.5 \times 1.25}{1.33 \times 21.6} = 7.51 \text{ in.}^2$$

These area requirements are for the full gross cross section of the members. Depending on the type of member and the method of connection, a reduced effective area or a net cross section may be critical in selection of the members. Once the members are selected, their actual full cross-section areas would be used for determination of the drift of the structure.

The type of member to be used for the braces depends on the problems of integrating the brace into the wall construction and connecting it to the steel frame. When the area requirements are higher than can be developed with small-diameter round rods, a common choice is that of single angles. If the angle legs are turned in opposite directions in a single X, they can be lapped and fastened at their intersection.

As discussed for the previous example, it may be desirable to oversize the brace members in order to reduce the stress and the accompanying drift.

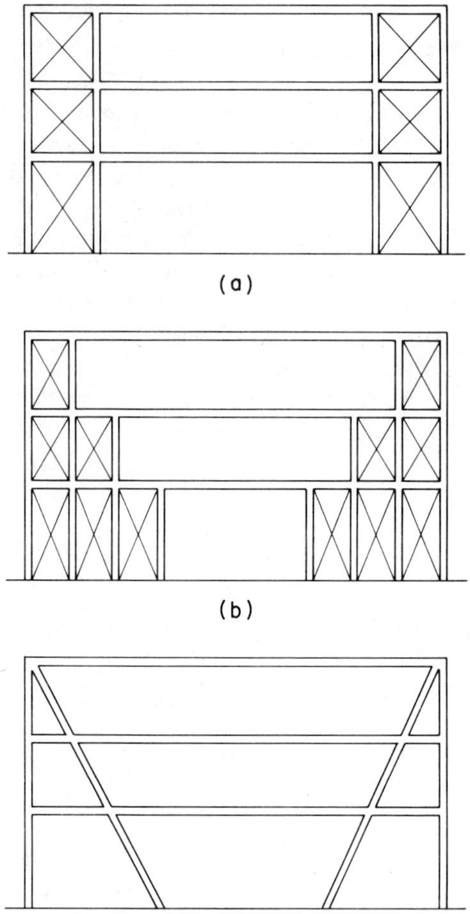

(a)

(b)

(c)

FIGURE 25.4. Trussed corner bracing.

25.3 THREE-STORY BUILDING WITH CORNER BENTS

The building discussed in Section 24.3 could be braced with corner-braced bents in place of the shear walls used in that example. Figure 25.4*a* shows the building elevation with the three-story bents at the building corners. If the bents are constant in width throughout their height (as shown in Fig. 25.4*a*), they do not, of course, accommodate the stepped window layout shown on the building elevation in Fig. 24.12. If the staggered width bent is required, it could be built as shown in Fig. 25.4*b*; however, the proliferation of members and connections seems prohibitive.

Figure 25.4*c* shows the possibility for a single large diagonal member at each corner that is used to constitute a truss with the frame members and the added diagonals. This scheme would accommodate the same elevation as that with the stepped shear wall. However, since the truss does not require a solid wall, there are additional possibilities for openings as well.

25.4 THREE-STORY BUILDING WITH PERIPHERAL BRACING

It would be possible to replace some, or conceivably all, of the shear walls in the example in Section 24.4 with braced frame bents. Figure 25.5 shows a partial elevation of the framing for the building in which the short masonry piers have been

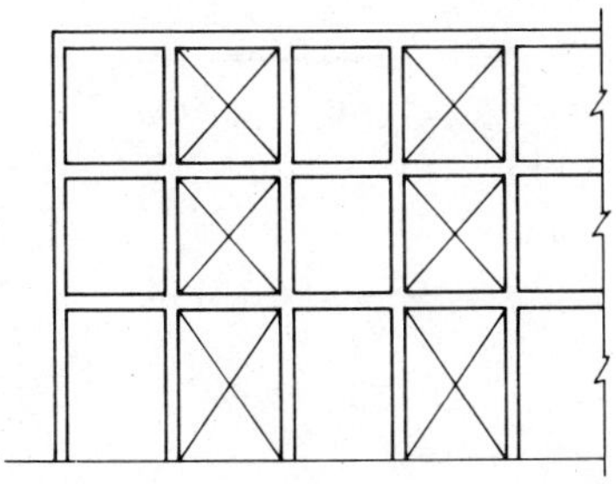

FIGURE 25.5. Peripheral bracing arrangement.

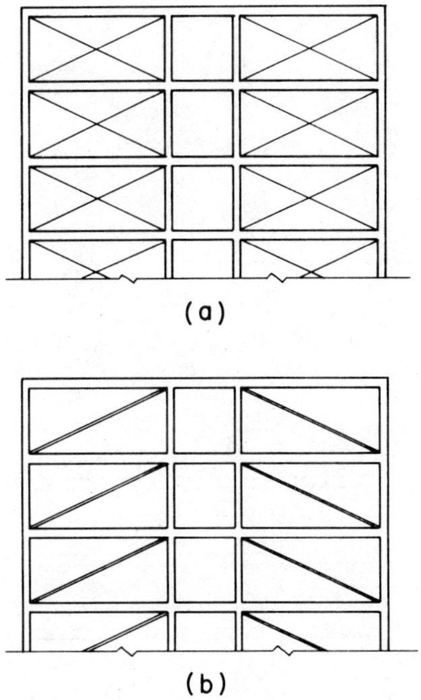

FIGURE 25.6. Single diagonal and X-bracing systems.

replaced with X-bracing. The end shear walls at the stairs and possibly even the interior shear walls above the basement level could also be replaced with braced bents. It is quite possible, however, that the shear walls may remain the system of choice for the bracing in the direction of the narrow plan dimension (north–south) even though the X-bracing is used on the long walls.

As with many other situations, the choice of one bracing scheme over another may be strongly influenced by considerations other than simply that for the most logical and efficient seismic bracing system. If the masonry walls are desirable for load bearing, for acoustic properties, or for thermal inertial mass, they have a head start against the X-bracing or a rigid frame. On the other hand the X-bracing permits a wide range of choice for the infill wall construction, or even allows for an all-glass wall if the X-braces are exposed.

25.5 TWELVE-STORY BUILDING

Truss bracing presents a possibility for the bracing of the building discussed in Section 24.5. A partial section of the upper portion of the structure for such a scheme is shown in Fig. 25.6*a*. Another alternative is for the single diagonal scheme shown in Fig. 25.6*b*. The latter scheme would require much larger members, but these may be desirable for their energy absorption, the reduced building drift, the out-of-plane bracing of the wall, and other possible reasons. The fact that there are half as many members and connections may be significant, even though the elements are heavier.

A third possibility for the truss bracing is the K-brace system shown in Fig. 25.7. Although increasingly popular for bracing of steel frames, the K-brace system may be a bit of an overkill for this particular structure. Because the K-brace layout permits the use of some space in the center portion of the bay, an alternative may be to use a single, central bay of bracing, as shown in Fig. 25.8. With a drop ceiling in the corridor, it is possible to facilitate the same basic plan as shown for the shear wall structure in Fig. 24.19.

With any of the bracing systems shown in Figs. 25.6 and 25.7—or indeed with the shear wall system discussed in Section 24.5—an economy measure that may be

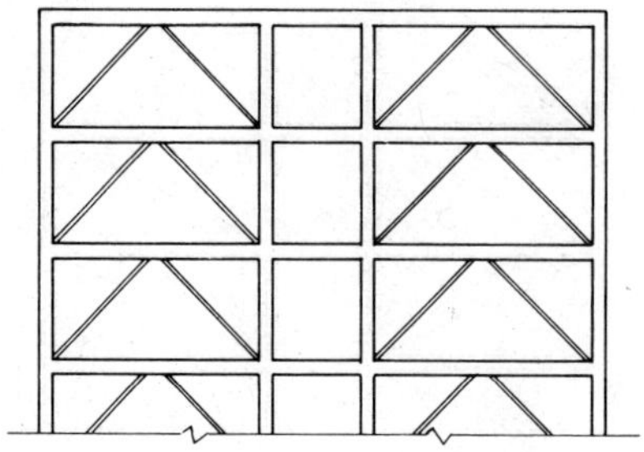

FIGURE 25.7. K-bracing—double-bay system.

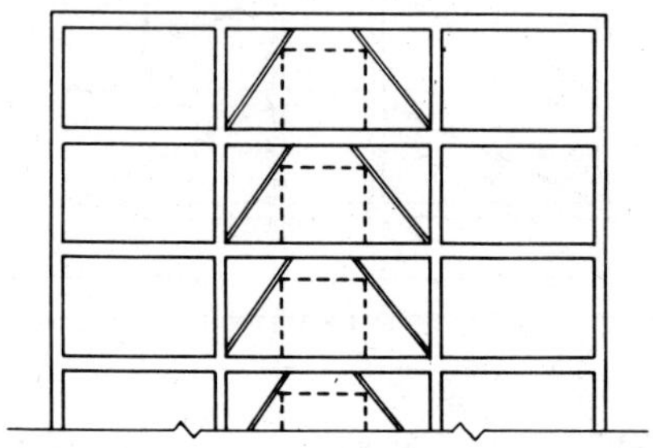

FIGURE 25.8. K-bracing—single-bay system.

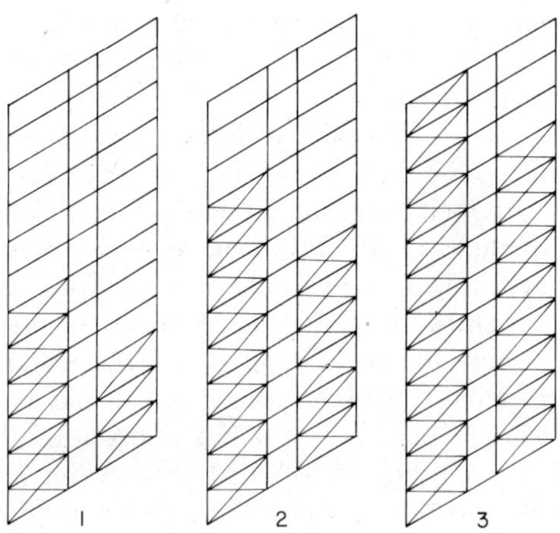

FIGURE 25.9. Stepped arrangement of trussed bracing system.

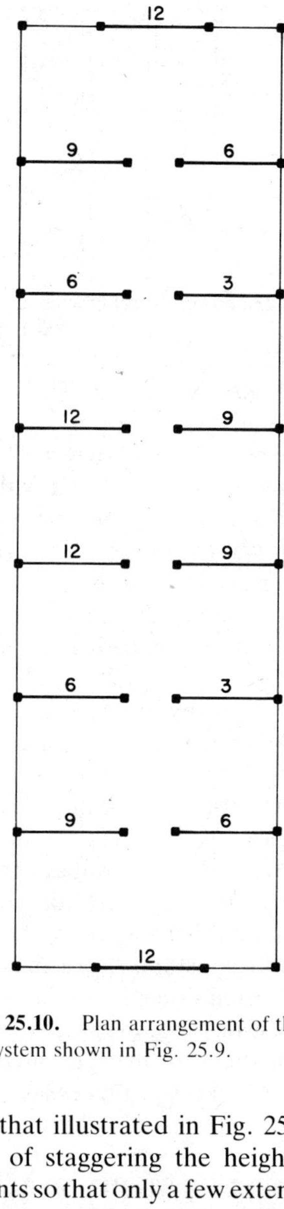

FIGURE 25.10. Plan arrangement of the stepped bracing system shown in Fig. 25.9.

used is that illustrated in Fig. 25.9. This consists of staggering the height of the truss bents so that only a few extend all the way to the top of the building. The isometric drawing in Fig. 25.9 indicates four heights of bents: 3 stories, 6 stories, 9 stories, and the full height of 12 stories. On the schematic plan of the building in Fig. 25.10 is shown a possible use of the various height bents.

CHAPTER TWENTY-SIX

Moment-Resistive Frames

For seismic resistance, rigid frames occur most often in the form of poured-in-place concrete frames or steel frames with heavy welded or bolted connections. Rigid frames are popular with many architects since they provide the least interference with planning and with wall openings. The building code also favors the rigid frame to some extent by assigning a value of K that is one-half that for a shear wall or braced frame. Other than cost factors, the single biggest drawback in the choice of a rigid frame is the magnitude of lateral drift that is normal for the system. Forming and sizing of members is often done with major consideration for their stiffness in resisting deflection, the primary purpose being to reduce drift.

26.1 ONE-STORY SINGLE-UNIT BENT

Figure 26.1 shows a simple one-story building in which the lateral resistive structure that resists force in the north–south direction consists of a series of single bent rigid frames. The bents occur at each end of the building and at 25-ft intervals throughout the length. With ordinary light roof and wall construction the total lateral seismic load to the roof diaphragm in the north–south direction will be approximately 20 kips or 200 lb/ft. If the roof diaphragm is flexible, this load is distributed to the bents on a peripheral basis. Thus, the bent loads are

$$\text{End bent: } 12.5 \times 200 = 2500 \text{ lb/bent}$$

$$\text{Intermediate bent: } 25 \times 200 = 5000 \text{ lb/bent}$$

If the roof diaphragm is considered to be reasonably rigid, it tends to distribute the loads to the bents in proportion to the bent sitffnesses. Thus, if they are of approximately equal stiffness, the load on each bent is

$$\frac{200 \times 100}{5} = 4000 \text{ lb/bent}$$

We assume the latter to be the case and design the bents for the 4000-lb load, applied as a concentrated load at the top of the bent.

Figure 26.2 shows the free-body diagram of the bent with the loads and reactions. The deflected shape under load is shown by the dashed line. Although this

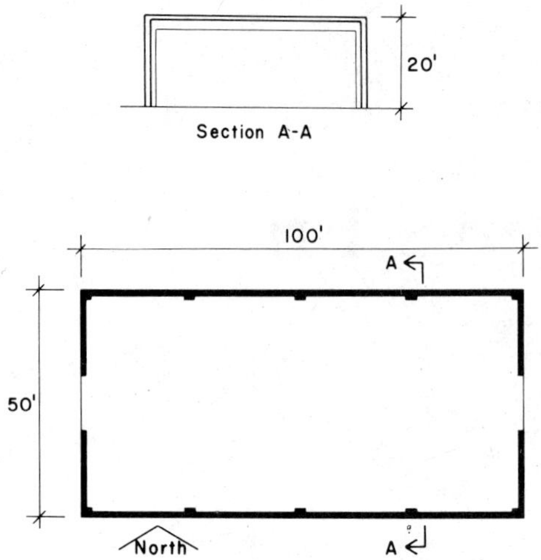

FIGURE 26.1

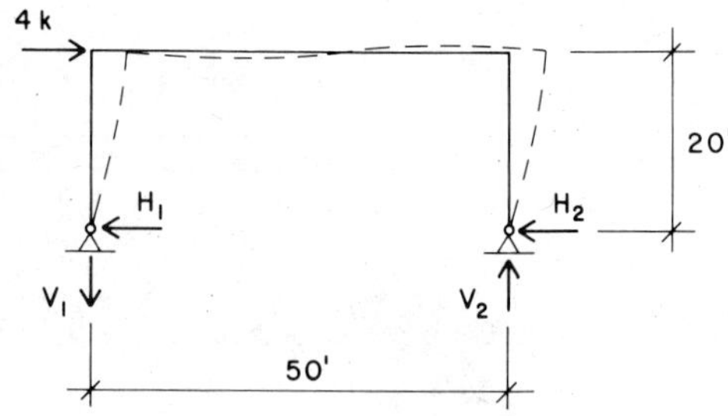

FIGURE 26.2. Loading of the bent.

problem is essentially indeterminate, if we assume the bent to be symmetrical, we may reasonably assume the two horizontal reactions to be equal. In any event the vertical reactions are statically determinate and may be determined as follows:

$$V_1 = V_2 = \frac{4 \times 20}{50} = 1.6 \text{ k}$$

On the basis of this analysis for the reactions, the distribution of internal forces is shown in the illustrations in Fig. 26.3. These forces must be combined with the forces caused by the gravity load to determine the critical design conditions for the bents.

Figure 26.4 shows the bent as loaded by a uniform load of 1000 lb/ft on top of the horizontal member. This effect is based on

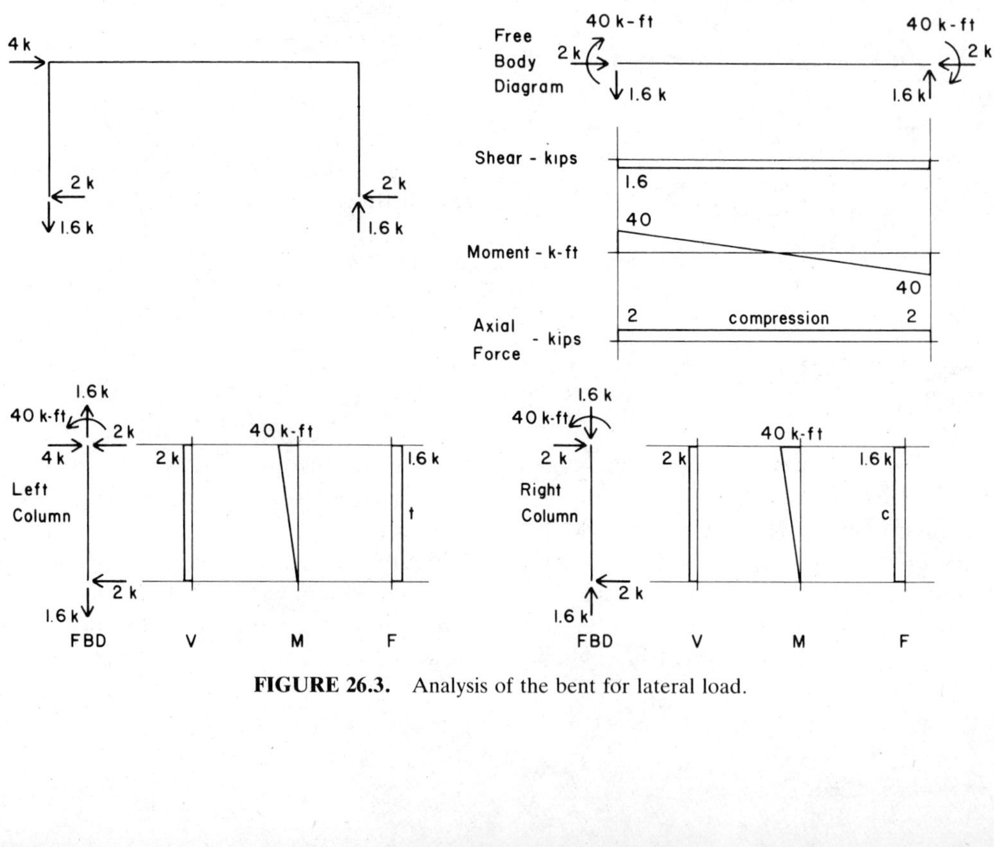

FIGURE 26.3. Analysis of the bent for lateral load.

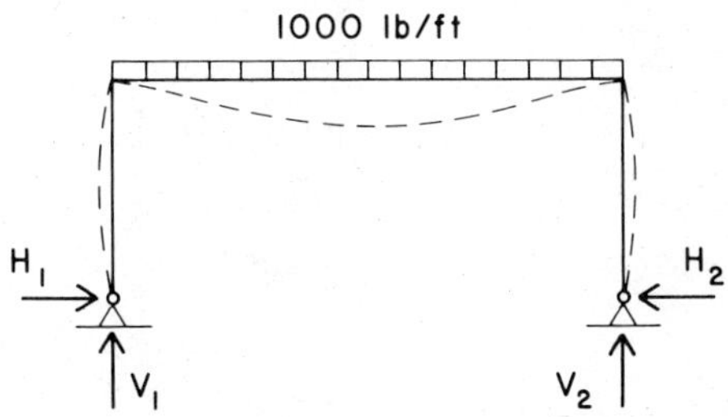

FIGURE 26.4. Gravity loading of the bent.

an assumption that the roof framing delivers an approximately uniform loading with a total dead plus live load of 40 psf as an average for the roof construction, including the weight of the horizontal bent member. As with the lateral load, the vertical reactions may be found on the basis of the bent symmetry to be

$$V_1 = V_2 = \frac{1 \times 50}{2} = 25 \text{ k each}$$

Determination of the horizontal reactions in this case, however, is indeterminate and must consider the relative stiffness (I/L) of the bent members. For the pin-based columns it will be found that the horizontal reactions will each be

$$H = \frac{wL^3 I_c}{8I_g h^2 + 12I_c hL}$$

In this calculation we may use the relative, rather than actual, values of the column and girder stiffness (I_c and I_g in the formula). If we assume the girder to be approximately 1.5 times as stiff as the column, the horizontal reaction will be

$$H = \frac{(1)(50)^3(1)}{8(1.5)(20)^2 + 12(1)(20)(50)}$$

$$= \frac{125{,}000}{4800 + 12{,}000} = 7.44 \text{ k}$$

With these values for the reactions the free-body diagrams and distribution of internal forces for the gravity loading are as shown in Fig. 26.5. These must next be combined with the previously determined lateral forces, as shown in Fig. 26.6. The design conditions for the individual bent members would be selected from the maximum values of Fig. 26.5 (gravity only) or three-quarters of the maximum values of the combined loading, as shown in Fig. 26.6. The adjustment of three-quarters for the comparison is based on the increased allowable stress for the combined forces that include the seismic load.

We may now proceed to make a preliminary design of the column and girder, based on these analyses. For the illustration we will design the bent in steel, using standard rolled sections, and then discuss other possibilities for its construction. For the column the critical condition is that of the leeward column, which must be designed for the compression plus bending. We assume the following for the first trial design:

A36 steel, rolled WF sections

Design for M = 148.8 k-ft, axial compression = 25 k

(Combined loading is not critical; no increase in allowable stresses.)

Assume the wall to brace the column continuously on its y axis.

We may now try a section by picking one from the S_x tables in the American Institute of Steel Construction Manual (Ref. 16) with a moment capacity slightly higher than 148.8 k-ft, based on the assumption that the axial load is not very important because of the relatively high stiffness on the unbraced x axis. We thus will try a W 16 × 57, with a listed moment capacity of 184 k-ft with F_y = 36 ksi. Checking this with formula 1.6–2 from the steel design specification of the AISC manual, we get

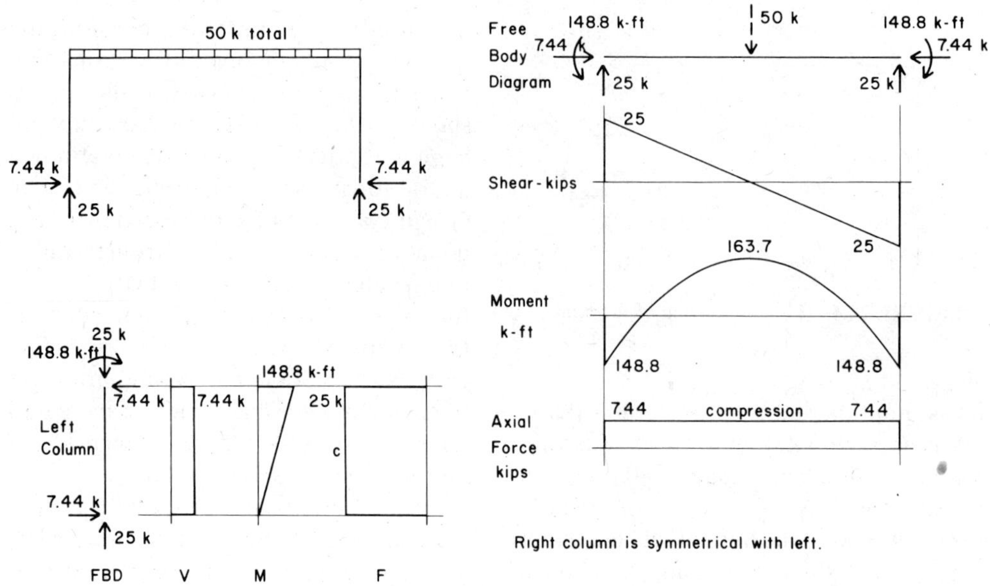

FIGURE 26.5. Analysis of the bent for gravity load.

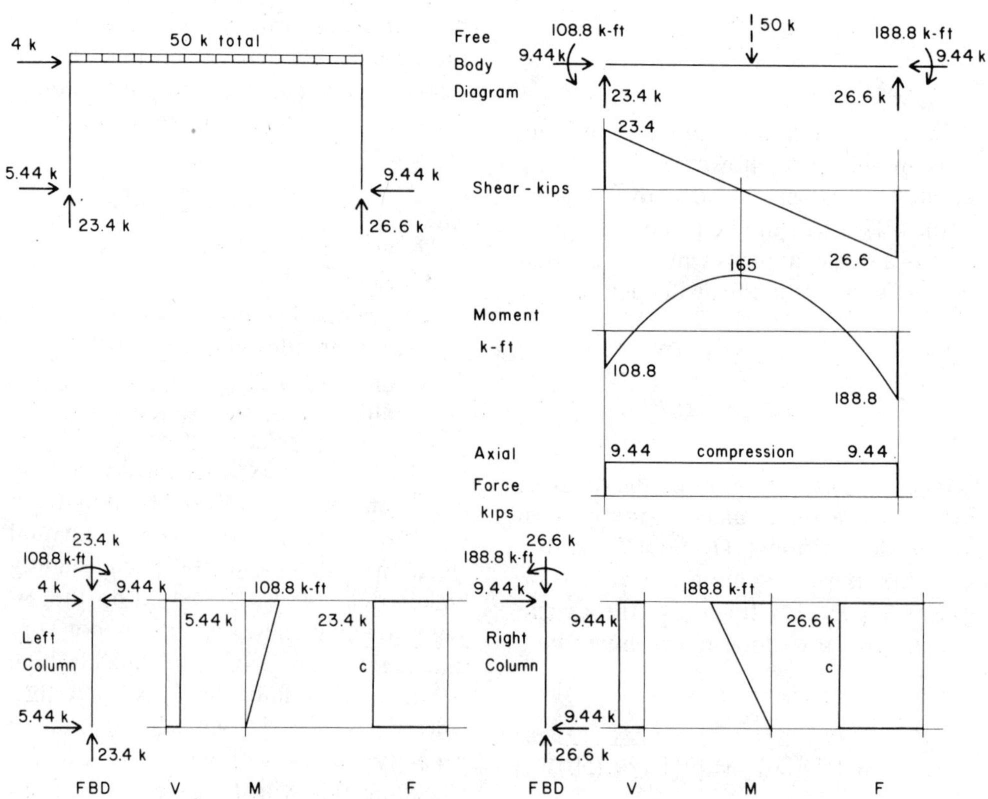

FIGURE 26.6. Analysis of the bent for the combined loads.

$$\frac{f_a}{F_a} + \frac{f_b}{F_b} \leqslant 1$$

where $f_a = \dfrac{25}{6.8} = 1.49$ ksi

$$\frac{KL}{r_x} = \frac{2(240)}{6.72} = 71.4$$

$F_a = 16.29$ ksi (from AISC Manual, p. 5-74)

$$f_b = \frac{M}{S} = \frac{(148.8 \times 12)}{2.2} = 19.36 \text{ ksi}$$

$F_b = 24$ ksi

Then

$$\frac{f_a}{F_a} + \frac{f_b}{F_b} = \frac{1.49}{16.29} + \frac{19.36}{24}$$
$$= 0.09 + 0.81 = 0.90$$

which indicates that the section is adequate on a stress basis.

There are two additional design considerations for the column that have some importance. One has to do with the connection of the column to the girder. If the connection is to be a fully welded connection, the flange widths of the two members should be reasonably matched. The other consideration has to do with the lateral deflection, or drift, of the bent, which is discussed later.

For the girder design we assume the following:

A36 steel, rolled WF section.

Design for M: 163.7 k-ft (positive moment at midspan).

Axial compression is negligible, based on the column design.

Assume roof framing braces the girder on 6-ft centers.

Once again using the S_x tables from the AISC manual, we try a W 21 × 50, with a listed moment capacity of 189 k-ft. This will make the f_b/F_b ratio in the combined stress formula

$$\frac{163.7}{189} = 0.866$$

Then $f_a = \dfrac{7.44}{14.7} = 0.506$ ksi

$$\frac{KL}{r_y} = \frac{1(72)}{1.30} = 55$$

$F_a = 17.90$ ksi

which clearly indicates that the combined stress is not critical.

Two deflection problems should be considered. The first is that of the vertical deflection of the girder, which is most critical for the end bent because the details of the end wall construction must tolerate at least the deflection due to the live load. If we assume the live load to be approximately half of the total load, this deflection will be something less than

$$\frac{5WL^3}{384EI} = \frac{5(25)(50)^3(1728)}{384(29{,}000)(984)} = 2.4 \text{ in.}$$

which is the simple beam deflection, based on no end moments.

The end moment will produce an upward deflection equal to $ML^2/8EI$. Using half of the calculated end moment for the live load deflection, this reduction amounts to

$$\frac{ML^2}{8EI} = \frac{(74.4)(50)^2(1728)}{8(29{,}000)(984)} = 1.4 \text{ in.}$$

resulting in a net midspan deflection of approximately 1 in.

The other deflection consideration is that of the lateral drift caused by wind, as mentioned previously. Again, this is most critical for the end wall construction. As shown in Fig. 26.7, the lateral deflection (Δ in the figure) may be calculated in two parts. The first part consists of the simple cantilever deflection of the column (t_1 in the figure). The second part is caused by the rotation at the top of the column (t_2 in the figure). The determination of this combined deflection is

$$\Delta = t_1 + t_2$$

$$= \frac{Hh^3}{3EI_c} + \frac{Hh^2L}{8EI_g}$$

$$= \frac{(2)(20)^3(1728)}{3(29{,}000)(758)} + \frac{(2)(20)^2(50)(1728)}{8(29{,}000)(984)}$$

$$= 0.41 + 0.30 = 0.71 \text{ in.}$$

Although these are live load deflections and consequently quite theoretical, the end wall construction, as well as any interior cross wall construction, should be designed to tolerate movements of this order of magnitude.

Connection of the column and girder rolled sections are done with some combination of bolting and welding. Figure 26.8 shows two possibilities, both based on the assumption that the bent cannot be transported from the fabricating shop to the site in one piece. In the upper part of

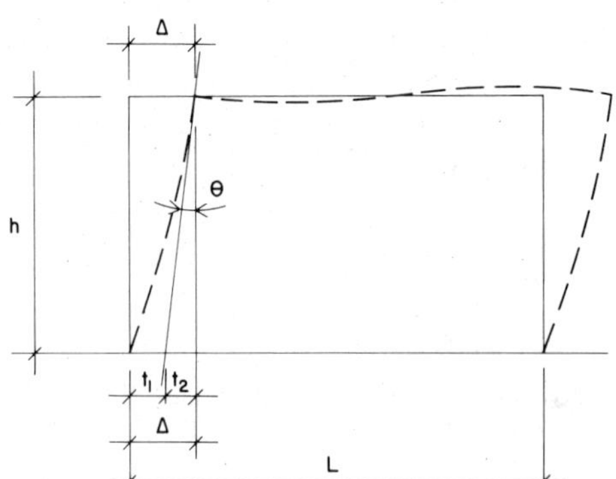

FIGURE 26.7. Deflection of the bent under lateral load.

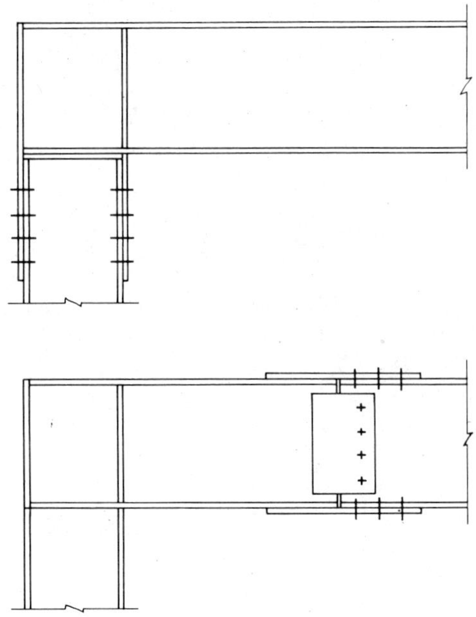

FIGURE 26.8. Framing details for the steel bent.

the figure welded plates extend beneath the girder and are bolted to each side of the column. In the lower part of the figure the column-to-girder connection is fully welded, and the field connection consists of a splice in the girder at the approximate location of the inflection point under gravity load. The web plates of this connection are designed for the girder shear and the flange plates are designed for the seismic moment plus some amount of the gravity moment because unbalanced live load would produce some moment at this theoretical zero moment point.

Although we do not illustrate their design, Fig. 26.9 shows details for four other possible constructions of the bents. Figure 26.9*a* shows a bent built up from flat plates of steel, producing the same basic I-shaped cross section for the members. An advantage of this construction, as well as of the others in the figure, is that the members may be tapered in length. This offers the possibility of designing the girder for the required depths at the midpoint and ends

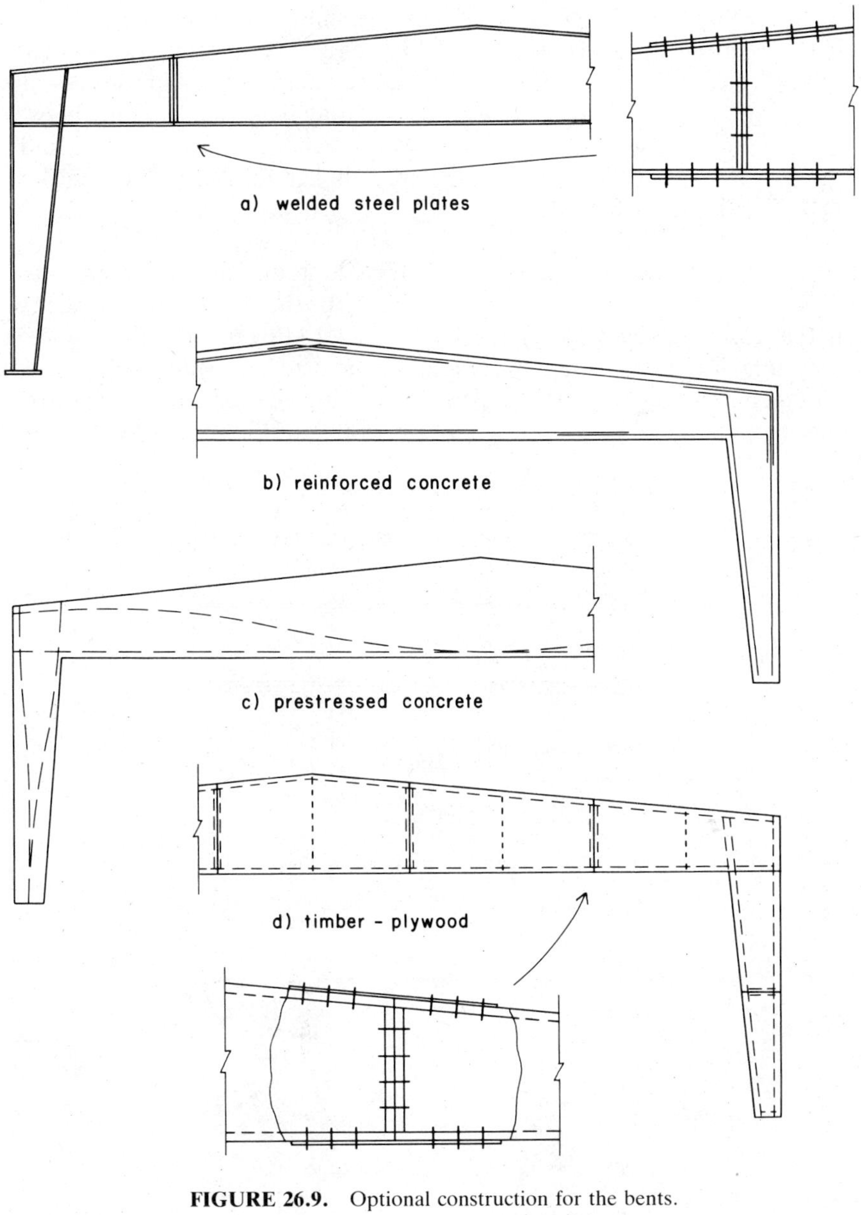

FIGURE 26.9. Optional construction for the bents.

and the possibility of providing for roof drainage while maintaining a flat bottom on the girder. And it offers the advantage of reducing the size of the column at the floor level where the moment capacity is not required.

Figures 26.9*b* and *c* illustrate construction of the bent in concrete with either conventional reinforcing or prestressing with steel cables. If the bents are cast flat on the site and tilted in place, they could be built in one piece.

Figure 26.9*d* shows the possibility of a plywood and timber box-type construction for the bent. Such bents would most likely be primarily shop-built, using field splices similar to those for the steel bents. The sketch in the figure shows a possibility for a joint placed near the girder inflection point.

26.2 THREE-STORY FRAME

Figure 26.10 shows a plan and section for a building similar in form to that in Section 24.2. In this case, however, the lateral resistive system for north–south forces consists of seven three-story, three-bay rigid bents.

Assuming the roof and floor diaphragms to be reasonably stiff, and the bents to be approximately equal in stiffness, the total load may be divided equally between the seven bents. The loading for an individual bent is determined as follows:

1. The total building weight is assumed to be the same as that found in Section 24.2, except for the absence of the heavy shear walls. We therefore assume the loads to be approximately 90% of those found in Table 24.1.

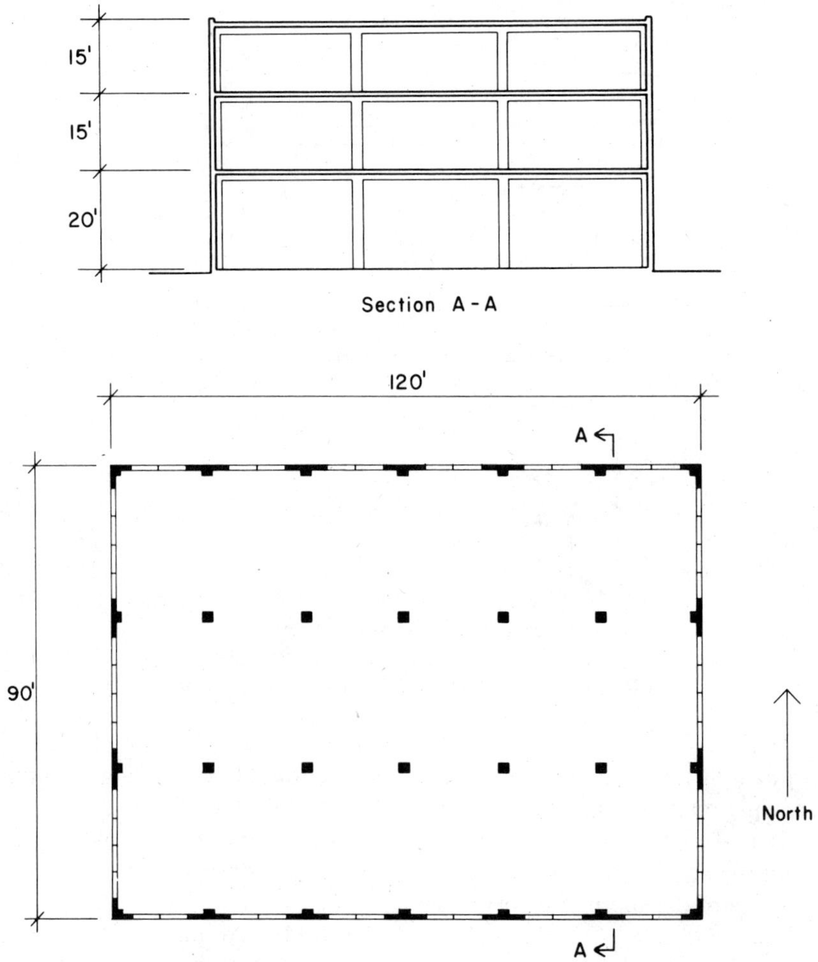

FIGURE 26.10

2. The K factor for the rigid frame (0.67) is one-half that for the shear wall (1.33), resulting in a further reduction of 50% of the values for the lateral loads.

The F_x values determined in Table 24.2 are thus used to find the bent loadings with the following adjustments:

$$\text{Bent Load} = F_x \times 0.90 \times 0.50 \times \tfrac{1}{7}$$
$$= 0.0643F_x$$

On this basis the bent loading will be as shown in Fig. 26.11. For an approximate analysis we assume the inflection points of both the columns and beams to be at the midpoints of the spans, except for the first-story columns. The internal shear forces are thus as shown in Fig. 26.12. The end moments in the members may then be determined from these shears.

These approximate shears and moments must be combined with the effects of gravity for the design of the beams and columns in the bents. The preliminary member sizes thus found may then be used for a more exact indeterminate analysis. They may also be used for a determination of the individual story drift (horizontal deflection), which is often a critical concern for the rigid frame.

An excellent discussion of the various problems of analysis and design of multistory framed structures may be found in *High-Rise Building Structures* (Ref. 28.)

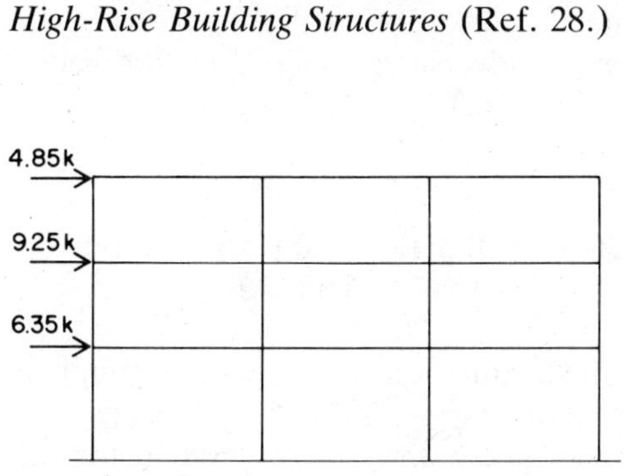

FIGURE 26.11. Loading of the three-story bent.

26.3 THREE-STORY BUILDING WITH PERIPHERAL BENTS

Figure 26.13 shows another rigid frame system for the building in the preceding section. In this case there are only two bents in each direction, constituted from the exterior spandrel beams and the exterior columns. This is yet another possibility for the building for which shear wall schemes were discussed in Sections 24.2 and 24.3 and braced frames were discussed in Sections 25.2 and 25.3.

The scheme shown here offers the distinct advantage of the greatest freedom in planning and in utilization of the exterior walls for windows. By freeing the interior columns from involvement in the lateral load resistance, it becomes possible to locate them without reference to the exterior framing. While some order may be proper, it is quite literally possible to locate them randomly with respect to the organization of the framing in the exterior wall. By contrast, the scheme in the preceding section requires an alignment of both interior and exterior columns.

There are two major structural advantages to this scheme. As with other peripheral systems (Sections 24.3 and 25.3), there is an optimal development of the torsional resistance of the whole system. The second advantage comes from the ability to increase the overall rigidity and strength of the frame by increasing the number of columns and the depth of the beams. It is usually much less critical to have exterior columns to plan around than to have a large number of interior columns, especially when the interior columns must be in a determined location. Increasing the number of the exterior columns reduces the shear on individual columns and also results in stiffer (shorter span) beams. The spandrel beams can also be stiffened by increasing their depth—a freedom not likely for interior beams because of the limits on the floor-to-floor

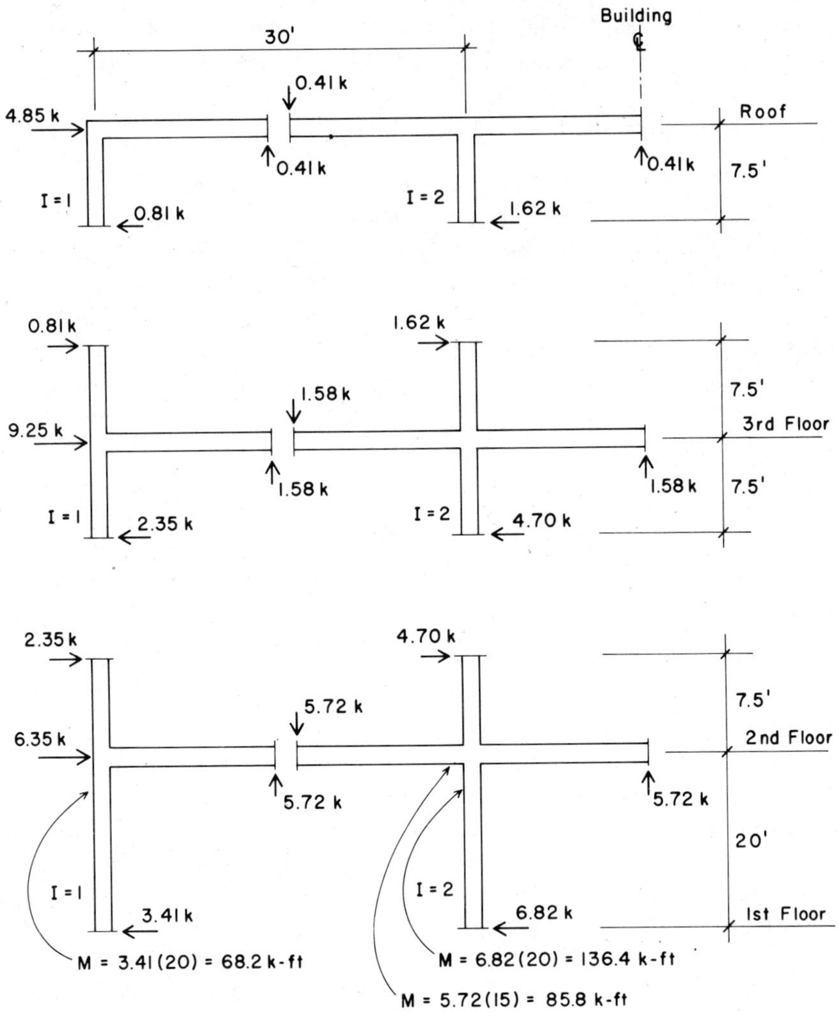

FIGURE 26.12. Approximate analysis of the bent.

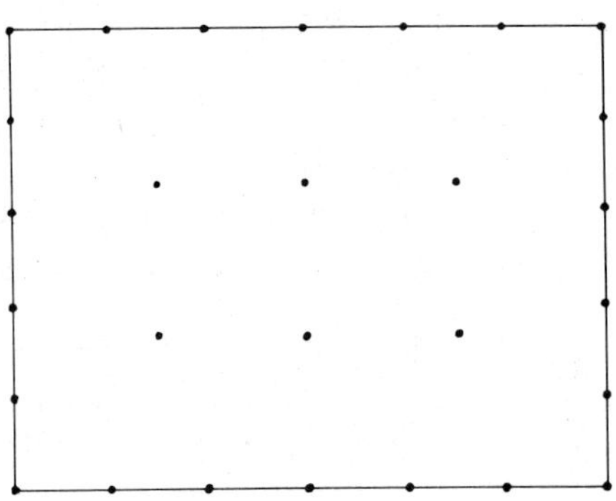

FIGURE 26.13

height. The stiffened frame offers the major advanatage of less drift (horizontal deflection.)

26.4 THREE-STORY BUILDING WITH MIXED SYSTEM

The mixture referred to in this example is that of interior and peripheral bents. This is an alternative bracing system for the building discussed in Section 24.4. The

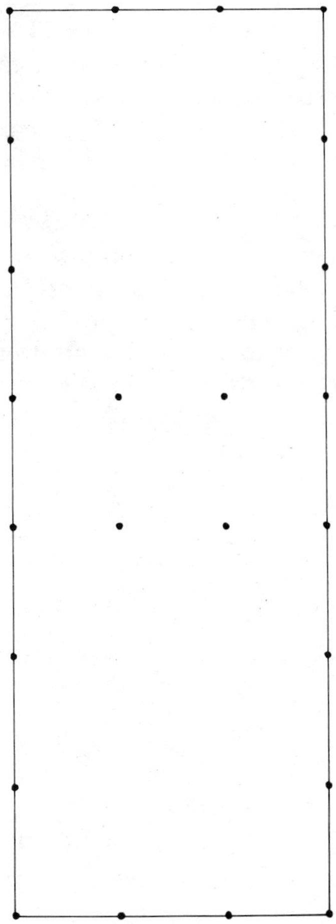

FIGURE 26.14

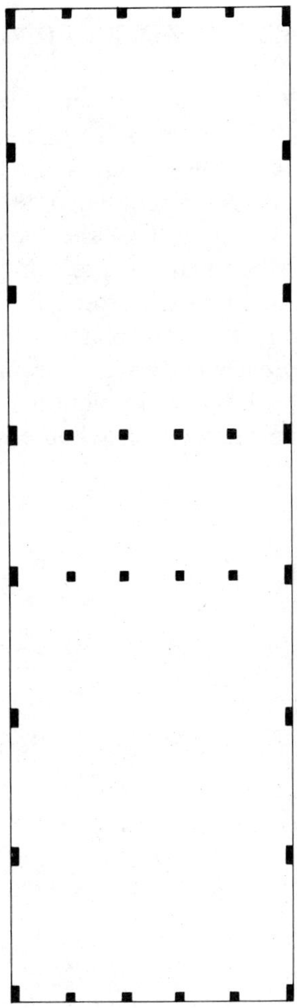

FIGURE 26.15

architectural plans for the building, developed for the all-shear wall system, are shown in Figs. 24.15 and 24.16.

The scheme shown in Fig. 26.14 utilizes peripheral bents in the long direction and a combination of the end bents plus two interior bents in the short direction. The proportions of the plan and the long dimension of the building make it unlikely that the interior bents could be omitted. In the scheme proposed, there are the same number of columns in the bents in the two directions (16).

With the absence of the shear walls or the X-braces from the exterior walls, it would be possible to have more windows than shown in the plan in Fig. 24.15. The planning of the building exterior, as well as the development of the stair, elevator, duct shaft, and restrooms layouts may quite likely be somewhat different if the specific choice is made for one lateral bracing system over the other, which is simply to say that neither decision should necessarily be made *first*, but that the total design schemes be considered as complete alternatives.

26.5 TWELVE-STORY FRAME

Figure 26.15 shows a scheme for a rigid frame system for the building discussed in Section 24.5. This scheme is similar to that for the three-story building in the preceding section. Because of the long narrow plan, it is necessary to use some interior bents in addition to the end bents in the narrow direction. The analysis and design of the system would have to be carried to a reasonable level of completion in order to determine whether this scheme is reasonable or not. If the load per bent, and the resulting forces and drifts, are excessive, it would be necessary to use more of the interior bents—possibly all of them as was proposed for the shear wall scheme in Section 24.5.

As with other buildings, the choice of the lateral bracing system has various architectural implications. In this example, if the scheme proposed in Fig. 26.15 is used, the apartment layouts would be much less constrained than they would be with the shear wall system.

CHAPTER TWENTY-SEVEN

Case Studies

The material in this chapter consists of excerpts from the design calculations for several buildings that were designed for various locations in the Los Angeles area. In most cases they were within the city limits of Los Angeles and thus designed for the city building code (essentially similar to the *UBC*). The illustrations consist mostly of direct reproduction of the handmade calculation sheets that were prepared for submission to the City of Los Angeles Building Department for review for the building permit. Bear in mind that these are "problem" buildings, for which the solutions are not very orderly, in most cases. The problems came from various sources, including site conditions, special requirements of the permit-granting agency, architectural design priorities, and client requests.

27.1 HILLSIDE HOUSE: MIXED WALL AND FRAME

This house is built on a steeply sloping site. Primarily because of the site, the plan is long and narrow. As often happens with one-story buildings of light wood frame construction, the wind load is greater than the seismic load. This is true for this structure, even though it is in seismic zone 4, and the wind pressure for design is only 15 psf. Although the calculations are not shown, it was determined that the wind load was approximately 20% higher than the seismic load.

The lateral resistive system consists of a horizontal plywood roof and vertical shear walls of wood frame. As the loading varies, the surfacing for shear ranges from $\frac{1}{2}$-in. gypsum board drywall to $\frac{3}{8}$-in. plywood. In the short direction there are a number of shear walls that are reasonably well dispersed along the plan. In the long direction, however, there is a lack of solid walls, especially at one end of the plan. Two steel rigid frames are therefore inserted, as shown in the plans on Figs. 27.1 and 27.2. At the time of the design and construction of this building, it was a policy of the city to permit the mixing of wood frame construction with steel rigid frames, on the condition that the maximum story drift for the frames be limited to 0.25 in. For average short story heights, this is about one-third of the usual limit ($0.005h$), the intent being to keep the bent stiffness within the range of that of ordinary wood framed shear walls.

Figure 27.1 presents the analysis for the wind loads developed by the roof diaphragm. These are distributed to the vertical bracing system on a peripheral basis. Wall surfacing for the wood framed shear walls is also indicated in the calculations. For this distribution the steel bents are treated on a par with the shear walls.

Figure 27.2 shows a similar analysis for the loads to the upper-floor diaphragm. There are additional walls at this level, but the steel bents are still used in the long direction.

Figure 27.3 shows an analysis for the combined lateral and gravity loads on the two-story bent on the downhill side. The end moments on the girders due to gravity load are assumed rather conservatively to be $wL^2/12$.

Figure 27.4 shows an analysis for the combined loads on the steel bent that is indicated on the uphill side of the plan in Fig. 27.1. Because of the site development, this is a single-story bent. The calculations indicate an initial choice of a W 8 × 17 for the column. Further on, the drift calculation indicates a drift considerably in excess of the limit of 0.25 in. This results in a change to a W 8 × 31 and a change to a fixed base for the column, which reduces the drift to an acceptable value.

27.2 BEACH HOUSE: MIXED WALLS AND FRAME

This is a large house, built on the beach in Malibu. The structure in this case is primarily a mixed system of wood frame shear walls and steel rigid bents, with the same conditions as described for the preceding examples. However, the steel frames are considerably more extensive, as shown in the photographs in Fig. 27.5. The steel frame is also used simply to achieve some long-span open-plan conditions, so there is a mix of rigid bents and simple post and beam steel framing.

The beach site consists of 10 to 15 ft of sand overlying a solid rock base that slopes toward the water. It is actually a mountainside site that happens to coincide with the ocean level in this millenium. The foundation consists of concrete piers that bear on the rock and support a reinforced concrete grade beam structure at ground level.

Figure 27.6 shows an analysis for the lateral load at the level of the roof diaphragm. Again, wind load slightly exceeds the seismic load, so it is used for the lateral loading for design.

Figure 27.7 shows an analysis for the lateral load at the level of the second floor. It may be observed that the plan has considerable open space, so the steel bents constitute a major portion of the vertical bracing system.

Figure 27.8 shows an analysis for the combined lateral and gravity loading on one of the bents. In this case the bent is three stories high and sits on the concrete grade level structure, as shown in the photographs in Fig. 27.5.

27.3 APARTMENT BUILDING: MIXED WALLS AND FRAME

This is a small, three-story apartment building utilizing wood frame construction as the basic structure. The plan allows for an adequate use of shear walls in the east–west direction, but not in the north–south direction. The solution here is similar to that in the preceding examples, with the use of steel rigid frame bents in conjunction with the wood-framed shear walls.

Plans for the layout of the vertical bracing elements are shown in the calculations in Figs. 27.9 and 27.10. This diaphragm distribution is further complicated by the large roof openings and the open space at the entries.

The loading for one of the three-story high bents is shown in Fig. 27.11, together with calculations for the preliminary sizing of the frame members. As with the bents in other cases, these first trial sizes are used for a lateral drift calculation, the results of which may require some revisions. In addition to the possibility of increasing the size of the columns, the column bases could be changed to a fixed condition to reduce drift. The latter, of course, must be added to the foundation design development for a determination of feasibility.

Where frames are actually built into walls, even where the walls are discontinuous, lateral deflection will most likely be considerably less than that calculated for the unrestrained frame.

27.4 BEACH HOUSE: PIERCED WOOD WALL

This is a small house with several interior levels, including a three-level roof. The wall elevation shown in Fig. 27.12 is that of one of the exterior walls that is used as a vertical bracing element, loaded at several levels by the roof and floor diaphragms.

The wall in this case represents a situation known affectionately as a "Swiss cheese wall" among structural designers. Openings pierce the wall in a haphazard way, constituting major discontinuities for the shear wall. Although there are several options, the design in this case makes use of the basic construction of the wood frame structure. Columns and beams are used to tie the fragments of shear wall together to form a cohesive whole.

The illustration shown in Fig. 27.12 was redrawn from the original calculations. It shows the loading for the wall and contains several notes relating to the use of ties, anchors, and various elements of the construction. This notation is frequently used on the calculations both for the information of the designer and reviewer and as instruction to the draftsperson who prepares the working drawings for construction.

Figure 27.13 shows the same information for another of the exterior walls. In this case the wall is only two stories high.

27.5 STORAGE WAREHOUSE: MASONRY WALL

This situation is similar to that in the preceding section in that the exterior wall with considerable openings is used as a vertical bracing element. The wall in this case is of reinforced concrete block masonry. As shown in the elevation in the calculations in Fig. 27.14, the only significant element that is usable is the single 6-ft-wide pier.

A distribution of the total force to all of the masonry piers, using a procedure essentially the same as that for the wall in the example in Section 23.4, is shown in the calculations. This analysis shows that approximately 85% of the total shear is absorbed by the large stiff pier. The design of this pier, as well as that of the smaller piers, is shown in the calculations in Fig. 27.15.

27.6 RETAIL STORE: MIXED WALLS AND TRUSSED BENT

As shown in the plan drawing in Fig. 27.16, this is a large single-space building with a clear span roof of 125 ft between exterior walls. Site and building planning conditions permit the use of considerable solid walls on three sides but indicate a need for considerable opening in the street side wall, a very common situation for commercial buildings in urban areas. The solution in this case, partly justified by the required depth for the long span roof, consists of a four-span steel trussed bent, as shown in the calculations in Fig. 27.17.

The analysis for seismic loading on the bent, and the calculations for the bent drift and the sizing of the bent members are shown in Figs. 27.17, 27.18, and 27.19. The bent construction is quite conventional, using steel wide flange members for the columns and the truss chords, and double angles for the diagonals.

Bent drift in this case is considerable, although still within the code limit. This situation is not the same as in the mixed systems in Section 27.1, 27.2, and 27.3, since the load sharing is not really affected by the relative stiffness of the vertical structure. There is therefore no special reason to restrict the bent drift, other than the usual concerns for distortion of the building construction.

ROOF DIAPH. LEVEL

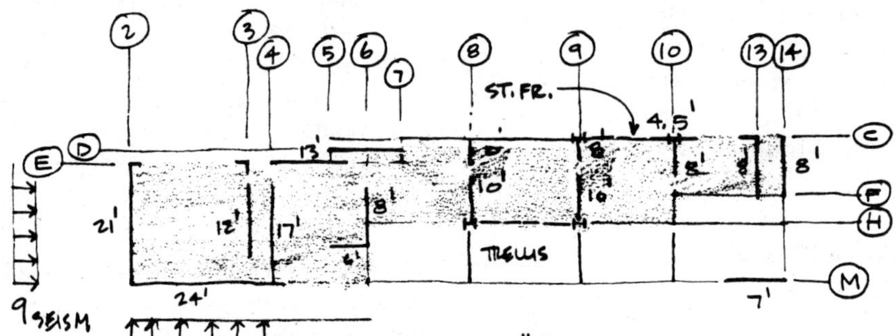

q_{WIND} = 15 PSF × 10/2 = 75 #/'

LAT ⊥ HILL: H/FT:

H(2) = 75 × 20/2 = 750#   750/21 = 35#/'   1/2" GB 5d @ 7" OC

H(3) = 75 × 24/2 = 900#   900/12 = 75#/'   " (100#/')

H(4) = 75 × 24/2 = 900#   900/17 = 53#/'   "

H(6) = 75 × 34/2 = 1275#   1275/8' = 160#/'   1/2" GB 5d @ 7 EA SIDE (2×100#/')

H(8) = 75 × 34/2 = 1275#   1275/18' = 71#/'   1/2" GB 5d @ 7 EA SIDE (2×100)

H(9) = 75 × 34/2 = 1275#   1275/18' = 71#/'   "

H(10) = 75 × 34/2 = 1275#   1275/8' = 160#/'   "

H(14) = 75 × 18/2 = 675#   675/8' = 84#/'   1/2" GB (100#/') OK

LAT ∥ TO HILL: SEISMIC GOVERNS

H_s = 25' × 42' × 15 PSF × .2 = 3150# q_s = 3150/21 = 150#/' (BETW (2) & (6))
(2)-(6)

H(E) = 150 × 21/2 = 1575#   H/FT = 1575/13' = 121#/'   3/8" PLY 8d @ 6 (240)

H(M) = 150 × 21/2 = 1575#   H/FT = 1575/24' = 65#/'   5/8" GB 6d @ 7 (125#/') OK

H_s (6)-(14) = 28 × 68 × 15 PSF × .2 = 5712#

H(C) = 5712/2 = 2852# - STEEL FRAME (8)(9)
H/FT DIAPH = 2852/18' = 158#/' OK (3/8" PLY 8d @ 6 165#/')

H(H) = 5712/2 = 2852# - STEEL FRAME LINE (H) (8)(9)
SEE PG ____

FIGURE 27.1

UPPER FLOOR DIAPHRAGM

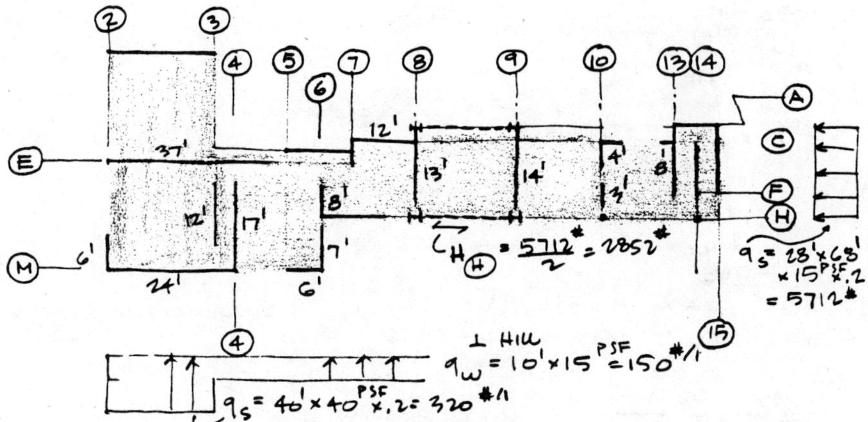

↑ $H_{(2)}$ = 320 × 20/2 + 750 (RF) = 3950# H/FT = 3950/6 = 658#/'

1/2" PLY 10d@2 (770×.9 = 693#/')

↑ $H_{(3)}$ = 3950 + 150 × 4/2 + 900 (RF) = 5150# H/FT = 5150/12 = 429#/'

1/2" PLY 10d@2 (693#/')

↑ $H_{(4)}$ = 150 × 22/2 + 900 (RF) = 2550# H/FT = 2550/17 = 150#/'

1/2" GB EA SIDE (2×100#/')

↑ $H_{(6)}$ = 34/2 × 150 + 1275 (RF) = 3825# H/FT = 3825/15 = 255#/'

3/8" PLY 8d@4 (320#/')

$H_{(10)}$ = 150 × 31/2 + 1275 = 3600#

H/FT = 3600/7' = 514#/'

1/2" PLY 10d@2 (693#/')

↑ $H_{(8)}$ = 34/2 × 150 + 1275 (RF) = 3825#

H/FT = 3825/14' = 273#/'

3/8" PLY 8d@4 (320#/')

$H_{(13)}$ = 150 × 18/2 + 0 = 1350#

H/FT = 1350/7' = 192#/'

1/2" GB EA SIDE (2×100#/' OK)

↑ $H_{(9)}$ = 150 × 34/2 + 1275 (RF) = 3825#

H/FT = 3825/14 = 273#/'

3/8" PLY 8d@4 (320#/')

$H_{(14)}$ = 150 × 4/2 = 300# + 675 (RF) = 975#

H/FT = 975/8' = 121#/' 3/8" PLY 8d@6 (220#/')

FIGURE 27.2

STEEL FRAME LINE (H) (8)&(9)

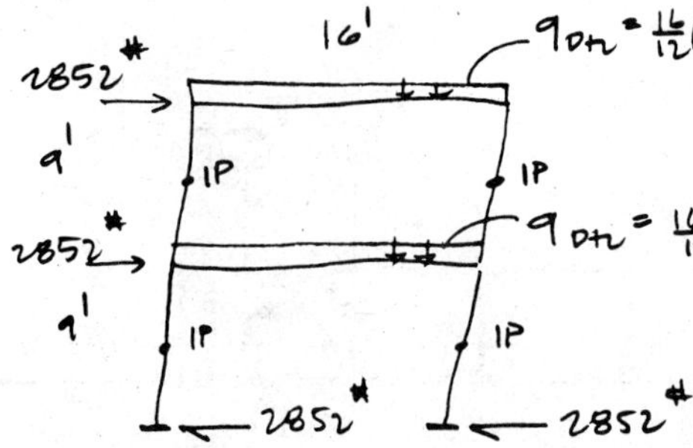

$w_{D+L} = \frac{16}{12}(20+10) + 50^{\#/'} = 90^{\#/'}$

$w_{D+L} = \frac{16}{12}(40+10) + 50^{\#/'}_{bm} = 117^{\#/'}$

DEAD + LIVE:

FEM: RF = $90 \times 16^2/12 = 1920^{\#-'}$
FL = $117 \times 16^2/12 = 2496^{\#-'}$
} D+L MOM @ GIRDER ENDS

SEISMIC MOM ON COL'S:

RF : $M_c = \frac{2852}{2} \times 5' = 7130^{\#-'}$

2ND FL : $M_{c_U} = \frac{2852}{2} \times 5' = 7130^{\#-'}$

$M_{c_L} = 2852 \times 5' = 14,260^{\#-'}$

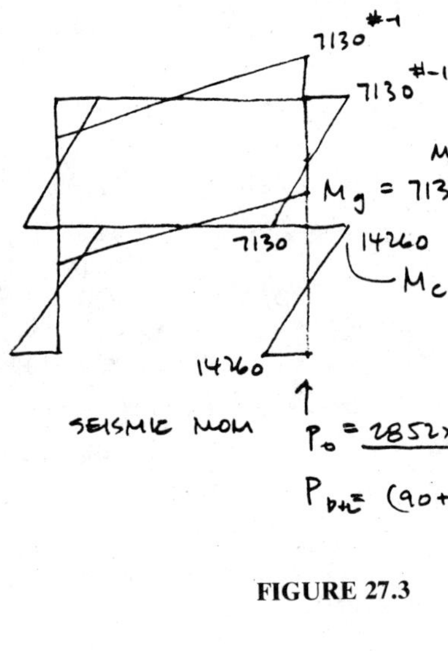

TOTAL MOMENTS;

$M_g = 7130 + 14260 + 2496 = 23,886^{\#-'}$ (M_{c_U}, M_{c_L}, FEM)

$M_c = 14260 + 2496 = 16,756^{\#-'}$ (ΣM_c, FEM)

SEISMIC MOM

$P_o = \frac{2852 \times 18 + 2852 \times 9}{16'} = 4812^{\#}$

$P_{D+L} = (90+117)\frac{16}{2} + 19 \times 50 = 2606^{\#}$

$\Sigma P = 7418^{\#}$

FIGURE 27.3

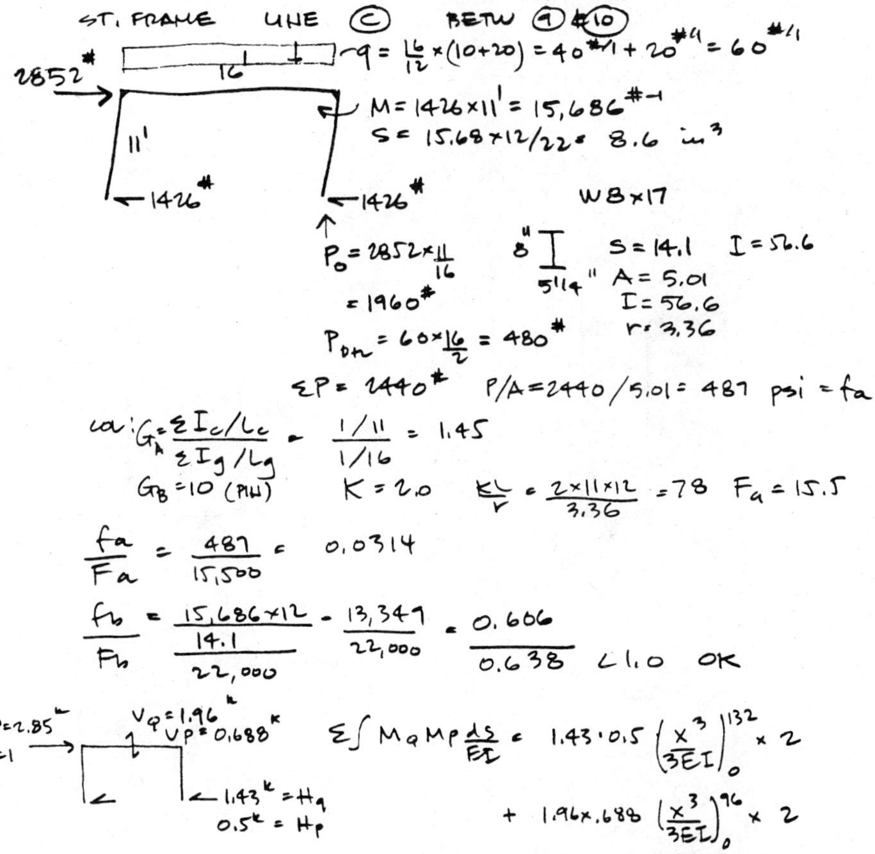

FIGURE 27.4

FIGURE 27.5. Building braced by three-dimensional steel rigid frame. Steel frame sits on a reinforced concrete grade structure that is supported on concrete piers.

LOADS TRIB TO RF DIAPH (WIND GOV)

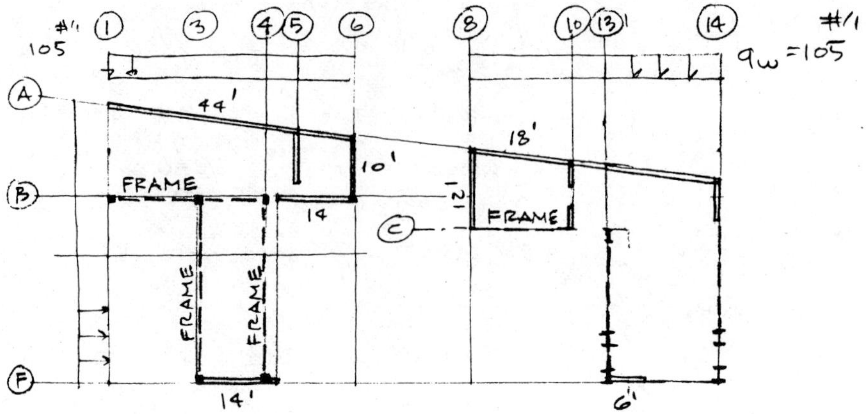

$q_{WIND} = \frac{14'}{2}(15^{PSF}) = 105^{\#/'}$

$H_{③} = 105 \times 44' \times \frac{7'}{12} = 2695^{\#}$

$H_{④} = 105 \times 44 - 2695 = 1925^{\#}$

①-⑥:

$H_{Ⓐ} = 105 \times \frac{16}{2} = 840^{\#}$

$H_{Ⓑ} = 105 \times \frac{48'}{2} = 2520^{\#}$

$H_{Ⓕ} = 105 \times \frac{33}{2} = 1733^{\#}$

⑧:

$H_{Ⓒ} = 105 \times \frac{38}{2} = 1995^{\#}$

$H_{Ⓕ} = 105 \times \frac{38}{2} = 1995^{\#}$

⑧-⑭

$H_{⑧} = 105 \times \frac{21}{2} = 1102^{\#}$

$H_{⑩} = 105 \times \frac{27}{2} = 1417^{\#}$

$H_{⑬} = 105 \times \frac{28}{2} = 1470^{\#}$

$H_{⑭} = 105 \times \frac{20}{2} = 1050^{\#}$ $H/FT = 1050/6' = 175^{\#/'}$ 3/8" PLY 8d @4 ($320^{\#/'}$)

$F_{OT} = 1050 \times 14/6 = 2460$

ST 6236 CAP = 2680

$H_{Ⓐ} = \frac{40}{2} \times 105 = 2100^{\#}$

$H_{Ⓕ} = \frac{40}{2} \times 105 = 2100^{\#}$

FIGURE 27.6

LOADS TRIB TO 2ND FL DIAPH.
(WIND GOV)

→ N

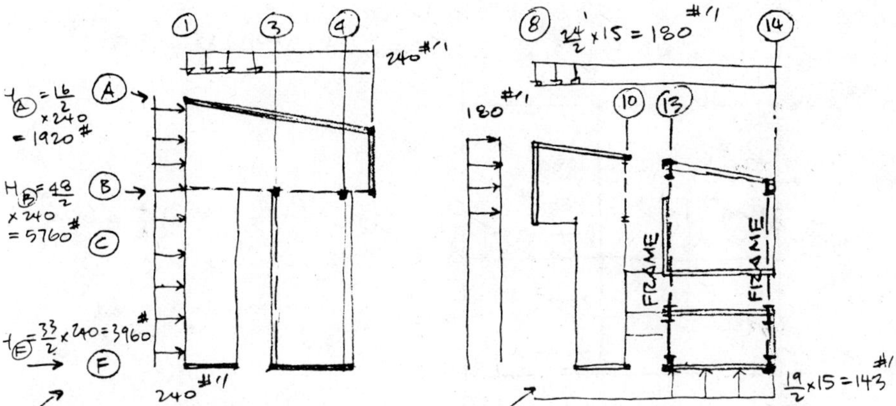

$A = \left(\frac{49' + 45}{2}\right) 33 - 6 \times 32 - 3 \times 32$

$= 1551 - 288 = 1263$ SF

$W_{FLOOR} = 1263 \times 20 + 14 \times 33 \times 25 = 36{,}810^{\#}$ (TILE)

$W_{WALLS} = (33 + 12 + 14 + 30) 12 \times 10^{PSF} + (46 + 44) 5^{PSF} \times 12 = 10{,}680 + 5400 = 16{,}080^{\#}$

$\Sigma W_D = 52.9^K$ SAY 53^K

$A = 45' \times \left(\frac{41 + 35}{2}\right) - 8 \times 32 - 6 \times 11$

$= 1710 - 322 = 1388$ SF

$W = 1388 \times 20 = 27{,}600$

$W_{WALLS} = (45' + 13' + 18' + 9') 12 \times 10^{PSF} + (24 + 16 + 27) 12 \times 5 = 10{,}200 + 4020 = 14{,}220^{\#}$

$\Sigma W_D = 41{,}820^{\#}$ SAY 42^K

$q_w = \frac{32'}{2} \times 15^{PSF} = 240^{\#/1}$

$H_{(1)} = \frac{16}{2} \times 240 = 1920^{\#} + 0$ (RF)

$H_{(3)} = 34 \times 240 \times \frac{17}{17} = 8160^{\#} + 2695^{\#}$ (RF) (FRAME)

$H_{(4)} = 18 \times 240 \times \frac{9}{12} = 3240^{\#} + 1925$ (RF) (FRAME)

(8)–(14) : 7

$H_{(A)} = 180 \times \frac{12}{2} = 1080^{\#}$

$H_{(C)} = 180 \times \frac{40}{2} = 3600^{\#}$

$H_{(F)} = 180 \times \frac{28}{2} = 2520^{\#}$

$H_{(8)} = 180 \times \frac{21}{2} = 1890^{\#} + 1102$ (RF) $= 2992^{\#}$

$H_{(10)} = 180 \times \frac{27}{2} = 2430^{\#} + 1417$ (RF) $= 3847^{\#}$

$H_{(13)} = 180 \times \frac{19}{2} = 1710^{\#} + 1000$ (RF) $= 2710^{\#}$

$H_{(14)} = 18 \times 9 \times \frac{180}{12} = 2243 + 1308$ (RF) $= 3551^{\#}$

FIGURE 27.7

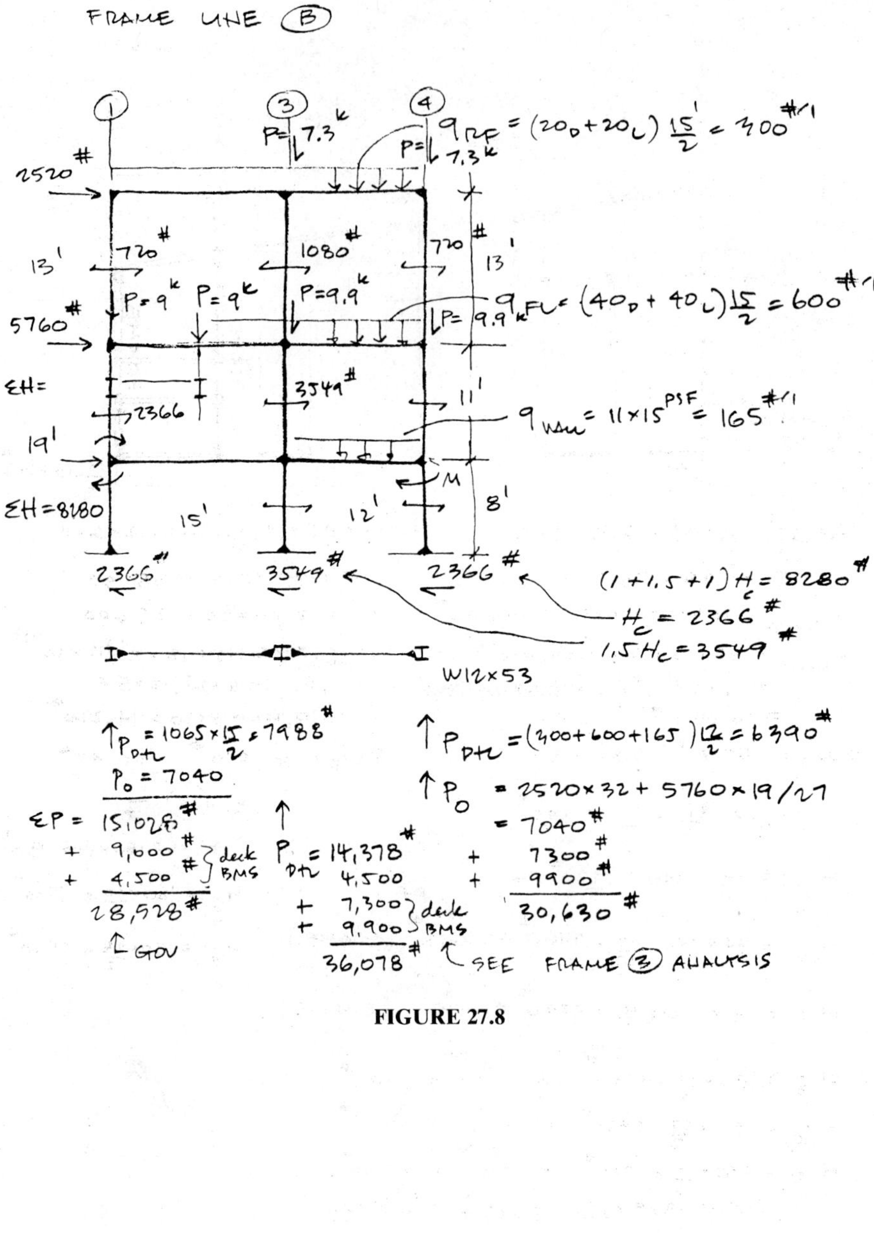
FRAME LINE (B)
(1)
(3)
(4)
P = 7.3^k
P = 7.3^k
q_RF = (20_D + 20_L) 15'/2 = 300 #/'
2520#
13'
720#
1080#
720#
13'
P = 9^k
P = 9^k
P = 9.9^k
P = 9.9^k
q_FL = (40_D + 40_L) 15/2 = 600 #/'
5760#
ΣH =
2366
3549#
11'
q_WALL = 11 × 15^PSF = 165 #/'
19'
M
ΣH = 8180
15'
12'
8'
2366#
3549#
2366#
(1 + 1.5 + 1) H_c = 8280#
H_c = 2366#
1.5 H_c = 3549#
W12×53
↑P_D+L = 1065 × 15/2 = 7988#
P_o = 7040
ΣP = 15,028#
+ 9,000#
+ 4,500#
deck
BMS
28,528#
GOV
↑P_D+L = 14,378#
4,500
+ 7,300
+ 9,900
deck
BMS
36,078#
↑P_D+L = (300 + 600 + 165) 12/2 = 6390#
↑P_O = 2520 × 32 + 5760 × 19/27
= 7040#
+ 7300#
+ 9900#
30,630#
SEE FRAME (3) ANALYSIS

FIGURE 27.8

3RD FL WALLS

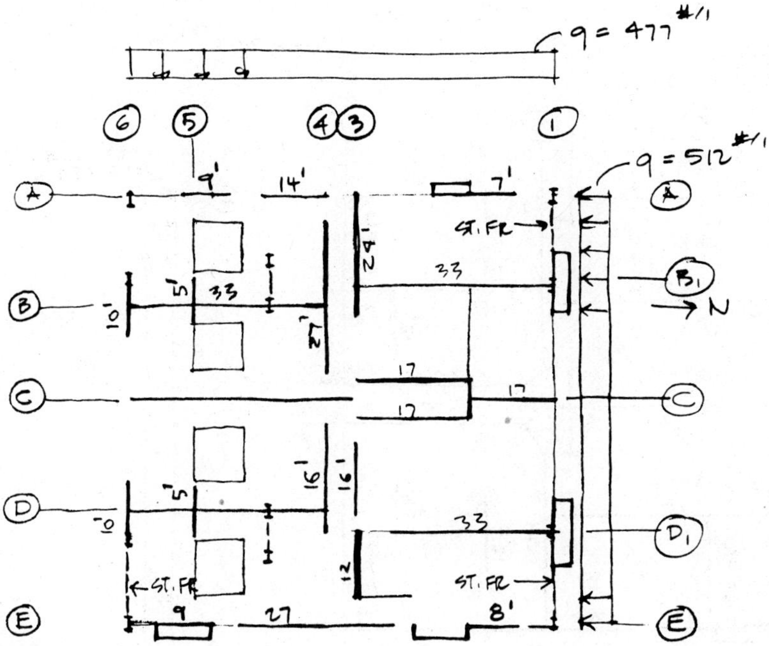

$H_{(A)} = 512 \times 17/2 = 4352^{\#}$ — $H/FT = 4352/30 = 145^{\#/'}$ 7/8" CP OK (180#/')

$H_{(B)+(B_1)} = 512 \times \frac{34}{2} = 8704^{\#}$ — $H/FT = 8704/66 = 131^{\#/'}$ 1/2" GB 5d@7 EA SIDE = 2×100 OK

$H_{(C)} = 512 \times 37/2 = 9472$ — $H/FT = 9472/66 = 144^{\#/'}$ "

$H_{(D)} = 512 \times 34/2 = 8704$ — $H/FT = 8704/66 = 132^{\#/'}$ "

$H_{(E)} = 512 \times 17/2 = 4352$ — $H/FT = 4352/44 = 98.9^{\#/'}$ 7/8" CP OK (180#/')

$H_{(6)} = 477 \times 34/2 = 8109 = H_{(1)}$ — $H/FT = 8109/18 = 450^{\#/'}$ 3/8" PLY 8d@2½ (480#/') 2 ST 6236 STRAPS (O.T.)

$F_{OT} = 4054 \times 9/10' = 3649^{\#}$

CAP 2 ST6236 = 2×2680 = 5360#

FIGURE 27.9

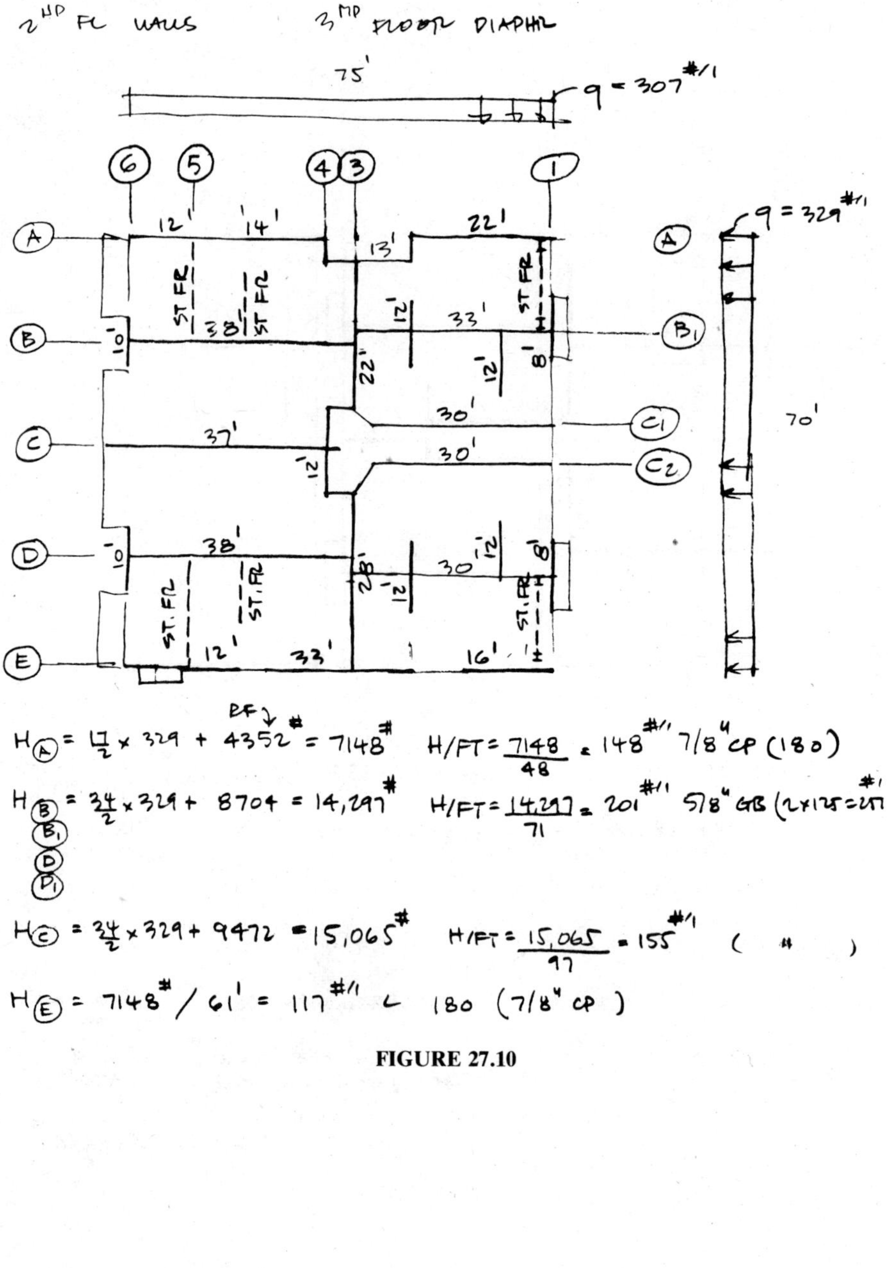

2ND FL WALLS
3RD FLOOR DIAPHR
75'
q = 307 #/'
q = 329 #/'
70'
ST FR
H(A) = 17/2 × 329 + 4352# = 7148#   H/FT = 7148/48 = 148 #/' 7/8" CP (180)
H(B)(B1)(D)(D1) = 34/2 × 329 + 8704 = 14,297#   H/FT = 14,297/71 = 201 #/' 5/8" GB (2×125=25
H(C) = 34/2 × 329 + 9472 = 15,065#   H/FT = 15,065/97 = 155 #/' (")
H(E) = 7148# / 61' = 117 #/' < 180 (7/8" CP)

FIGURE 27.10

2 FRAMES ON EACH LINE, BUT: ASSUME EACH FRAME WILL RECEIVE 60% OF TOTAL LOAD TRIBUTARY TO THAT LINE TO ACCOUNT FOR IMPERFECT RIGIDITY MATCHING, ETC.

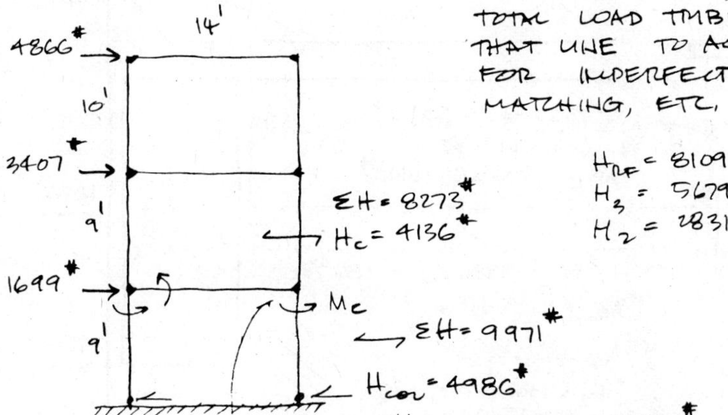

$H_{RF} = 8109 \times 0.6 = 4866^{\#}$

$H_3 = 5679 \times 0.6 = 3407^{\#}$

$H_2 = 2831 \times 0.6 = 1699^{\#}$

$P_0 = \frac{\Sigma M_0}{14} = \frac{212{,}865}{14} = 15{,}204^{\#}$

$P_{DL} = 7 \times 28 \times 10 + 3 \times 7 \times 50 = 3010^{\#}$

$M_g = 67.2^{k'}$

$M_c = 4136 \times 9 \times .6 = 22{,}337^{\#\text{-}'}$

$M_c = 4986 \times 9' = 44{,}874^{\#\text{-}'}$

$\Sigma P_{DL+0} = 18{,}214^{\#}$

GIRDER: $-M = 67.2^{k'}$

$S = 67.2 \times 12/22 = 36.65$

W12×31 (6½" wide) I = 239

W10×39 (8") I = 210

COL: $G_A = 10$ (PIN)

TRY W8×28: I = 97, r = 3.45, A = 8.22, S = 24.3

$G_B = \frac{\Sigma I_c/L_c}{\Sigma I_g/L_g} = \frac{2 \times 97/9}{210 \times 1/14} = \frac{97}{15} = 6.47$ (AISC FIG 1.8.2)

$K = 2.7$ $KL/r = 2.7 \times 9 \times 12/3.45 = 84.52$

$F_a = 14.79$

$f_a = 18{,}214/8.22 = 2213$ psi $\frac{f_a}{F_a} = \frac{2213}{14{,}790} = 0.1496$

$f_b = 44{,}874 \times 12/24.3 = 22{,}160$ psi $\frac{f_b}{F_b} = \frac{22{,}160}{22{,}000} = 1.0072$

$\Sigma \frac{f}{F} = 1.156 < 1.33$ OK

W12×32

W8×28 COL OK W12×31 ✓ OR W10×39 BM OK

FIGURE 27.11

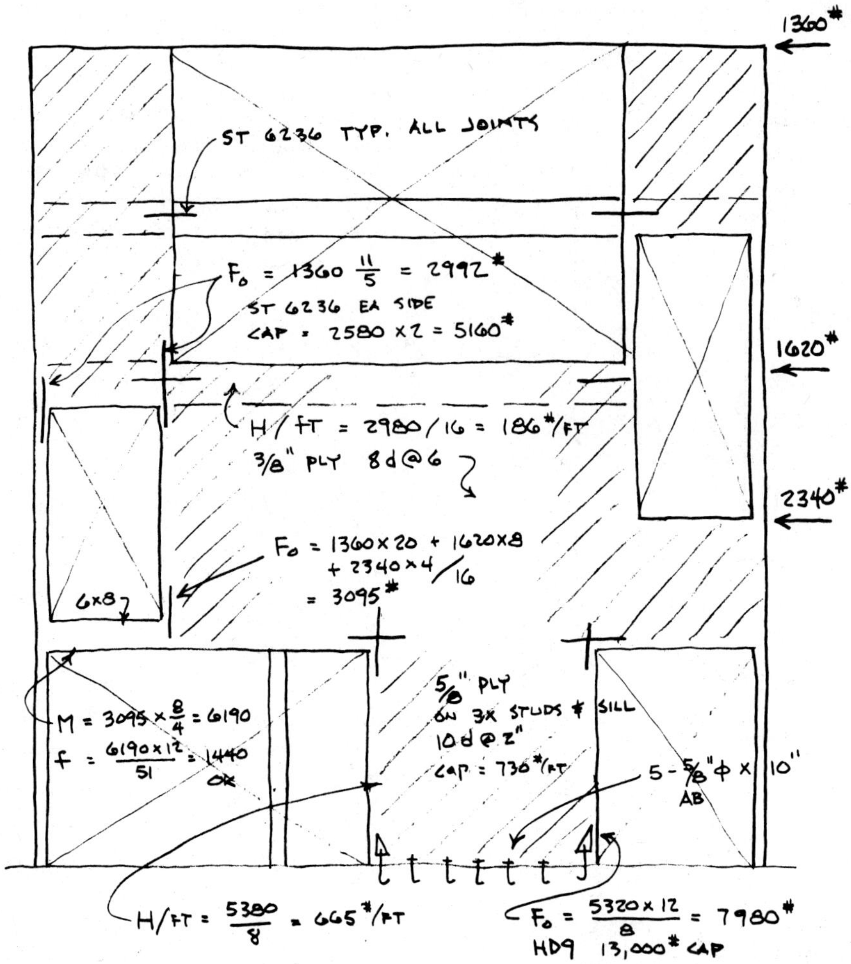

FIGURE 27.12

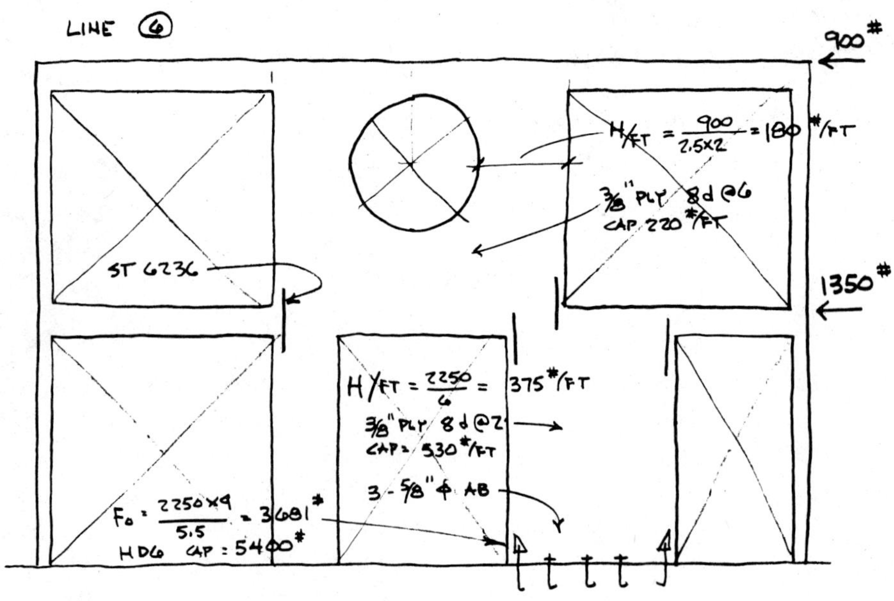

FIGURE 27.13

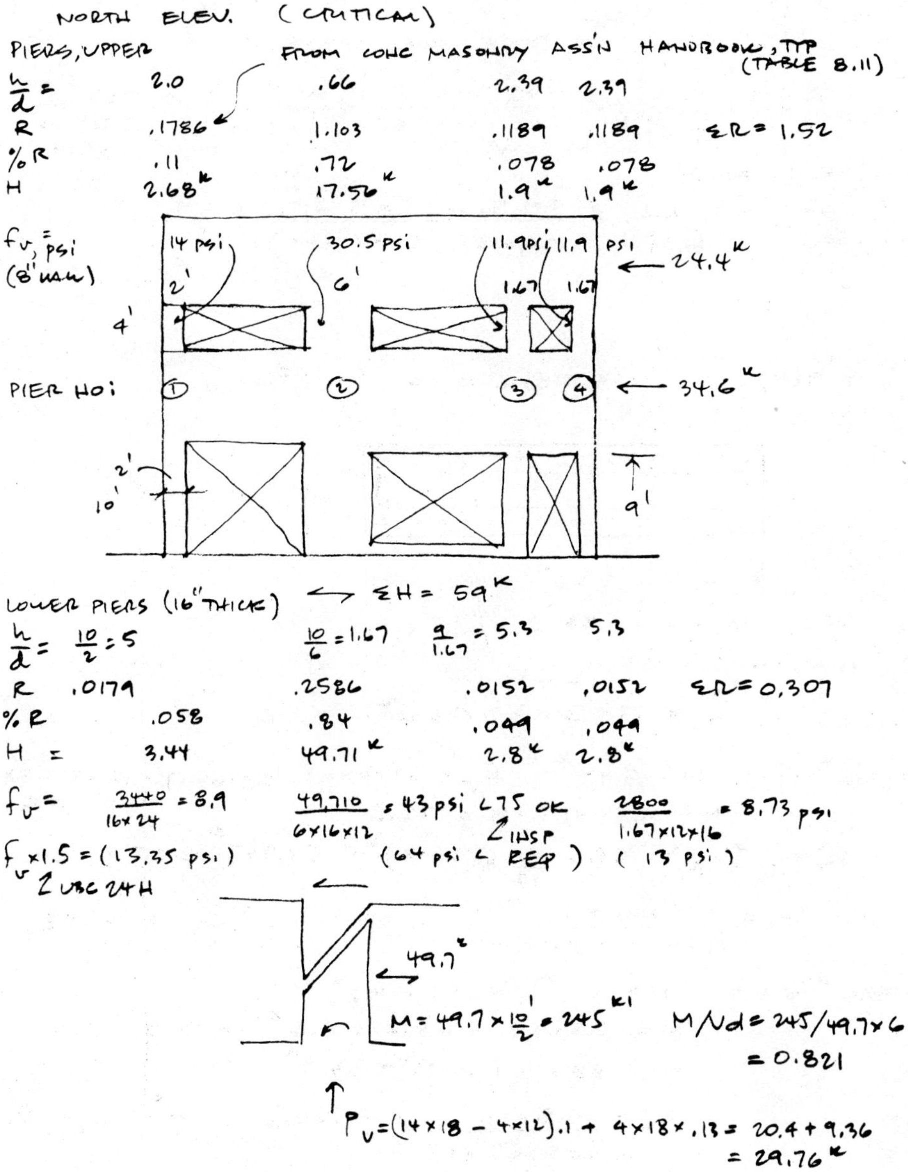
NORTH ELEV. (CRITICAL)
PIERS, UPPER
FROM CONC MASONRY ASS'N HANDBOOK, TYP (TABLE 8.11)
h/d = 2.0 .66 2.39 2.39
R .1786 1.103 .1189 .1189 ΣR = 1.52
%R .11 .72 .078 .078
H 2.68k 17.56k 1.9k 1.9k
fv, psi (8" wall)
14 psi
30.5 psi
11.9 psi
11.9 psi
24.4k
2'
6'
1.67
1.67
4'
PIER NO:
1
2
3
4
34.6k
2'
10'
9'
LOWER PIERS (16" THICK)
ΣH = 59k
h/d = 10/2 = 5 10/6 = 1.67 9/1.67 = 5.3 5.3
R .0179 .2586 .0152 .0152 ΣR = 0.307
%R .058 .84 .049 .049
H = 3.44 49.71k 2.8k 2.8k
fv = 3440/16×24 = 8.9
49,710/6×16×12 = 43 psi < 75 OK
2800/1.67×12×16 = 8.73 psi
fv ×1.5 = (13.35 psi) < UBC 24H
< INSP (64 psi < REQ)
(13 psi)
49.7k
M = 49.7 × 10/2 = 245k'
M/Vd = 245/49.7×6 = 0.821
Pv = (14×18 − 4×12).1 + 4×18×.13 = 20.4 + 9.36 = 29.76k

FIGURE 27.14

PIER # ② 1ST FLOOR

SHEAR REINF TO TAKE ALL SHEAR

$$A_v = \frac{V s}{f_v j d} = \frac{49{,}700 \times 1.5 \times 8''}{20{,}000 \times .87 \times (6 \times 12 - 4)} = 0.504 \text{ in}^2$$

@ 8" oc HORIZ

VERT BARS

$$A_s = \frac{M}{f_s j d} = \frac{49{,}700 \times 1.5 \times 5' \times 12}{20{,}000 \times .87 \times (72-4)} = 3.7 \text{ in}^2$$

4 #9 (V)

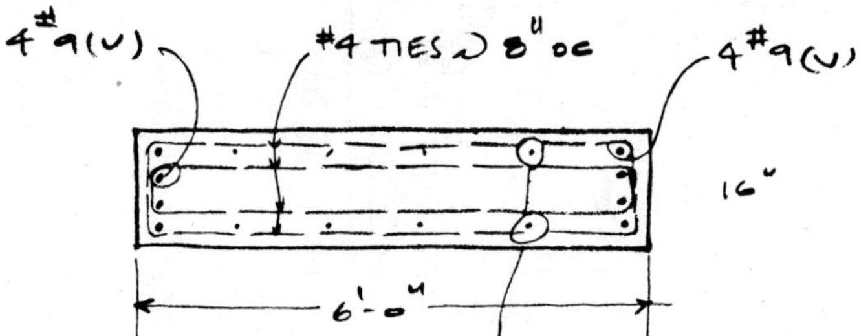

MIN A_s: .002 × 16 × 12 = .384

.0007 × 16 × 12 = .134

2 #4 @ 16

$A_s = .2 \times \frac{12}{16} \times 2 = 0.3 \text{ in}^2$

PIER # ①: $H = 3.44^k$ $f_v = 3440 \times 1.5 / 2 \times 16 \times 12 = 13.35 < 34$ psi OK

$M = 3.44 \times 1.5 \times 5' = 25.8^{k'}$

$M_R = 88 \times \frac{16}{12} \times (24-3)^2 = 51{,}743^{\#\text{-}'} > 25.8$ OK

$$A_s = \frac{25{,}800 \times 12}{(24-3)\, 20{,}000 \times .87} = 0.84 \qquad 2\ \#6 \quad A_s = .88$$

PIERS # 3 & 4 $H = 2.8^k (\times 1.5) = 4.2^k$

$M = (2.8 \times 1.5)\ 4.5' \times 12 = 226^{k''} = 18.8^{k'}$

$M_R = 88 \times \frac{16}{12} (20-3)^2 = 33.9^{k'}$

$$A_s = \frac{18.8 \times 12}{20 \times .87 \times 17} = 0.764 \text{ in}^2 \qquad 4\ \#5 \quad A_s = 1.24$$

$$f_v = \frac{4200}{16 \times 20 \times .87} = 15 \text{ psi} < 34 \text{ OK}$$

7

FIGURE 27.15

SEISMIC FORCES GOVERN OVER WIND

$H_{WIND} = 14' \times 15^{PSF} \times 400' = 84{,}000^{\#}$

$< V_s = 284^K$ (SEE NEXT PAGE)

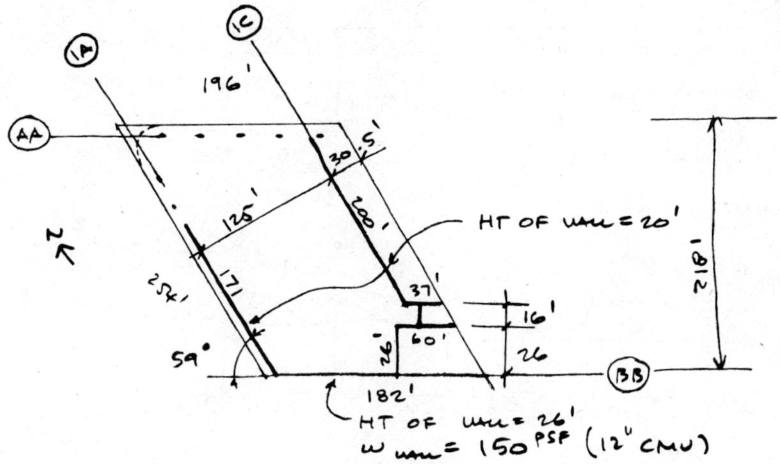

1" = 100

TOTAL ROOF AREA = $196 \times 254 \times .86 = 42{,}814$ SF

DEAD WT ROOF:

RFG =	5.6^{PSF}
5/8" PLY	1.8
2x4 @ 16"	1.1
4x16 @ 8'	1.67 "
TG 142 @ 24'	5.92 "
INSUL	1.00 "
CEIL (T-BAR)	1.80 "
MISC	1.00 "
	20.89 PSF SAY 21 PSF

$W_{D\ ROOF} = 21^{PSF} \times 42{,}814^{SF} = 894.4^K$

W_{WALLS} TRIB TO N-S DIR: (N & S WALLS)

W TRIB TO RF: $27' \times 120^{PSF} \times \frac{13}{20} = 2106^{\#/'} \times 182' = 383{,}29$

W_{WALLS} TRIB TO E-W DIR: (E & W WALLS)

$20' \times \frac{1}{2} \times 2 \times (120 + 5) = 2500^{\#/'} \times 254' = 635{,}000^{\#}$

(5 = CLADDING)

FIGURE 27.16

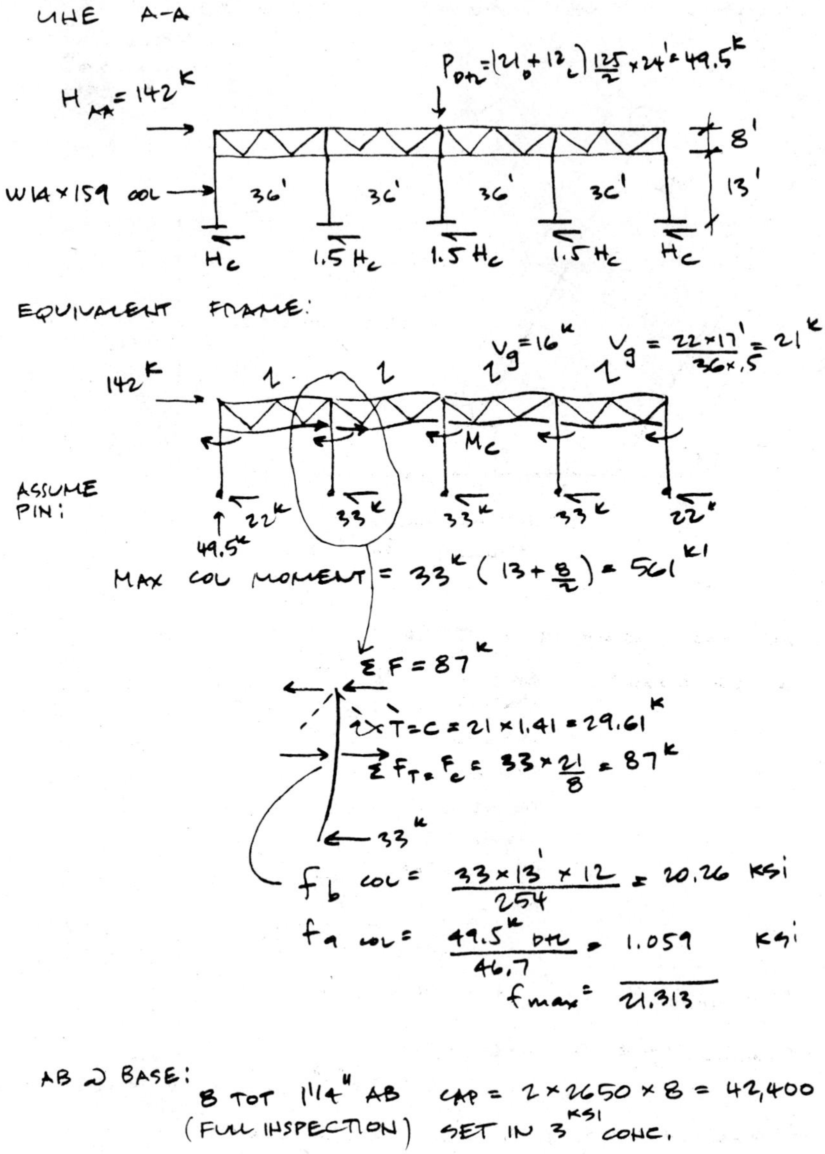
LINE A-A
H = 142 K
P D+L = (21 D + 12 L) 125/2 × 24' = 49.5 K
W14×159 COL
36'
36'
36'
36'
8'
13'
Hc
1.5 Hc
1.5 Hc
1.5 Hc
Hc
EQUIVALENT FRAME:
142 K
Vg = 16 K
Vg = 22×17'/36×.5 = 21 K
Mc
ASSUME PIN:
22 K
33 K
33 K
33 K
22 K
49.5 K
MAX COL MOMENT = 33 K (13 + 8/2) = 561 K'
ΣF = 87 K
2×T=C = 21×1.41 = 29.61 K
ΣFT = Fc = 33 × 21/8 = 87 K
33 K
fb col = 33×13'×12 / 254 = 20.26 KSI
fa col = 49.5 K D+L / 46.7 = 1.059 KSI
fmax = 21.313
AB @ BASE:
8 TOT 1 1/4" AB CAP = 2×2650×8 = 42,400
(FULL INSPECTION) SET IN 3 KSI CONC.

FIGURE 27.17

CHECK COL FOR COMBINED STRESSES

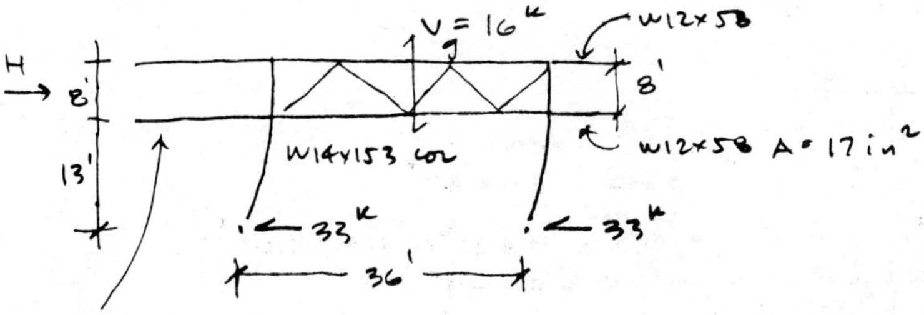

$I_{girder} \cong 0.9\, I_{TRUSS} = \Sigma A d^2 = 17 \times (4\times 12)^2 \times 2 = 78{,}336\ in^4$

$I_{col} = 1900\ in^4 \quad G_A = \dfrac{\Sigma I_c/L_c}{\Sigma I_g/L_g} = \dfrac{1900/13}{78{,}336/36} = \dfrac{146}{2176} = .067$

COL X-AXIS:

$G_B = 10$ (PIN) AISC

$K = 1.7 \quad \dfrac{KL}{r_x} = \dfrac{1.7\times 13\times 12}{6.38} = 41.56$

COL Y-AXIS

$K = 1.0 \quad L = 8' + 13' = 21'$

$\dfrac{KL}{r_y} = \dfrac{21\times 12}{4.00} = 63 \leftarrow$ GOVERNS

$F_a = 17.14$

$\dfrac{f_a}{F_a} = \dfrac{1.059}{17.14} = 0.061$

$\dfrac{f_b}{F_b} = \dfrac{\dfrac{33\times 13\times 12}{254}}{22} = \dfrac{20.26}{22} = 0.92$

$\Sigma \dfrac{f}{F} = 0.981 < 1.33$ COL OK

CHECK Δ_H FRAME:

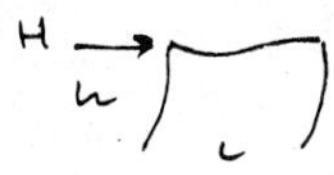

$\Delta = \dfrac{Hh^3}{3EI_c} + \dfrac{Hh^2L}{8EI_g}$

$= \dfrac{33\times(13\times 12)^3}{3\times 30\times 10^3\times 1900} + \dfrac{33\times(13\times 12)^2(36\times 12)}{8\times 30\times 10^3\times 78{,}336\times .9} =$

$= 0.73 + 0.01845 = 0.748 < .005h = .005\times 13\times 12 = .78''$

CHECK P-D EFFECT:

$\Delta = .78$

$M_\Delta = 49.5^k \times .78'' = 38.61^{k''}$

$f_{b_\Delta} = \dfrac{38.61^{k''}}{254 \sim S_x} = 0.15$ KSI VERY SMALL

$\Sigma \dfrac{f_a}{F_a} + \Sigma \dfrac{f_b}{F_b} = \dfrac{1.059}{17.14} + \dfrac{20.26 + .15}{22} = .0617 + .927 = .988 < 1.33$

W14×159 COL **OK**

FIGURE 27.18

CHECK TRUSS - COL CONNECTION:

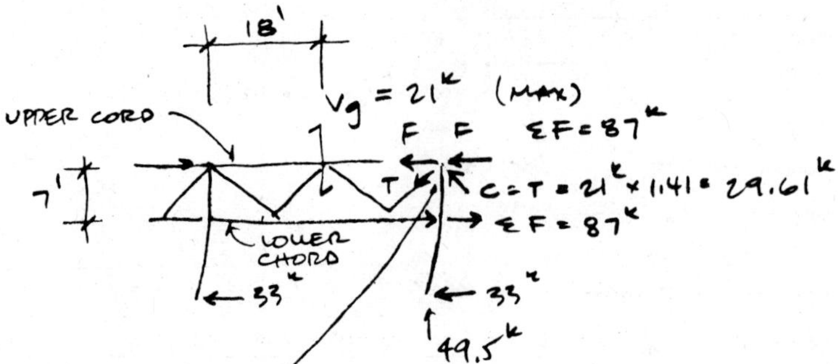

DET $\left(\frac{10}{56}\right)$: DIAGONAL CHORD MEMBER : MAX FORCE $= 29.61^k$

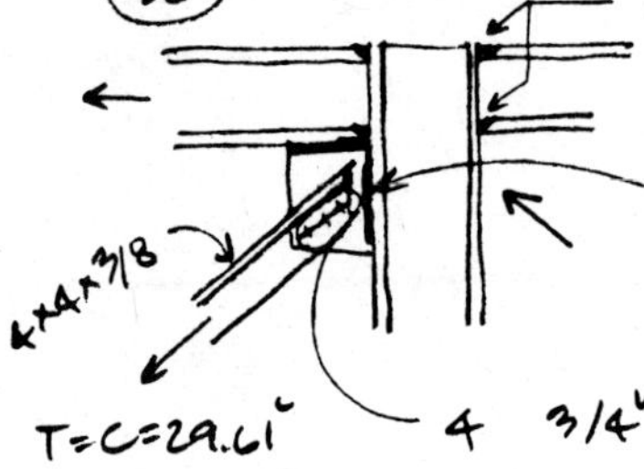

$f_w = 44^k / 10'' \times .648 \times 2 = 3.39^{ksi} < 14.4^{ksi}$ FLANGE CONN OK

$f_{WELD} = \dfrac{29.61^k}{12 \times .375 \times .707 \times 2 \times 2} = 2.3^{ksi} < 14.4$ ALL

4 3/4" ϕ MB CAP = $8.8 \times 4 = 35.2 > 29.6$
DS

USE 4 7/8" ϕ MB CAP $= 4 \times 12 = 48^k$

ALLOW COMPR LOAD ON STRUT:
JL 4×4×3/8

L_U MAX = 12.5' $P_{ALL} = 61^k$ ($L_U = 12'$)
$P_{ALL} = 46^k$ ($L_U = 14'$) OK $> 29.61^k$

4×4×3/8" STRUTS OK

LOWER CHORD: $F_{CH} = 44^k$ WEB OF W12×58 WELDED TO CONN PL (3/8")

T $f_{weld} = \dfrac{44{,}000^{\#}}{9.5 \times .375} = 12{,}350$ psi $< 22{,}000$ OK

FIGURE 27.19

PART FIVE

DESIGN DATA AND AIDS

CHAPTER TWENTY-EIGHT

Capacities of Wood-Framed Diaphragms

The material in this chapter consists of reprinted excerpts from the *UBC*. References to these tables can be found in Section 25 of the *UBC*, which is reprinted in Chapter 33.

28.1 PLYWOOD DIAPHRAGMS

Tables 28.1 and 28.2 give values for the shear capacities of plywood diaphragms, consisting of plywood panels that are nailed to wood framing systems. These tables are reprints of Tables 25−J and 25−K from the 1982 edition of the *Uniform Building Code* (Ref. 2) and have been reproduced with permission of the publisher. This data is based on the assumption that the entire wood frame construction is in conformance with all the applicable requirements of Chapter 25 of the *UBC*.

28.2 MISCELLANEOUS WOOD-FRAMED DIAPHRAGMS

Tables 28.3 and 28.4 give values for the diaphragm shear capacities of various materials that are used for surfacing on light wood frames. These tables are reprinted from the 1982 edition of the *Uniform Building Code* (Ref. 2) with permission of the publisher.

Reference should be made to the various provisions of Chapters 25 and 47 of the *UBC* pertaining to wood frame construction and the installation requirements for the surfacing materials described in the tables.

TABLE 28.1. Allowable Shear in 16/ft for Horizontal Plywood Diaphragms with Framing of Douglas Fir-Larch or Southern Pine[1] (Table 25–J of the 1982 UBC)

PLYWOOD GRADE	Common Nail Size	Minimum Nominal Penetration in Framing (in Inches)	Minimum Nominal Plywood Thickness (in Inches)	Minimum Nominal Width of Framing Member (in Inches)	BLOCKED DIAPHRAGMS: Nail spacing at diaphragm boundaries (all cases), at continuous panel edges parallel to load (Cases 3 and 4) and at all panel edges (Cases 5 and 6)				UNBLOCKED DIAPHRAGM: Nails spaced 6" max. at supported end	
					6	4	2½[2]	2[2]	Load perpendicular to unblocked edges and continuous panel joints (Case 1)	Other configurations (Cases 2, 3 & 4)
					Nail spacing at other plywood panel edges					
					6	6	4	3		
STRUCTURAL I	6d	1¼	5/16	2	185	250	375	420	165	125
				3	210	280	420	475	185	140
	8d	1½	⅜	2	270	360	530	600	240	180
				3	300	400	600	675	265	200
	10d	1⅝	½	2	320	425	640[2]	730[2]	285	215
				3[2]	360	480	720	820	320	240
C-D, C-C, STRUCTURAL II and other grades covered in U.B.C. Standard No. 25-9	6d	1¼	5/16	2	170	225	335	380	150	110
				3	190	250	380	430	170	125
			⅜	2	185	250	375	420	165	125
				3	210	280	420	475	185	140
	8d	1½	⅜	2	240	320	480	545	215	160
				3	270	360	540	610	240	180
			½	2	270	360	530	600	240	180
				3	300	400	600	675	265	200
	10d	1⅝	½	2	290	385	575[2]	655[2]	255	190
				3[2]	325	430	650	735	290	215
			⅝	2	320	425	640[2]	730[2]	285	215
				3[2]	360	480	720	820	320	240

[1]These values are for short-time loads due to wind or earthquake and must be reduced 25 percent for normal loading. Space nails 10 inches on center for floors and 12 inches on center for roofs along intermediate framing members.

Allowable shear values for nails in framing members of other species set forth in Table No. 25-17-J of U.B.C. Standards shall be calculated for all grades by multiplying the values for nails in STRUCTURAL I by the following factors: Group III, 0.82 and Group IV, 0.65.

[2]Framing shall be 3-inch nominal or wider and nails shall be staggered where nails are spaced 2 inches or 2½ inches on center, and where 10d nails having penetration into framing of more than 1⅝ inches are spaced 3 inches on center.

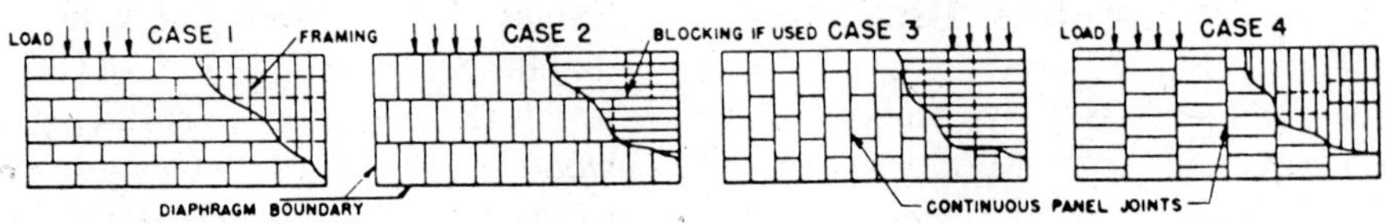

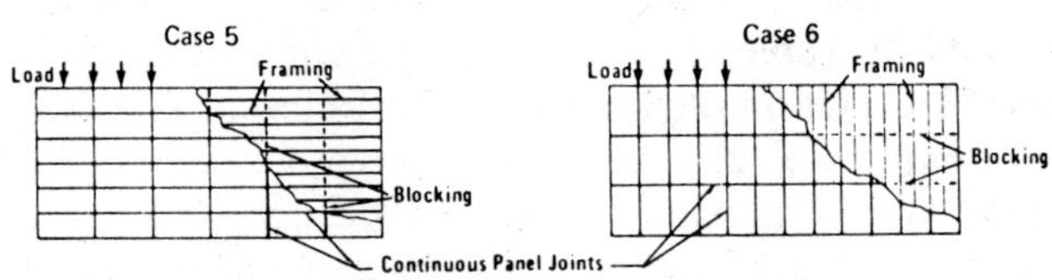

NOTE: Framing may be located in either direction for blocked diaphragms.

TABLE 28.2. Allowable Shear for Wind or Seismic Forces in 16/ft for Plywood Shear Walls with Framing of Douglas Fir-Larch or Southern Pine[1] (Table 25–K of the 1982 UBC)

PLYWOOD GRADE	MINIMUM NOMINAL PLYWOOD THICKNESS (Inches)	MINIMUM NAIL PENETRATION IN FRAMING (Inches)	NAIL SIZE (Common or Galvanized Box)	PLYWOOD APPLIED DIRECT TO FRAMING — Nail Spacing at Plywood Panel Edges: 6	4	3	2	NAIL SIZE (Common or Galvanized Box)	PLYWOOD APPLIED OVER ½-INCH GYPSUM SHEATHING — Nail Spacing at Plywood Panel Edges: 6	4	3	2
	5/16	1¼	6d	200	300	390	510	8d	200	300	390	510
STRUCTURAL I	3/8	1½	8d	230[3]	360[3]	460[3]	610[3]	10d	280	430	550[2]	730[2]
	½	1⅝	10d	340	510	665[2]	870[2]	—	—	—	—	—
C-D, C-C STRUCTURAL II and other grades covered in U.B.C. Standard No. 25-9.	5/16	1¼	6d	180	270	350	450	8d	180	270	350	450
	3/8	1½	8d	220[3]	320[3]	410[3]	530[3]	10d	260	380	490[2]	640[2]
	½	1⅝	10d	310	460	600[2]	770[2]	—	—	—	—	—
			NAIL SIZE (Galvanized Casing)					NAIL SIZE (Galvanized Casing)				
Plywood panel siding in grades covered in U.B.C. Standard No. 25-9	5/16	1¼	6d	140	210	275	360	8d	140	210	275	360
	3/8	1½	8d	130[3]	200[3]	260[3]	340[3]	10d	160	240	310[2]	410

[1]All panel edges backed with 2-inch nominal or wider framing. Plywood installed either horizontally or vertically. Space nails at 6 inches on center along intermediate framing members for 3/8-inch plywood installed with face grain parallel to studs spaced 24 inches on center and 12 inches on center for other conditions and plywood thicknesses. These values are for short-time loads due to wind or earthquake and must be reduced 25 percent for normal loading.

Allowable shear values for nails in framing members of other species set forth in Table No. 25-17-J of U.B.C. Standards shall be calculated for all grades by multiplying the values for common and galvanized box nails in STRUCTURAL I and galvanized casing nails in other grades by the following factors: Group III, 0.82 and Group IV, 0.65.

[2]Framing shall be 3-inch nominal or wider and nails shall be staggered where nails are spaced 2 inches on center, and where 10d nails having penetration into framing of more than 1⅝ inches are spaced 3 inches on center.

[3]The values for 3/8-inch-thick plywood applied direct to framing may be increased 20 percent, provided studs are spaced a maximum of 16 inches on center or plywood is applied with face grain across studs or if the plywood thickness is increased to ½ inch or greater.

TABLE 28.3. Allowable Shears for Wind or Seismic Loading on Vertical Diaphragms of Fiberboard Sheathing Board Construction for Type V Construction Only[1] (Table 25–P of the 1982 UBC)

SIZE AND APPLICATION	NAIL SIZE	SHEAR VALUE 3-INCH NAIL SPACING AROUND PERIMETER AND 6-INCH AT INTERMEDIATE POINTS
1. ½″ x 4′ x 8′	No. 11 gauge galvanized roofing nail 1½″ long, 7/16″ head	125[2]
2. 25/32″ x 4′ x 8′	No. 11 gauge galvanized roofing nail 1¾″ long, 7/16″ head	175

[1]Fiberboard sheathing diaphragms shall not be used to brace concrete or masonry walls.

[2]The shear value may be 175 for ½-inch x 4-foot x 8-foot fiberboard nail-base sheathing.

TABLE 28.4 Allowable Shear for Wind or Seismic Forces in 16/ft for Vertical Diaphragms of Lath and Plaster or Gypsum Board Frame Wall Assemblies[1] (Table 47–I of the 1982 UBC)

TYPE OF MATERIAL	THICKNESS OF MATERIAL	WALL CONSTRUCTION	NAIL SPACING[2] MAXIMUM (In Inches)	SHEAR VALUE	MINIMUM NAIL SIZE
1. Expanded metal, or woven wire lath and portland cement plaster	⅞"	Unblocked	6	180	No. 11 gauge, 1½" long, 7/16" head No. 16 gauge staple, ⅞" legs
2. Gypsum lath, plain or perforated	⅜" Lath and ½" Plaster	Unblocked	5	100	No. 13 gauge, 1⅛" long, 19/64" head, plasterboard blued nail
3. Gypsum sheathing board	½" x 2′ x 8′	Unblocked	4	75	No. 11 gauge, 1¾" long, 7/16" head, diamond-point, galvanized
	½" x 4′ ½" x 4′	Blocked Unblocked	4 7	175 100	
4. Gypsum wallboard or veneer base	½"	Unblocked	7	100	5d cooler nails
			4	125	
		Blocked	7	125	
			4	150	
	⅝"	Blocked	4	175	6d cooler nails
		Blocked Two-ply	Base ply 9 Face ply 7	250	Base ply—6d cooler nails Face ply—8d cooler nails

[1]These vertical diaphragms shall not be used to resist loads imposed by masonry or concrete construction. See Section 4713 (b). Values are for short-time loading due to wind or earthquake and must be reduced 25 percent for normal loading.

[2]Applies to nailing at all studs, top and bottom plates and blocking.

CHAPTER TWENTY-NINE

Steel Deck Diaphragms

When used for roofs, formed sheet steel decks are often utilized as diaphragms. Diaphragm shear capacity of a particular installation depends on the type of deck, the thickness (gage) of the steel, the fastening between adjacent deck units, the fastening of the deck to the supports, and the span of the deck (spacing of supports). In general, design of steel decks for diaphragm actions is based on the *Steel Deck Institute Diaphragm Design Manual* (Ref. 17).

When used for floors, steel decks are covered with a substantial amount of structural concrete fill. The relationship between the steel deck and the concrete fill varies. In some cases the fill is essentially inert, and the steel deck is assumed to perform the required structural tasks—for both gravity and lateral loads. When this is the case, the deck is designed in the same manner as for a roof deck, except that the concrete contributes considerably to deflection resistance, making flexibility of less concern. Another possibility is that which occurs in composite construction in which the concrete is used structurally, the deck units being bonded to the concrete for interaction as in the manner of steel reinforcing rods.

Although there are industry standards, steel decks occur in wide variety as products of individual manufacturers. For the design of a particular building, it is best to ascertain the products commonly available in the region, and to obtain information from the suppliers or the manufacturers of the products. This information will usually include data regarding the structural capacity of the deck units for various installation conditions. The material that follows is a sample of diaphragm design information obtained from one manufacturer—Inryco Inc. It is used in the design example in section 23.5. It was obtained from *Inryco Lateral Diaphragm Data Manual 20−2* and is reprinted with permission of the publishers, Inryco Inc., Building Panels Division, P. O. Box 393, Milwaukee, WI, 53201. We emphasize that this material is given for illustration only, and should not be used for any design work unless it is verified by the manufacturer.

1½″ B (24″)

ROOF DECK OR ACOUSTIDECK®

Painted and Galvanized

3 Puddle Welds

Side Seams

Button-Punched 36″ O.C.

GAGE		SPANS 5′ - 0″	6′ - 0″	7′ - 0″	8′ - 0″	9′ - 0″	10′ - 0″
22	q	**280**	**240**	**210**	**185**	**170**	**155**
	F	23.84 + 224.5R	27.87 + 187.0R	31.83 + 160.3R	35.73 + 140.3R	39.56 + 124.7R	43.32 + 112.2R
20	q	**445**	**375**	**330**	**295**	**270**	**245**
	F	16.88 + 129.7R	19.88 + 108.1R	22.91 + 92.6R	25.94 + 81.0R	28.97 + 72.0R	32.01 + 64.8R
18	q	**790**	**665**	**580**	**515**	**465**	**425**
	F	10.44 + 54.9R	12.38 + 45.8R	14.38 + 39.2R	16.43 + 34.3R	18.52 + 30.5R	20.66 + 27.5R
16	q	**1115**	**940**	**815**	**725**	**650**	**590**
	F	7.23 + 28.1R	8.60 + 23.4R	10.03 + 20.0R	11.52 + 17.5R	13.06 + 15.6R	14.65 + 14.0R

q = diaphragm shear load in lb/ft

FIGURE 29.1.

CHAPTER THIRTY

Capacities of Framing and Anchorage Devices

Putting together a whole structure—especially one of wood, steel, or precast concrete—involves the fastening of a lot of separate parts. The material in this chapter consists of illustrations and load capacity data for some widely used fastening elements for wood framed structures.

30.1 HOLD-DOWN DEVICES FOR WOOD FRAMES

Wood-framed shear walls must often be anchored for uplift resistance due to overturning effects. The most common device used for this is the hold down (also called tie down or holdown), which is bolted to a wood post at the end of the wall and tied to an anchor bolt imbedded in concrete. The material that follows has been abstracted from the product catalog of the Simpson Co., P.O. Box 1568, San Leandro, CA, 94577, with their permission. The material presented is a small sampling of the whole product line; it is given for illustration only and should not be used for any design work without verification by the manufacturer. It is also advisable to verify that any data provided and used for design work is acceptable to the code-enforcing agency that grants building permits.

30.2 FRAMING DEVICES FOR WOOD STRUCTURES

Light wood frames (stud–rafter–joist) are often assembled by using a number of framing connectors made of light-gage sheet steel that are nailed in place. These are used primarily for ease and speed of the carpentry work, but offer the side benefit of providing considerable tying action. They are of particular significance in developing the various load transfers required for the resistance of lateral loads.

The material that follows is a small sampling of some typical framing anchors, used for holding rafters or light wood trusses to their supporting wood struc-

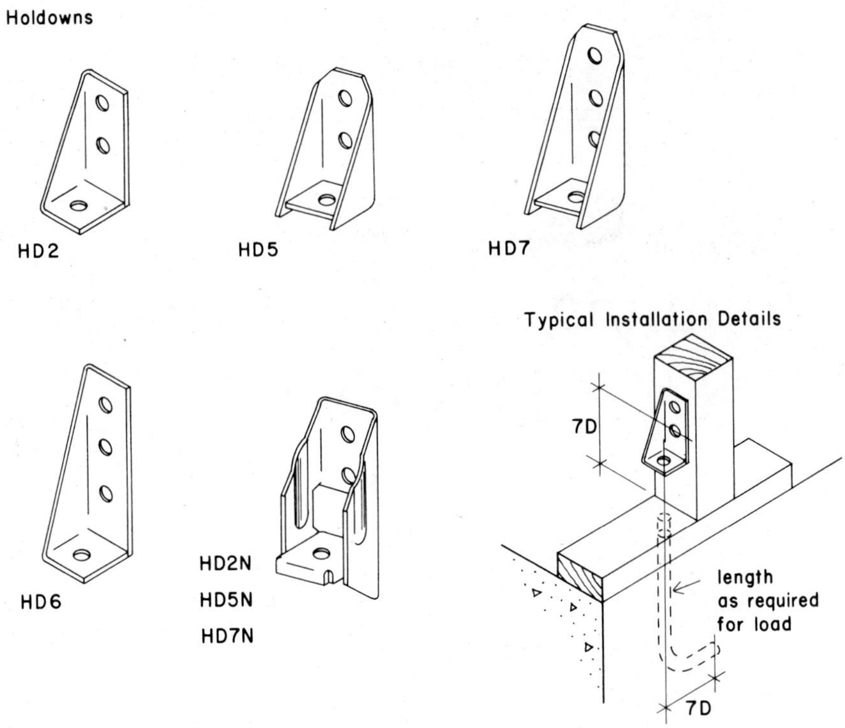

No.	Bolts		Design Load - in lbs (Normal Loading) with stud sizes shown		
	base	stud	1 9/16	2 9/16	3 9/16
HD2 HD2N	5/8	2 - 5/8	2450	2520	2520
HD5 HD5N	3/4	2 - 3/4	3375	3610	3610
HD6	1	3 - 3/4	5060	5410	5410
HD7	1	3 - 7/8	6350	6500	6500
	1 1/8	3 - 7/8	6350	7380	7380
HD7N	1	2 - 1	6350	6500	6500

FIGURE 30.1.

Seismic and Hurricane Anchors
for trusses and rafters

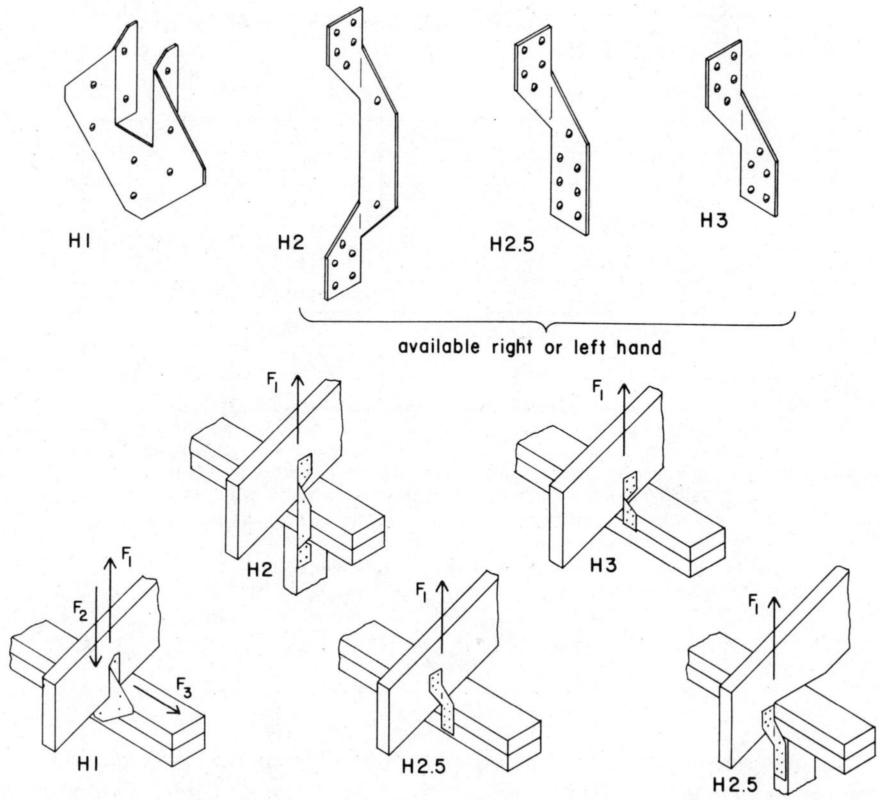

No.	Fasteners			Load Capacity - lb		
	to rafters	to plates	to studs	F_1	F_2	F_3
H1	4 - 8d	4 - 8d	—	520	390	520
H2	5 - 8d	—	5 - 8d	370	—	—
H2.5	5 - 8d	4 to 6 - 8d	4 to 6 - 8d	370	—	—
H3	4 - 8d	4 - 8d	—	305	—	—

FIGURE 30.2.

TABLE 30.1 Allowable Shear and Tension on Bolts Embedded in Concrete (in lb)—UBC Table 26–G[1,2]

DIAMETER (In Inches)	MINIMUM[3] EMBEDMENT (In Inches)	MINIMUM CONCRETE STRENGTH (In psi)		
		SHEAR[4]		TENSION[5]
		2000	3000	2000 to 5000
1/4	2 1/2	500	500	200
3/8	3	1100	1100	500
1/2	4	2000	2000	950
5/8	4	2750	3000	1500
3/4	5	2940	3560	2250
7/8	6	3580	4150	3200
1	7	3580	4150	3200
1 1/8	8	3580	4500	3200
1 1/4	9	3580	5300	3200

[1]Values are for natural stone aggregate concrete and bolts of at least A307 quality. Bolts shall have a standard bolt head or an equal deformity in the embedded portion.

[2]Values are based upon a bolt spacing of 12 diameters with a minimum edge distance of 6 diameters. Such spacing and edge distance may be reduced 50 percent with an equal reductión in value. Use linear interpolation for intermediate spacings and edge margins.

[3]An additional 2 inches of embedment shall be provided for anchor bolts located in the top of columns for buildings located in Seismic Zones Nos. 2, 3 and 4.

[4]Values shown are for work with or without special inspection.

[5]Values shown are for work without special inspection. Where special inspection is provided values may be increased 100 percent.

Abstracted from the *Uniform Building Code* (Ref. 2), with permission of the publisher.

tures. This material has been abstracted from the product catalog of the Simpson Co., P.O. Box 1568, San Leandro, CA, 94577, with their permission. This material is presented for illustration only and should not be used for any design work without verification by the manufacturer.

30.3 ANCHOR BOLTS

Steel bolts embedded in concrete or masonry are used for a variety of situations—a common one being the fastening of sill plates for wood stud walls. Table 30.1 is reprinted from the 1982 edition of the *Uniform Building Code*, with permission of the publisher. It gives values for both shear and tension loadings for bolts embedded in concrete with design strength of 2000 or 3000 psi. The reason for the very low strength of 2000 psi is that most local codes permit the omission of concrete testing if this value is used for design.

With reinforced masonry construction using hollow concrete blocks, anchor bolts are embedded in the concrete fill in the block voids. This is essentially similar to the situation of casting them into a concrete member, so that these values are sometimes also used for masonry.

CHAPTER THIRTY-ONE

Stiffness Factors for Masonry Piers

The following tables present factors that are used for the evaluation of relative stiffness of masonry piers. This relates to the determination of load distribution for a series of interacting piers. Use of this data is illustrated in the design examples in Sections 23.4 and 27.5. Table data is reprinted from the *Concrete Masonry Design Manual* (Ref. 20), with permission of the publishers.

TABLE 31.1 Distribution of Horizontal Forces Along a Cantilevered Masonry Wall—Average Rigidities

h/d	Rc	h/d	Rc	h/d	Rc	h/d	Rc	h/d	Rc	h/d	Rc
9.90	.0006	5.20	.0043	1.85	.0810	1.38	.1706	0.91	.4352	0.45	1.4582
9.80	.0007	5.10	.0046	1.84	.0821	1.37	.1737	0.90	.4452	0.44	1.5054
9.70	.0007	5.00	.0049	1.83	.0833	1.36	.1768	0.89	.4554	0.43	1.5547
9.60	.0007	4.90	.0052	1.82	.0845	1.35	.1800	0.88	.4659	0.42	1.6063
9.50	.0007	4.80	.0055	1.81	.0858	1.34	.1832	0.87	.4767	0.41	1.6604
9.40	.0007	4.70	.0058	1.80	.0870	1.33	.1866	0.86	.4899	0.40	1.7170
9.30	.0008	4.60	.0062	1.79	.0883	1.32	.1900	0.85	.4994	0.39	1.7765
9.20	.0008	4.50	.0066	1.78	.0896	1.31	.1935	0.84	.5112	0.38	1.8380
9.10	.0008	4.40	.0071	1.77	.0909	1.30	.1970	0.83	.5233	0.37	1.9098
9.00	.0008	4.30	.0076	1.76	.0923	1.29	.2007	0.82	.5359	0.36	1.9738
8.90	.0009	4.20	.0081	1.75	.0937	1.28	.2044	0.81	.5488	0.35	2.0467
8.80	.0009	4.10	.0087	1.74	.0951	1.27	.2083	0.80	.5621	0.34	2.1237
8.70	.0009	4.00	.0093	1.73	.0965	1.26	.2122	0.79	.5758	0.33	2.2051
8.60	.0010	3.90	.0100	1.72	.0980	1.25	.2162	0.78	.5899	0.32	2.2913
8.50	.0010	3.80	.0108	1.71	.0995	1.24	.2203	0.77	.6044	0.31	2.3828
8.40	.0010	3.70	.0117	1.70	.1010	1.23	.2245	0.76	.6194	0.30	2.4802
8.30	.0011	3.60	.0127	1.69	.1026	1.22	.2289	0.75	.6349	0.29	2.5838
8.20	.0012	3.50	.0137	1.68	.1041	1.21	.2333	0.74	.6509	0.28	2.6945
8.10	.0012	3.40	.0149	1.67	.1058	1.20	.2378	0.73	.6674	0.27	2.8130
8.00	.0012	3.30	.0163	1.66	.1074	1.19	.2425	0.72	.6844	0.26	2.9401
7.90	.0013	3.20	.0178	1.65	.1091	1.18	.2472	0.71	.7019	0.25	3.0769
7.80	.0013	3.10	.0195	1.64	.1108	1.17	.2521	0.70	.7200	0.24	3.2246
7.70	.0014	3.00	.0214	1.63	.1125	1.16	.2571	0.69	.7388	0.23	3.3845
7.60	.0014	2.90	.0235	1.62	.1143	1.15	.2622	0.68	.7581	0.22	3.5583
7.50	.0015	2.80	.0260	1.61	.1162	1.14	.2675	0.67	.7781	0.21	3.7479
7.40	.0015	2.70	.0288	1.60	.1180	1.13	.2729	0.66	.7987	0.20	3.9557
7.30	.0016	2.60	.0320	1.59	.1199	1.12	.2784	0.65	.8201	.195	4.0673
7.20	.0017	2.50	.0357	1.58	.1218	1.11	.2841	0.64	.8422	.190	4.1845
7.10	.0017	2.40	.0400	1.57	.1238	1.10	.2899	0.63	.8650	.185	4.3079
7.00	.0018	2.30	.0450	1.56	.1258	1.09	.2959	0.62	.8886	.180	4.4379
6.90	.0019	2.20	.0508	1.55	.1279	1.08	.3020	0.61	.9131	.175	4.5751
6.80	.0020	2.10	.0577	1.54	.1300	1.07	.3083	0.60	.9384	.170	4.7201
6.70	.0020	2.00	.0658	1.53	.1322	1.06	.3147	0.59	.9647	.165	4.8736
6.60	.0021	1.99	.0667	1.52	.1344	1.05	.3213	0.58	.9919	.160	5.0364
6.50	.0022	1.98	.0676	1.51	.1366	1.04	.3281	0.57	1.0201	.155	5.2095
6.40	.0023	1.97	.0685	1.50	.1389	1.03	.3351	0.56	1.0493	.150	5.3937
6.30	.0025	1.96	.0694	1.49	.1412	1.02	.3422	0.55	1.0797	.145	5.5904
6.20	.0026	1.95	.0704	1.48	.1436	1.01	.3496	0.54	1.1112	.140	5.8008
6.10	.0027	1.94	.0714	1.47	.1461	1.00	.3571	0.53	1.1439	.135	6.0261
6.00	.0028	1.93	.0724	1.46	.1486	0.99	.3649	0.52	1.1779	.130	6.2696
5.90	.0030	1.92	.0734	1.45	.1511	0.98	.3729	0.51	1.2132	.125	6.5306
5.80	.0031	1.91	.0744	1.44	.1537	0.97	.3811	0.50	1.2500	.120	6.8136
5.70	.0033	1.90	.0754	1.43	.1564	0.96	.3895	0.49	1.2883	.115	7.1208
5.60	.0035	1.89	.0765	1.42	.1591	0.95	.3981	0.48	1.3281	.110	7.4555
5.50	.0037	1.88	.0776	1.41	.1619	0.94	.4070	0.47	1.3696	.105	7.8215
5.40	.0039	1.87	.0787	1.40	.1647	0.93	.4162	0.46	1.4130	.100	8.2237
5.30	.0041	1.86	.0798	1.39	.1676	0.92	.4255				

TABLE 31.2 Distribution of Horizontal Forces Along a Fixed Masonry Wall—Average Rigidities

h/d	R_f	h/d	R_f	h/d	R_f	h/d	R_f	h/d	R_f	h/d	R_f
9.90	.0025	5.20	.0160	1.85	.2104	1.38	.3694	0.91	.7177	0.45	1.736
9.80	.0026	5.10	.0169	1.84	.2128	1.37	.3742	0.90	.7291	0.44	1.779
9.70	.0027	5.00	.0179	1.83	.2152	1.36	.3790	0.89	.7407	0.43	1.825
9.60	.0027	4.90	.0189	1.82	.2176	1.35	.3840	0.88	.7527	0.42	1.874
9.50	.0028	4.80	.0200	1.81	.2201	1.34	.3890	0.87	.7649	0.41	1.924
9.40	.0029	4.70	.0212	1.80	.2226	1.33	.3942	0.86	.7773	0.40	1.978
9.30	.0030	4.60	.0225	1.79	.2251	1.32	.3994	0.85	.7901	0.39	2.034
9.20	.0031	4.50	.0239	1.78	.2277	1.31	.4047	0.84	.8031	0.38	2.092
9.10	.0032	4.40	.0254	1.77	.2303	1.30	.4100	0.83	.8165	0.37	2.154
9.00	.0033	4.30	.0271	1.76	.2330	1.29	.4155	0.82	.8302	0.36	2.219
8.90	.0034	4.20	.0288	1.75	.2356	1.28	.4211	0.81	.8442	0.35	2.287
8.80	.0035	4.10	.0308	1.74	.2384	1.27	.4267	0.80	.8585	0.34	2.360
8.70	.0037	4.00	.0329	1.73	.2411	1.26	.4324	0.79	0.873	0.33	2.437
8.60	.0038	3.90	.0352	1.72	.2439	1.25	.4384	0.78	0.888	0.32	2.518
8.50	.0039	3.80	.0377	1.71	.2468	1.24	.4443	0.77	0.904	0.31	2.605
8.40	.0040	3.70	.0405	1.70	.2497	1.23	.4504	0.76	0.920	0.30	2.697
8.30	.0042	3.60	.0435	1.69	.2526	1.22	.4566	0.75	0.936	0.29	2.795
8.20	.0043	3.50	.0468	1.68	.2556	1.21	.4628	0.74	0.952	0.28	2.900
8.10	.0045	3.40	.0505	1.67	.2586	1.20	.4692	0.73	0.969	0.27	3.013
8.00	.0047	3.30	.0545	1.66	.2617	1.19	.4757	0.72	0.987	0.26	3.135
7.90	.0048	3.20	.0590	1.65	.2648	1.18	.4823	0.71	1.005	0.25	3.265
7.80	.0050	3.10	.0640	1.64	.2679	1.17	.4891	0.70	1.023	0.24	3.407
7.70	.0052	3.00	.0694	1.63	.2711	1.16	.4959	0.69	1.042	0.23	3.560
7.60	.0054	2.90	.0756	1.62	.2744	1.15	.5029	0.68	1.062	0.22	3.728
7.50	.0056	2.80	.0824	1.61	.2777	1.14	.5100	0.67	1.082	0.21	3.911
7.40	.0058	2.70	.0900	1.60	.2811	1.13	.5173	0.66	1.103	0.20	4.112
7.30	.0061	2.60	.0985	1.59	.2844	1.12	.5247	0.65	1.124	.195	4.220
7.20	.0063	2.50	.1081	1.58	.2879	1.11	.5322	0.64	1.146	.190	4.334
7.10	.0065	2.40	.1189	1.57	.2914	1.10	.5398	0.63	1.168	.185	4.454
7.00	.0069	2.30	.1311	1.56	.2949	1.09	.5476	0.62	1.191	.180	4.580
6.90	.0072	2.20	.1449	1.55	.2985	1.08	.5556	0.61	1.216	.175	4.714
6.80	.0075	2.10	.1607	1.54	.3022	1.07	.5637	0.60	1.240	.170	4.855
6.70	.0078	2.00	.1786	1.53	.3059	1.06	.5719	0.59	1.266	.165	5.005
6.60	.0081	1.99	.1805	1.52	.3097	1.05	.5804	0.58	1.292	.160	5.164
6.50	.0085	1.98	.1824	1.51	.3136	1.04	.5889	0.57	1.319	.155	5.334
6.40	.0089	1.97	.1844	1.50	.3175	1.03	.5977	0.56	1.347	.150	5.514
6.30	.0093	1.96	.1864	1.49	.3214	1.02	.6066	0.55	1.376	.145	5.707
6.20	.0097	1.95	.1885	1.48	.3245	1.01	.6157	0.54	1.407	.140	5.914
6.10	.0102	1.94	.1905	1.47	.3295	1.00	.6250	0.53	1.438	.135	6.136
6.00	.0107	1.93	.1926	1.46	.3337	0.99	.6344	0.52	1.470	.130	6.374
5.90	.0112	1.92	.1947	1.45	.3379	0.98	.6441	0.51	1.504	.125	6.632
5.80	.0118	1.91	.1969	1.44	.3422	0.97	.6540	0.50	1.539	.120	6.911
5.70	.0124	1.90	.1991	1.43	.3465	0.96	.6641	0.49	1.575	.115	7.215
5.60	.0130	1.89	.2013	1.42	.3510	0.95	.6743	0.48	1.612	.110	7.545
5.50	.0137	1.88	.2035	1.41	.3555	0.94	.6848	0.47	1.651	.105	7.908
5.40	.0144	1.87	.2058	1.40	.3600	0.93	.6955	0.46	1.692	.100	8.306
5.30	.0152	1.86	.2081	1.39	.3647	0.92	.7065				

CHAPTER THIRTY-TWO

Weight of Elements of Building Construction

This chapter contains material that will assist in the determination of the weight of various parts of the building as is required in the process of finding the lateral inertial force. Table 32.1 contains data for the determination of the weight of specific materials, where a unique assemblage must be evaluated. Table 32.2 presents the average weights of commonly used elements of construction. For simpler and faster approximations of weights in preliminary calculations, Table 32.3 presents average weights of ordinary total assemblages of typical construction.

TABLE 32.1. Weight of Materials

Material	Weight lb/ft^3	Specific Gravity[a]
Soil		
Top soil		
Dry, loose	75	1.21
Moist, packed	100	1.61
Sand, gravel		
Dry, Loose	90–100	1.53
Moist, packed	110–120	1.85
Clay		
Dry	65	1.05
Wet	100–110	1.69
Wood		
Cedar	22	0.35
Fir, northern	35	0.56
Fir, white	25	0.40
Hemlock	30	0.48
Oak, red	40	0.65
Pine, white	25	0.40
Pine, yellow, long-leaf	45	0.73
Poplar	30	0.48

TABLE 32.1. *(Continued)*

Material	Weight lb/ft³	Specific Gravity[a]
Redwood	25	0.40
Spruce	25	0.40
Metal		
Aluminum	165	2.66
Copper	556	8.97
Lead	710	11.45
Steel	490	7.90
Masonry		
Brick		
Finished, pressed	130–150	2.26
Common, soft	110–120	1.85
Concrete block		
Normal weight, voids empty	90	1.45
Lightweight, voids empty	65	1.05
Lightweight, voids 25% filled	85	1.37
Lightweight, voids 100% filled	140	2.26
Concrete		
Normal, gravel	145	2.34
(with average reinforcing)	150	2.42
Lightweight		
Structural	110	1.77
Masonry units	90	1.45
Insulating	30–40	0.56
Miscellaneous		
Asphalt paving	100	1.61
Glass	160	2.58
Plaster		
Cement, stucco	120	1.94
Gypsum	40	0.65
Water	62	1.00

[a] Average weight divided by weight of water at 62 lb/ft³.

TABLE 32.2. Weight of Elements of Building Construction

	lb/ft²	kN/m²
Roofs		
3-ply ready roofing (roll, composition)	1	0.05
3-ply felt and gravel	5.5	0.26
5-ply felt and gravel	6.5	0.31
Shingles		
Wood	2	0.10
Asphalt	2–3	0.10–0.15

(continued)

TABLE 32.2. *(Continued)*

	lb/ft^2	kN/m^2
Clay tile	9–12	0.43–0.58
Concrete tile	8–12	0.38–0.58
Slate, $\frac{1}{4}$ in.	10	0.48
Fiber glass	2–3	0.10–0.15
Aluminum	1	0.05
Steel	2	0.10
Insulation		
Fiber glass batts	0.5	0.025
Rigid foam plastic	1.5	0.075
Foamed concrete, mineral aggregate	2.5/in.	0.0047/mm
Wood rafters		
2 × 6 at 24 in.	1.0	0.05
2 × 8 at 24 in.	1.4	0.07
2 × 10 at 24 in.	1.7	0.08
2 × 12 at 24 in.	2.1	0.10
Steel deck, painted		
22 ga	1.6	0.08
20 ga	2.0	0.10
18 ga	2.6	0.13
Skylight		
Glass with steel frame	6–10	0.29–0.48
Plastic with aluminum frame	3–6	0.15–0.29
Plywood or softwood board sheathing	3.0/in.	0.0057/mm
Ceilings		
Suspended steel channels	1	0.05
Lath		
Steel mesh	0.5	0.025
Gypsum board, $\frac{1}{2}$ in.	2	0.10
Fiber tile	1	0.05
Dry wall, gypsum board, $\frac{1}{2}$ in.	2.5	0.12
Plaster		
Gypsum, acoustic	5	0.24
Cement	8.5	0.41
Suspended lighting and air distribution Systems, average	3	0.15
Floors		
Hardwood, $\frac{1}{2}$ in.	2.5	0.12
Vinyl tile, $\frac{1}{8}$ in.	1.5	0.07
Asphalt mastic	12/in.	0.023/mm
Ceramic tile		
$\frac{3}{4}$ in.	10	0.48
Thin set	5	0.24
Fiberboard underlay, $\frac{5}{8}$ in.	3	0.15
Carpet and pad, average	3	0.15
Timber deck	2.5/in.	0.0047/mm
Steel deck, stone concrete fill, average	35–40	1.68–1.92
Concrete deck, stone aggregate	12.5/in.	0.024/mm

TABLE 32.2. *(Continued)*

	lb/ft²	kN/m²
Wood joists		
2 × 8 at 16 in.	2.1	0.10
2 × 10 at 16 in.	2.6	0.13
2 × 12 at 16 in.	3.2	0.16
Lightweight concrete fill	8.0/in.	0.015/mm
Walls		
2 × 4 studs at 16 in., average	2	0.10
Steel studs at 16 in., average	4	0.20
Lath, plaster; see Ceilings		
Gypsum dry wall, $\frac{5}{8}$ in. single	2.5	0.12
Stucco, $\frac{7}{8}$ in., or wire and paper or felt	10	0.48
Windows, average, glazing + frame		
small pane, single glazing, wood or metal frame	5	0.24
Large pane, single glazing, wood or metal frame	8	0.38
Increase for double glazing	2–3	0.10–0.15
Curtain walls, manufactured units	10–15	0.48–0.72
Brick veneer		
4 in., mortar joints	40	1.92
$\frac{1}{2}$ in., mastic	10	0.48
Concrete block		
Lightweight, unreinforced—4 in.	20	0.96
6 in.	25	1.20
8 in.	30	1.44
Heavy, reinforced, grouted—6 in.	45	2.15
8 in.	60	2.87
12 in.	85	4.07

TABLE 32.3. Average Values of Total Dead Weight for Typical Building Construction

Description of Construction	Weight[a] lb/ft²
Roofs	
1. Light wood frame, shingles, no ceiling	5
2. Light wood frame, 3-ply felt + gravel, gypsum drywall ceiling	12
3. Medium span wood or steel trusses, plywood deck, 3-ply felt + gravel, no ceiling	15
4. Steel open-web joists, metal deck, lightweight concrete insulating fill, 3-ply felt + gravel, suspended drywall ceiling	18–25
5. Glue-laminated wood girders, timber purlins,	25–30

(continued)

TABLE 32.3. *(Continued)*

Description of Construction	Weight[a] lb/ft^2
plywood deck, clay or concrete tile shingles, no ceiling	
6. Steel beams, otherwise same as 5	22–30
7. Precast concrete plank, 3-ply felt + gravel, 2-in. concrete fill, no ceiling,	
6-in.-thick units	50
8-in.-thick units	65
10-in.-thick units	80
8. Precast concrete double-tees, 2-in concrete fill, 3-ply felt + gravel, no ceiling	80–100
9. Poured-in-place concrete slab and beams, insulating concrete fill, 3-ply felt + gravel, suspended drywall ceiling	100–125
Floors	
10. Wood joist, plywood deck, carpet, drywall ceiling	10
11. Steel- or wood-trussed joists, plywood deck, 2-in. concrete fill, carpet, drywall ceiling	30
12. Steel beam, steel deck, concrete fill, suspended lay-in tile ceiling	60–70
13. Precast concrete plank, 2-in. concrete fill, carpet, suspended drywall ceiling,	
6-in.-thick units	50
8-in.-thick units	65
10-in.-thick units	80
14. Poured-in-place concrete slab and beams, 2-in. concrete fill, suspended lay-in tile ceiling	100–125
Walls	
15. Exterior, wood studs, stucco, drywall	15–20
16. Exterior, wood studs, wood siding, drywall	10–15
17. Exterior, brick veneer, wood studs, plywood, drywall (real bricks)	60
18. Exterior, fake brick (thin tile), plywood, wood studs, drywall	25
19. Exterior, reinforced concrete block, 8-in. units,, 25% voids filled, average reinforcing, insulation, furred-out drywall	60
20. Steel frame, aluminum + glass windowall (curtain wall), 25% windows—small size	25
21. Windows, wood or metal frame, small panes of glass (5 to 10 ft^2/window)	5
22. Interior, wood stud, drywall both sides	8
23. Interior, wood stud, cement plaster both sides	20

[a]Average weight per sq ft of surface.

CHAPTER THIRTY-THREE

Design Guides

33.1 SEISMIC DESIGN REQUIREMENTS OF THE 1982 *UBC*

The material that follows is reprinted from the 1982 edition of the *Uniform Building Code* (Ref. 2), with the permission of the publisher. Some of the tables from this publication are reprinted elsewhere in the book and are therefore not included here. These consist of the following:

1. Tables 25−J and 25−K, shear capacities of plywood diaphragms, reprinted in Chapter 28.
2. Table 25−P, shear capacity of fiberboard diaphragms, and Table 47−I, shear capacity of plastered and drywall diaphragms, reprinted in Chapter 28.
3. Table 26−G, shear and tension capacity of anchor bolts in concrete, reprinted in Chapter 30.

Earthquake Regulations

Sec. 2312. (a) **General.** Every building or structure and every portion thereof shall be designed and constructed to resist stresses produced by lateral forces as provided in this section. Stresses shall be calculated as the effect of a force applied horizontally at each floor or roof level above the base. The force shall be assumed to come from any horizontal direction.

Structural concepts other than set forth in this section may be approved by the building official when evidence is submitted showing that equivalent ductility and energy absorption are provided.

Where prescribed wind loads produce higher stresses, such loads shall be used in lieu of the loads resulting from earthquake forces.

(b) **Definitions.** The following definitions apply only to the provisions of this section:

BASE is the level at which the earthquake motions are considered to be imparted to the structure or the level at which the structure as a dynamic vibrator is supported.

BOX SYSTEM is a structural system without a complete vertical load-carrying space frame. In this system the required lateral forces are resisted by shear walls or braced frames as hereinafter defined.

BRACED FRAME is a truss system or its equivalent which is provided to resist lateral forces in the frame system and in which the members are subjected primarily to axial stresses.

DUCTILE MOMENT-RESISTING SPACE FRAME is a moment-resisting space frame complying with the requirements for a ductile moment-resisting space frame as given in Section 2312 (j).

ESSENTIAL FACILITIES—See Section 2312 (k).

LATERAL FORCE-RESISTING SYSTEM is that part of the structural system assigned to resist the lateral forces prescribed in Section 2312 (d).

MOMENT-RESISTING SPACE FRAME is a vertical load-carrying space frame in which the members and joints are capable of resisting forces primarily by flexure.

SHEAR WALL is a wall designed to resist lateral forces parallel to the wall.

SPACE FRAME is a three-dimensional structural system without bearing walls, composed of interconnected members laterally supported so as to function as a complete self-contained unit with or without the aid of horizontal diaphragms or floor-bracing systems.

VERTICAL LOAD-CARRYING SPACE FRAME is a space frame designed to carry all vertical loads.

(c) **Symbols and Notations.** The following symbols and notations apply only to the provisions of this section:

C = Numerical coefficient as specified in Section 2312 (d).

C_p = Numerical coefficient as specified in Section 2312 (g) and as set forth in Table No. 23-J.

D = The dimension of the structure, in feet, in a direction parallel to the applied forces.

δ_i = Deflection at level i relative to the base, due to applied lateral forces, Σf_i, for use in Formula (12-3).

$F_i F_n F_x$ = Laterial force applied to level i, n or x, respectively.

F_p = Lateral forces on a part of the structure and in the direction under consideration.

F_t = That portion of V considered concentrated at the top of the structure in addition to F_n.

f_i = Distributed portion of a total lateral force at level i for use in Formula (12-3).

g = Acceleration due to gravity.

$h_i h_n h_x$ = Height in feet above the base to level i, n or x respectively.

I = Occupancy Importance Factor as set forth in Table No. 23-K.

K = Numerical coefficient as set forth in Table No. 23-I.

Level i

I = Level of the structure referred to by the subscript i.

i = 1 designates the first level above the base.

Level n = That level which is uppermost in the main portion of the structure.

Level x = That level which is under design consideration.

x = 1 designates the first level above the base.

N = The total number of stories above the base to level n.

S = Numerical coefficient for site-structure resonance.

T = Fundamental elastic period of vibration of the building or structure in seconds in the direction under consideration.

T_s = Characteristic site period.

V = The total lateral force or shear at the base.

W = The total dead load as defined in Section 2302 including the partition loading specified in Section 2304 (d) where applicable.

EXCEPTION: W shall be equal to the total dead load plus 25 percent of the floor live load in storage and warehouse occupancies. Where the design snow load is 30 psf or less, no part need be included in the value of W. Where the snow load is greater than 30 psf, the snow load shall be included; however, where the snow load duration warrants, the building official may allow the snow load to be reduced up to 75 percent.

$w_i w_x$ = That portion of W which is located at or is assigned to level i or x respectively.

W_p = The weight of a portion of a structure or nonstructural component.

Z = Numerical coefficient dependent upon the zone as determined by Figures No. 1, No. 2 and No. 3 in this chapter. For locations in Zone No. 1, $Z = 3/16$. For locations in Zone No. 2, $Z = 3/8$. For locations in Zone No. 3, $Z = 3/4$. For locations in Zone No. 4, $Z = 1$.

(d) **Minimum Earthquake Forces for Structures.** Except as provided in Section 2312 (g) and (i), every structure shall be designed and constructed to resist minimum total lateral seismic forces assumed to act nonconcurrently in the direction of each of the main axes of the structure in accordance with the following formula:

$$V = ZIKCSW \quad \text{(12-1)}$$

The value of K shall be not less than that set forth in Table No. 23-I. The value of C and S are as indicated hereafter except that the product of CS need not exceed 0.14.

The value of C shall be determined in accordance with the following formula:

$$C = \frac{1}{15\sqrt{T}} \quad \text{(12-2)}$$

The value of C need not exceed 0.12.

The period T shall be established using the structural properties and deformational characteristics of the resisting elements in a properly substantiated analysis such as the following formula:

$$T = 2\pi\sqrt{\left(\sum_{i=1}^{n} w_i\delta_i^2\right) \div \left(g\sum_{i=1}^{n} f_i\delta_i\right)} \quad \text{.................. (12-3)}$$

where the values of f_i represent any lateral force distributed approximately in accordance with the principles of Formulas (12-5), (12-6) and (12-7) or any other rational distribution. The elastic deflections, δ_i, shall be calculated using the applied lateral forces, f_i.

In the absence of a determination as indicated above, the value of T for buildings may be determined by the following formula:

$$T = \frac{0.05h_n}{\sqrt{D}} \quad \text{...................... (12-3A)}$$

Or in buildings in which the lateral force-resisting system consists of ductile moment-resisting space frames capable of resisting 100 percent of the required lateral forces and such system is not enclosed by or adjoined by more rigid elements tending to prevent the frame from resisting lateral forces:

$$T = 0.10N \quad \text{...................... (12-3B)}$$

The value of S shall be determined by the following formulas, but shall be not less than 1.0:

for T/T_s = 1.0 or less $S = 1.0 + \frac{T}{T_s} - 0.5\left[\frac{T}{T_s}\right]^2$ .... (12-4)

for T/T_s greater than 1.0 or less $S = 1.2 + 0.6\frac{T}{T_s} - 0.3\left[\frac{T}{T_s}\right]^2$ ... (12-4A)

WHERE:

T in Formulas (12-4) and (12-4A) shall be established by a properly substantiated analysis but T shall be not less than 0.3 second.

The range of values of T_s may be established from properly substantiated geotechnical data, in accordance with U.B.C. Standard No. 23-1, except that T_s shall not be taken as less than 0.5 second nor more than 2.5 seconds. T_s shall be that value within the range of site periods, as determined above, that is nearest to T.

When T_s is not properly established, the value of S shall be 1.5.

EXCEPTION: Where T has been established by a properly substantiated analysis and exceeds 2.5 seconds, the value of S may be determined by assuming a value of 2.5 seconds for T_s.

(e) **Distribution of Lateral Forces. 1. Structures having regular shapes or framing systems.** The total lateral force V shall be distributed over the height of the structure in accordance with Formulas (12-5), (12-6) and (12-7).

$$V = F_t + \sum_{i=1}^{n} F_i \quad \text{..................... (12-5)}$$

The concentrated force at the top shall be determined according to the following formula:

$$F_t = 0.07TV \quad \text{...................... (12-6)}$$

F_t need not exceed $0.25V$ and may be considered as 0 where T is 0.7 second or less. The remaining portion of the total base shear V shall be distributed over the height of the structure including level n according to the following formula:

$$F_x = \frac{(V - F_t)\, w_x h_x}{\sum_{i=1}^{n} w_i h_i} \qquad \text{(12-7)}$$

At each level designated as x, the force F_x shall be applied over the area of the building in accordance with the mass distribution on that level.

2. **Setbacks.** Buildings having setbacks wherein the plan dimension of the tower in each direction is at least 75 percent of the corresponding plan dimension of the lower part may be considered as uniform buildings without setbacks, provided other irregularities as defined in this section do not exist.

3. **Structures having irregular shapes or framing systems.** The distribution of the lateral forces in structures which have highly irregular shapes, large differences in lateral resistance or stiffness between adjacent stories, or other unusual structural features, shall be determined considering the dynamic characteristics of the structure.

4. **Accidental torsion.** In addition to the requirements of Section 2303 (b) 2, where the vertical resisting elements depend on diaphragm action for shear distribution at any level, the shear-resisting elements shall be capable of resisting a torsional moment assumed to be equivalent to the story shear acting with an eccentricity of not less than 5 percent of the maximum building dimension at that level.

(f) **Overturning.** At any level the incremental changes of the design overturning moment, in the story under consideration, shall be distributed to the various resisting elements in the same proportion as the distribution of the shears in the resisting system. Where other vertical members are provided which are capable of partially resisting the overturning moments, a redistribution may be made to these members if framing members of sufficient strength and stiffness to transmit the required loads are provided.

Where a vertical resisting element is discontinuous, the overturning moment carried by the lowest story of that element shall be carried down as loads to the foundation.

(g) **Lateral Force on Elements of Structures and Nonstructural Components.** Parts or portions of structures, nonstructural components and their anchorage to the main structural system shall be designed for lateral forces in accordance with the following formula:

$$F_p = ZIC_pW_p \qquad \text{(12-8)}$$

The values of C_p are set forth in Table No. 23-J. The value of the I coefficient shall be the value used for the building.

> **EXCEPTIONS:** 1. The value of I for panel connectors shall be as given in Section 2312 (j) 3 C.
>
> 2. The value of I for anchorage of machinery and equipment required for life safety systems shall be 1.5.

The distribution of these forces shall be according to the gravity loads pertaining thereto.

For applicable forces on diaphragms and connections for exterior panels, refer to Sections 2312 (j) 2 C and 2312 (j) 3 C.

(h) **Drift and Building Separations.** Lateral deflections or drift of a story relative to its adjacent stories shall not exceed 0.005 times the story height unless it can be demonstrated that greater drift can be tolerated. The displacement calculated from the application of the required lateral forces shall be multiplied by $(1.0/K)$ to obtain the drift. The ratio $(1.0/K)$ shall be not less than 1.0.

All portions of structures shall be designed and constructed to act as an integral unit in resisting horizontal forces unless separated structurally by a distance sufficient to avoid contact under deflection from seismic action or wind forces.

(i) **Alternate Determination and Distribution of Seismic Forces.** Nothing in Section 2312 shall be deemed to prohibit the submission of properly substantiated technical data for establishing the lateral forces and distribution by dynamic analyses. In such analyses the dynamic characteristics of the structure must be considered.

(j) **Structural Systems.** 1. **Ductility requirements.** A. All buildings designed with a horizontal force factor $K = 0.67$ or 0.80 shall have ductile moment-resisting space frames.

B. Buildings more than 160 feet in height shall have ductile moment-resisting space frames capable of resisting not less than 25 percent of the required seismic forces for the structure as a whole.

> **EXCEPTION:** Buildings more than 160 feet in height in Seismic Zones Nos. 1 and 2 may have concrete shear walls designed in accordance with Section 2627 or braced frames designed in conformance with Section 2312 (j) 1 G of this code in lieu of a ductile moment-resisting space frame, provided a K value of 1.00 or 1.33 is utilized in the design.

C. In Seismic Zones No. 2, No. 3 and No. 4 all concrete space frames required by design to be part of the lateral force-resisting system and all concrete frames located in the perimeter line of vertical support shall be ductile moment-resisting space frames.

> **EXCEPTION:** Frames in the perimeter line of the vertical support of buildings designed with shear walls taking 100 percent of the design lateral forces need only conform with Section 2312 (j) 1 D.

D. In Seismic Zones No. 2, No. 3 and No. 4 all framing elements not required by design to be part of the lateral force-resisting system shall be investigated and shown to be adequate for vertical load-carrying capacity and induced moment due to $3/K$ times the distortions resulting from the code-required lateral forces. The rigidity of other elements shall be considered in accordance with Section 2303 (b) 1.

E. Moment-resisting space frames and ductile moment-resisting space frames may be enclosed by or adjoined by more rigid elements which would tend to prevent the space frame from resisting lateral forces where it can be shown that the action or failure of the more rigid elements will not impair the vertical and lateral load-resisting ability of the space frame.

F. Necessary ductility for a ductile moment-resisting space frame shall be provided by a frame of structural steel with moment-resisting connections (complying with Section 2722 for buildings in Seismic Zones No. 3 and No. 4 or Section 2723 for buildings in Seismic Zones No. 1 and No. 2) or by a reinforced concrete frame (complying with Section 2625 for buildings in Seismic Zones No. 3 and No. 4 or Section 2626 for buildings in Seismic Zones No. 1 and No. 2).

> **EXCEPTION:** Buildings with ductile moment-resisting space frames in Seismic Zones No. 1 and No. 2 having an importance factor I greater than 1.0 shall comply with Section 2625 or 2722.

G. In Seismic Zones No. 3 and No. 4 and for buildings having an importance factor I greater than 1.0 located in Seismic Zone No. 2, all members in braced frames shall be designed for 1.25 times the force determined in accordance with Section 2312 (d). Connections shall be designed to develop the full capacity of the members or shall be based on the above forces without the one-third increase usually permitted for stresses resulting from earthquake forces.

Braced frames in buildings shall be composed of axially loaded bracing members of A36, A441, A500 Grades B and C, A501, A572 (Grades 42, 45, 50 and 55) or A588 structural steel, or reinforced concrete members conforming to the requirements of Section 2627.

H. Reinforced concrete shear walls for all buildings shall conform to the requirements of Section 2627.

I. In structures where $K = 0.67$ and $K = 0.80$, the special ductility requirements for structural steel or reinforced concrete specified in Section 2312 (j) 1 F, shall apply to all structural elements below the base which are required to transmit to the foundation the forces resulting from lateral loads.

2. **Design requirements.** A. **Minor alterations.** Minor structural alterations may be made in existing buildings and other structures, but the resistance to lateral forces shall be not less than before such alterations were made, unless the building as altered meets the requirements of this section.

B. **Reinforced masonry or concrete.** All elements within structures located in Seismic Zones No. 2, No. 3 and No. 4 which are of masonry or concrete shall be reinforced so as to qualify as reinforced masonry or concrete under the provisions of Chapters 24 and 26. Principal reinforcement in masonry shall be spaced 2 feet maximum on center in buildings using a moment-resisting space frame.

C. **Diaphragms.** Floor and roof diaphragms and collectors shall be designed to resist the forces determined in accordance with the following formula:

$$F_{px} = \frac{\sum_{i=x}^{n} F_i}{\sum_{i=x}^{n} w_i} w_{px} \qquad \text{(12-9)}$$

WHERE:

F_i = the lateral force applied to level i.

w_i = the portion of W at level i.

w_{px} = the weight of the diaphragm and the elements tributary thereto at level x, including 25 percent of the floor live load in storage and warehouse occupancies.

The force F_{px} determined from Formula (12-9) need not exceed $0.30ZIw_{px}$.

When the diaphragm is required to transfer lateral forces from the vertical resisting elements above the diaphragm to other vertical resisting elements below the diaphragm due to offsets in the placement of the elements or to changes in stiffness in the vertical elements, these forces shall be added to those determined from Formula (12-9).

However, in no case shall lateral force on the diaphragm be less than $0.14ZIw_{px}$.

Diaphragms supporting concrete or masonry walls shall have continuous ties between diaphragm chords to distribute, into the diaphragm, the anchorage forces specified in this chapter. Added chords may be used to form subdiaphragms to transmit the anchorage forces to the main cross ties. Diaphragm deformations shall be considered in the design of the supported walls. See Section 2312 (j) 3 A for special anchorage requirements of wood diaphragms.

3. **Special requirements.** A. **Wood diaphragms providing lateral support for concrete or masonry walls.** Where wood diaphragms are used to laterally support concrete or masonry walls the anchorage shall conform to Section 2310. In Zones No. 2, No. 3 and No. 4 anchorage shall not be accomplished by use of toenails or nails subjected to withdrawal; nor shall wood framing be used in cross-grain bending or cross-grain tension.

B. **Pile caps and caissons.** Individual pile caps and caissons of every building or structure shall be interconnected by ties, each of which can carry by tension and compression a minimum horizontal force equal to 10 percent of the larger pile cap or caisson loading, unless it can be demonstrated that equivalent restraint can be provided by other approved methods.

C. **Exterior elements.** Precast or prefabricated nonbearing, nonshear wall panels or similar elements which are attached to or enclose the exterior shall be designed to resist the forces determined from Formula (12-8) and shall accommodate movements of the structure resulting from lateral forces or temperature changes. The concrete panels or other similar elements shall be supported bŷ means of cast-in-place concrete or mechanical connections and fasteners in accordance with the following provisions:

Connections and panel joints shall allow for a relative movement between stories of not less than two times story drift caused by wind or $(3.0/K)$ times the calculated elastic story displacement caused by required seismic forces, or ½ inch, whichever is greater. Connections to permit movement in the plane of the panel for story drift shall be properly designed sliding connections using slotted or oversized holes or may be connections which permit movement by bending of steel or other connections providing equivalent sliding and ductility capacity.

Bodies of connectors shall have sufficient ductility and rotation capacity so as to preclude fracture of the concrete or brittle failures at or near welds.

The body of the connector shall be designed for one and one-third times the force determined by Formula (12-8). Fasteners attaching the connector to the panel or the structure such as bolts, inserts, welds, dowels, etc., shall be designed to ensure ductile behavior of the connector or shall be designed for four times the load determined from Formula (12-8).

Fasteners embedded in concrete shall be attached to or hooked around reinforcing steel or otherwise terminated so as to effectively transfer forces to the reinforcing steel.

The value of the coefficient I shall be 1.0 for the entire connector assembly in Formula (12-8).

(k) **Essential Facilities.** Essential facilities are those structures or buildings which must be safe and usable for emergency purposes after an earthquake in order to preserve the health and safety of the general public. Such facilities shall include but not be limited to:

1. Hospitals and other medical facilities having surgery or emergency treatment areas.
2. Fire and police stations.
3. Municipal government disaster operation and communication centers deemed to be vital in emergencies.

The design and detailing of equipment which must remain in place and be functional following a major earthquake shall be based upon the requirements of Section 2312 (g) and Table No. 23-J. In addition, their design and detailing shall consider effects induced by structure drifts of not less than $(2.0/K)$ times the story drift caused by required seismic forces nor less than the story drift caused by wind. Special consideration shall also be given to relative movements at separation joints.

(l) **Earthquake-recording Instrumentations.** For earthquake-recording instrumentations see Appendix, Section 2312 (l).

TABLE NO. 23-I—HORIZONTAL FORCE FACTOR *K* FOR BUILDINGS OR OTHER STRUCTURES[1]

TYPE OR ARRANGEMENT OF RESISTING ELEMENTS	VALUE[2] OF *K*
1. All building framing systems except as hereinafter classified	1.00
2. Buildings with a box system as specified in Section 2312 (b) **EXCEPTION:** Buildings not more than three stories in height with stud wall framing and using plywood horizontal diaphragms and plywood vertical shear panels for the lateral force system may use $K = 1.0$.	1.33
3. Buildings with a dual bracing system consisting of a ductile moment-resisting space frame and shear walls or braced frames using the following design criteria: a. The frames and shear walls or braced frames shall resist the total lateral force in accordance with their relative rigidities considering the interaction of the shear walls and frames b. The shear walls or braced frames acting independently of the ductile moment-resisting portions of the space frame shall resist the total required lateral forces c. The ductile moment-resisting space frame shall have the capacity to resist not less than 25 percent of the required lateral force	0.80
4. Buildings with a ductile moment-resisting space frame designed in accordance with the following criteria: The ductile moment-resisting space frame shall have the capacity to resist the total required lateral force	0.67
5. Elevated tanks plus full contents, on four or more cross-braced legs and not supported by a building	2.5[3]
6. Structures other than buildings and other than those set forth in Table No. 23-J	2.00

[1]Where wind load as specified in Section 2311 would produce higher stresses, this load shall be used in lieu of the loads resulting from earthquake forces.

[2]See Figures Nos. 1, 2 and 3 in this chapter and definition of Z as specified in Section 2312 (c).

[3]The minimum value of KC shall be 0.12 and the maximum value of KC need not exceed 0.25.

The tower shall be designed for an accidential torsion of 5 percent as specified in Section 2312 (e) 4. Elevated tanks which are supported by buildings or do not conform to type or arrangement of supporting elements as described above shall be designed in accordance with Section 2312 (g) using $C_p = .3$.

TABLE NO. 23-J—HORIZONTAL FORCE FACTOR C_p FOR ELEMENTS OF STRUCTURES AND NONSTRUCTURAL COMPONENTS

PART OR PORTION OF BUILDINGS	DIRECTION OF HORIZONTAL FORCE	VALUE OF C_p[1]
1. Exterior bearing and nonbearing walls, interior bearing walls and partitions, interior nonbearing walls and partitions—see also Section 2312 (j) 3 C. Masonry or concrete fences over 6 feet high	Normal to flat surface	0.3[6]
2. Cantilever elements: a. Parapets	Normal to flat surfaces	0.8
b. Chimneys or stacks	Any direction	
3. Exterior and interior ornamentations and appendages	Any direction	0.8
4. When connected to, part of, or housed within a building: a. Penthouses, anchorage and supports for chimneys, stacks and tanks, including contents b. Storage racks with upper storage level at more than 8 feet in height, plus contents c. All equipment or machinery	Any direction	0.3[2 3]
5. Suspended ceiling framing systems (applies to Seismic Zones Nos. 2, 3 and 4 only)—see also Section 4701 (e)	Any direction	0.3[4 7]
6. Connections for prefabricated structural elements other than walls, with force applied at center of gravity of assembly	Any direction	0.3[5]

[1]C_p for elements laterally self-supported only at the ground level may be two thirds of value shown.

[2]W_p for storage racks shall be the weight of the racks plus contents. The value of C_p for racks over two storage support levels in height shall be 0.24 for the levels below the top two levels. In lieu of the tabulated values steel storage racks may be designed in accordance with U.B.C. Standard No. 27-11.

Where a number of storage rack units are interconnected so that there are a minimum of four vertical elements in each direction on each column line designed to resist horizontal forces, the design coefficients may be as for a building with K values from Table No. 23-I, $CS = 0.2$ for use in the formula $V = ZIKCSW$ and W equal to the total dead load plus 50 percent of the rack-rated capacity. Where the design and rack configurations are in accordance with this paragraph, the design provisions in U.B.C. Standard No. 27-11 do not apply.

[3]For flexible and flexibly mounted equipment and machinery, the appropriate values of C_p shall be determined with consideration given to both the dynamic properties of the equipment and machinery and to the building or structure in which it is placed but shall be not less than the listed values. The design of the equipment and machinery and their anchorage is an integral part of the design and specification of such equipment and masonry.

For essential facilities and life safety systems, the design and detailing of equipment which must remain in place and be functional following a major earthquake shall consider drifts in accordance with Section 2312 (k).

[4]Ceiling weight shall include all light fixtures and other equipment which is laterally supported by the ceiling. For purposes of determining the lateral force, a ceiling weight of not less than 4 pounds per square foot shall be used.

[5]The force shall be resisted by positive anchorage and not by friction.

[6]See also Section 2309 (b) for minimum load and deflection criteria for interior partitions.

[7]Does not apply to ceilings constructed of lath and plaster or gypsum board screw or nail attached to suspended members that support a ceiling at one level extending from wall to wall.

TABLE NO. 23-K—VALUES FOR OCCUPANCY IMPORTANCE FACTOR I

TYPE OF OCCUPANCY	I
Essential facilities[1]	1.5
Any building where the primary occupancy is for assembly use for more than 300 persons (in one room)	1.25
All others	1.0

[1]See Section 2312 (k) for definition and additional requirements for essential facilities.

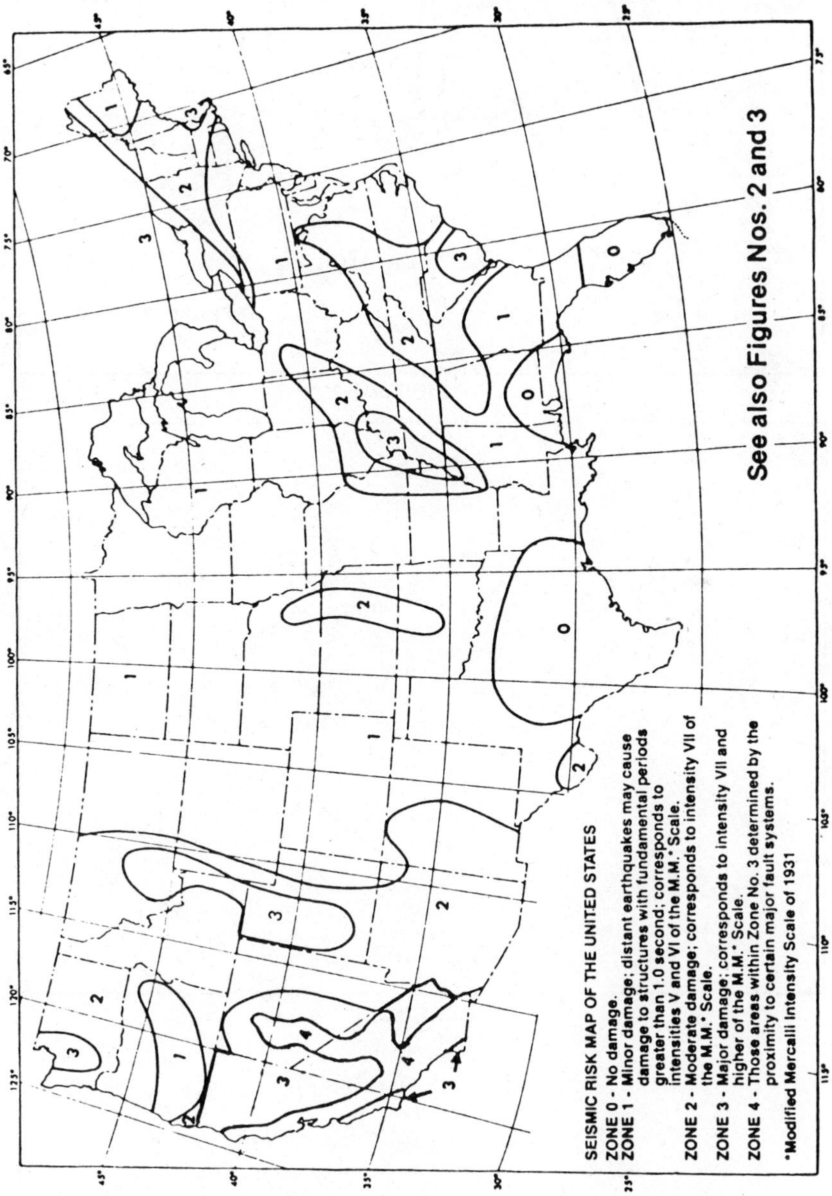

FIGURE NO. 1—SEISMIC ZONE MAP OF THE UNITED STATES
For areas outside of the United States, see Appendix Chapter 23

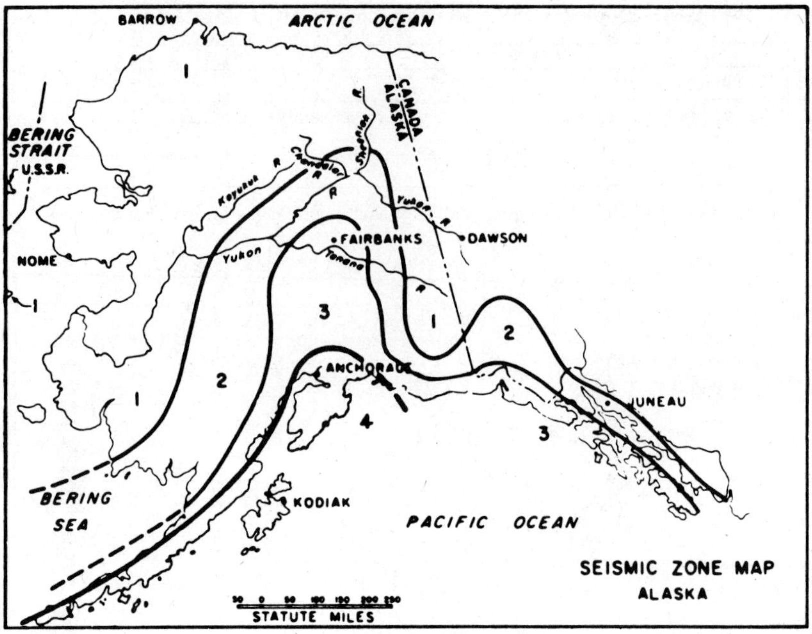

FIGURE NO. 2

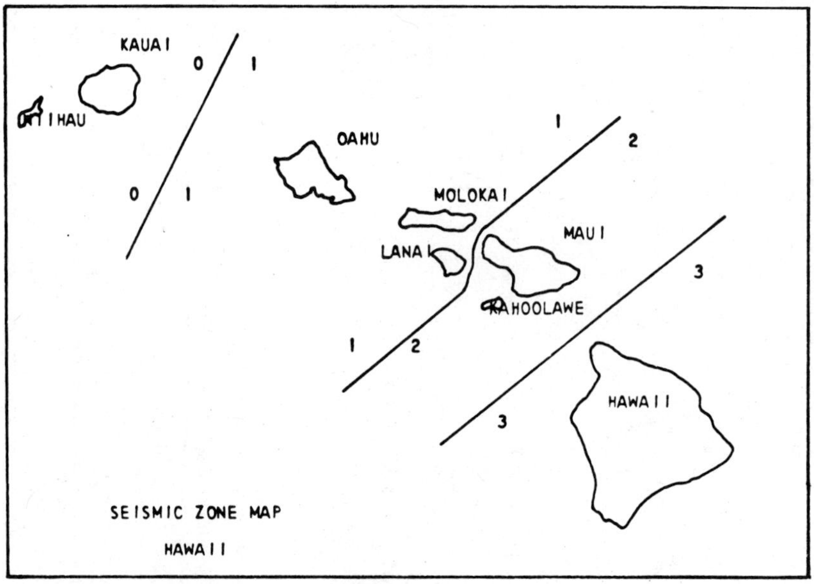

FIGURE NO. 3

TABLE NO. 24-G—ALLOWABLE SHEAR ON BOLTS FOR ALL MASONRY EXCEPT UNBURNED CLAY UNITS

DIAMETER OF BOLT (Inches)	EMBEDMENT[1] (Inches)	SOLID MASONRY (Shear in Pounds)	GROUTED MASONRY (Shear in Pounds)
1/2	4	350	550
5/8	4	500	750
3/4	5	750	1100
7/8	6	1000	1500
1	7	1250	1850[2]
1 1/8	8	1500	2250[2]

[1]An additional 2 inches of embedment shall be provided for anchor bolts located in the top of columns for buildings located in Seismic Zones Nos. 2, 3 and 4.

[2]Permitted only with not less than 2500 pounds per square inch units.

TABLE NO. 24-H—MAXIMUM WORKING STRESSES IN POUNDS PER SQUARE INCH FOR REINFORCED SOLID AND HOLLOW UNIT MASONRY[1]

TYPE OF STRESS	SPECIAL INSPECTION REQUIRED: Yes	SPECIAL INSPECTION REQUIRED: No
1. Compression, axial: Walls	See Section 2417	One half of the values permitted under Section 2417
2. Compression, axial: Columns	See Section 2417	One-half of the values permitted under Section 2417
3. Compression, flexural	$0.33f'_m$ but not to exceed 900	$0.166f'_m$ but not to exceed 450
4. Shear:		
a. No shear reinforcement:		
Flexural[2]	$1.1\sqrt{f'_m}$ 50 Max.	25
Shear walls[3]		
$M/Vd \geq 1$[4]	$.9\sqrt{f'_m}$ 34 Max.	17
$M/Vd = 0$[4]	$2.0\sqrt{f'_m}$ 50 Max.	25
b. Reinforcing taking all shear:		
Flexural	$3.0\sqrt{f'_m}$ 150 Max.	75
Shear walls[3]		
$M/Vd \geq 1$[4]	$1.5\sqrt{f'_m}$ 75 Max.	35
$M/Vd = 0$[4]	$2.0\sqrt{f'_m}$ 120 Max.	60
5. Modulus of elasticity[5]	$1000f'_m$ but not to exceed 3,000,000	$500f'_m$ but not to exceed 1,500,000
6. Modulus of rigidity[5]	$400f'_m$ but not to exceed 1,200,000	$200f'_m$ but not to exceed 600,000
7. Bearing on full area[6]	$0.25f'_m$ but not to exceed 900	$0.125f'_m$ but not to exceed 450
8. Bearing on one-third area or less[6]	$0.30f'_m$ but not to exceed 1200	$0.15f'_m$ but not to exceed 600
9. Bond—Plain bars	60	30
10. Bond—Deformed bars	140	100

[1]Stresses for hollow unit masonry are based on net section.

[2]Web reinforcement shall be provided to carry the entire shear in excess of 20 pounds per square inch whenever there is required negative reinforcement and for a distance of one sixteenth the clear span beyond the point of inflection.

[3]When calculating shear or diagonal tension stresses, shear walls which resist seismic forces shall be designed to resist 1.5 times the forces required by Section 2312 (d).

[4]M is the maximum bending moment occurring simultaneously with the shear load V at the section under consideration. Interpolate by straight line for M/Vd values between 0 and 1.

[5]Where determinations involve rigidity considerations in combination with other materials or where deflections are involved, the moduli of elasticity and rigidity under columns entitled "yes" for special inspection shall be used.

[6]This increase shall be permitted only when the least distance between the edges of the loaded and unloaded areas is a minimum of one fourth of the parallel side dimension of the loaded area. The allowable bearing stress on a reasonably concentric area greater than one third, but less than the full area, shall be interpolated between the values given.

Wood Diaphragms

Sec. 2513. (a) **General.** Lumber and plywood diaphragms may be used to resist horizontal forces in horizontal and vertical distributing or resisting elements, provided the deflection in the plane of the diaphragm, as determined by calculations, tests or analogies drawn therefrom, does not exceed the permissible deflection of attached distributing or resisting elements. See U.B.C. Standard No. 25-9 for a method of calculating the deflection of a blocked plywood diaphragm.

Permissible deflection shall be that deflection up to which the diaphragm and any attached distributing or resisting element will maintain its structural integrity under assumed load conditions, i.e., continue to support assumed loads without danger to occupants of the structure.

Connections and anchorages capable of resisting the design forces shall be provided between the diaphragms and the resisting elements. Openings in diaphragms which materially affect their strength shall be fully detailed on the plans and shall have their edges adequately reinforced to transfer all shearing stresses.

Size and shape of diaphragms shall be limited as set forth in Table No. 25-I.

In buildings of wood frame construction where rotation is provided for, the depth of the diaphragm normal to the open side shall not exceed 25 feet nor two thirds the diaphragm width, whichever is the smaller depth. Straight sheathing shall not be permitted to resist shears in diaphragms acting in rotation.

> **EXCEPTIONS:** 1. One-story, wood-framed structures with the depth normal to the open side not greater than 25 feet may have a depth equal to the width.
>
> 2. Where calculations show that diaphragm deflections can be tolerated, the depth normal to the open end may be increased to a depth-to-width ratio not greater than 1½:1 for diagonal sheathing or 2:1 for special diagonal sheathed or plywood diaphragms.

In masonry or concrete buildings, lumber and plywood diaphragms shall not be considered as transmitting lateral forces by rotation.

Diaphragm sheathing nails or other approved sheathing connectors shall be driven flush but shall not fracture the surface of the sheathing.

(b) **Diagonally Sheathed Diaphragms.** 1. **Conventional construction.** Such lumber diaphragms shall be made up of 1-inch nominal sheathing boards laid at an angle of approximately 45 degrees to supports. Sheathing boards shall be directly nailed to each intermediate bearing member with not less than two 8d nails for 1-inch by 6-inch nominal boards and three 8d nails for boards 8 inches or wider; and, in addition, three 8d nails and four 8d nails shall be used for 6-inch and 8-inch boards, respectively, at the diaphragm boundaries. End joints in adjacent boards shall be separated by at least one joist or stud space, and there shall be at least two boards between joints on the same support. Boundary members at edges of diaphragms shall be designed to resist direct tensile or compressive chord stresses and shall be adequately tied together at corners.

Conventional lumber diaphragms of Douglas fir-larch or southern pine may be used to resist shear due to wind or seismic forces not exceeding 300 pounds per lineal foot of width. The allowable strength shall be adjusted by the factors 0.82 and 0.65 where nails are used with sheathing and framing of Group III or IV wood species as listed in Table No. 25-17-J of U.B.C. Standard No. 25-17.

2. **Special construction.** Special diagonally sheathed diaphragms shall conform to conventional construction and, in addition, shall have all elements designed in conformance with the provisions of this code.

Each chord or portion thereof may be considered as a beam loaded with a uniform load per foot equal to 50 percent of the unit shear due to diaphragm action. The load shall be assumed as acting normal to the chord, in the plane of the diaphragm and either toward or away from the diaphragm. The span of the chord, or portion thereof, shall be the distance between structural members of the diaphragm, such as the joists, studs and blocking, which serve to transfer the assumed load to the sheathing.

Special diagonally sheathed diaphragms shall include conventional diaphragms sheathed with two layers of diagonal sheathing at 90 degrees to each other and on the same face of the supporting members.

Special diagonally sheathed diaphragms of Douglas fir-larch or southern pine may be used to resist shears due to wind or seismic loads, provided such shears do not stress the nails beyond their allowable safe lateral strength and do not exceed 600 pounds per lineal foot of width. The allowable strength shall be adjusted by the factors 0.82 and 0.65 where nails are used with sheathing and framing of Group III or IV wood species as listed in Table No. 25-17-J of U.B.C. Standard No. 25-17.

(c) **Plywood Diaphragms.** Horizontal and vertical diaphragms sheathed with plywood may be used to resist horizontal forces not exceeding those set forth in Table No. 25-J for horizontal diaphragms and Table No. 25-K for vertical diaphragms, or may be calculated by principles of mechanics without limitation by using values of nail strength and plywood shear values as specified elsewhere in this code. Plywood for horizontal diaphragms shall be as set forth in Table No. 25-S for corresponding joist spacing and loads. Plywood in shear walls shall be at least 5/16 inch thick for studs spaced 16 inches on center and 3/8 inch thick where studs are spaced 24 inches on center.

Maximum spans for plywood subfloor underlayment shall be as set forth in Table No. 25-T. Plywood used for horizontal and vertical diaphragms shall conform to U.B.C. Standard No. 25-9.

All boundary members shall be proportioned and spliced where necessary to transmit direct stresses. Framing members shall be at least 2-inch nominal in the dimension to which the plywood is attached. In general, panel edges shall bear on the framing members and butt along their center lines. Nails shall be placed not less than 3/8 inch in from the panel edge, shall be spaced not more than 6 inches on center along panel edge bearings, and shall be firmly driven into the framing members. No unblocked panels less than 12 inches wide shall be used.

Fiberboard Sheathing Diaphragms

Sec. 2514. Wood stud walls sheathed with fiberboard sheathing complying with U.B.C. Standard No. 25-24 may be used to resist horizontal forces not exceeding those set forth in Table No. 25-P. The fiberboard sheathing, 4 feet by 8 feet, shall be applied vertically to wood studs not less than 2-inch nominal in thickness spaced 16 inches on center. Nailing shown in Table No. 25-P shall be provided at the perimeter of the sheathing board and at intermediate studs. Blocking not less than 2-inch nominal in thickness shall be provided at horizontal joints when wall height exceeds length of sheathing panel, and sheathing shall be fastened to the blocking with nails sized as shown in Table No. 25-P spaced 3 inches on centers each side of joint. Nails shall be spaced not less than 3/8 inch from edges and ends of sheathing. Marginal studs of shear walls or shear-resisting elements shall be adequately anchored at top and bottom and designed to resist all forces. The maximum height-width ratio shall be 1½:1.

Wood Combined with Masonry or Concrete

Sec. 2515.

(b) **Horizontal Force.** Wood members shall not be used to resist horizontal forces contributed by masonry or concrete construction in buildings over one story in height.

> **EXCEPTION:** Wood floor and roof members may be used in horizontal trusses and diaphragms to resist horizontal forces imposed by wind, earthquake or earth pressure, provided such forces are not resisted by rotation of the truss or diaphragm.

2517 (g) **Wall Framing.**

3. **Bracing.** All exterior walls and main cross-stud partitions shall be effectively and thoroughly braced at each end, or as near thereto as possible, and at least every 25 feet of length by one of the following methods:

A. Nominal 1-inch by 4-inch continuous diagonal braces let into top and bottom plates and intervening studs, placed at an angle not more than 60 degrees nor less than 45 degrees from the horizontal, and attached to the framing in conformance with Table No. 25-Q.

B. Wood boards of ⅝-inch net minimum thickness applied diagonally on studs spaced not over 24 inches on center.

C. Plywood sheathing with a thickness not less than 5/16 inch for 16-inch stud spacing and not less than ⅜ inch for 24-inch stud spacing in accordance with Tables No. 25-M and No. 25-N.

D. Fiberboard sheathing 4-foot by 8-foot panels not less than ½ inch thick applied vertically on studs spaced not over 16 inches on center when installed in accordance with Section 2514 and Table No. 25-P.

E. Gypsum sheathing panels not less than ½ inch thick on studs spaced not over 16 inches on center when installed in accordance with Table No. 47-I.

F. Particleboard Exterior Type 2-M-1 sheathing panels not less than ⅜ inch thick on studs spaced not more than 16 inches on center.

G. Gypsum wallboard not less than ½ inch thick on studs spaced not over 24 inches on center when installed in accordance with Table No. 47-I.

H. Portland cement plaster on studs spaced 16 inches on center installed in accordance with Table No. 47-I.

I. Hardboard panel siding when installed in accordance with Section 2516 (g) 6 and Table No. 25-O.

For methods B, C, D, E, F, G, H and I, the braced panel must be at least 48 inches in width, covering three stud spaces where studs are spaced 16 inches apart and covering two stud spaces where studs are spaced 24 inches apart.

Solid sheathing of one of the materials specified in Items B through F, gypsum wallboard in Item G applied to supports at 16 inches on center, portland cement plaster in Item H, or hardboard panel siding in Item I, shall be applied to the exterior walls of the first story of all wood framed buildings three stories in height. In Seismic Zones Nos. 3 and 4 such braced wall sections shall be located at each end, or as near thereto as possible, and shall comprise at least 40 percent of the linear length of the wall.

Solid sheathing of one of the materials specified in Items B through F, gypsum wallboard in Item G applied to supports at 16 inches on center, portland cement plaster in Item H, or hardboard panel siding in Item I, shall be applied on either face of the exterior walls of the first story of all wood framed, two-story buildings and the second story of three-story buildings located in Seismic Zones No. 3 and No. 4. Braced wall sections shall be located at each end or as near thereto as possible and comprise at least 25 percent of the linear length of the wall.

All vertical joints of panel sheathing shall occur over studs. Horizontal joints shall occur over blocking equal in size to the studding except where waived by the installation requirements for the specific sheathing materials.

TABLE NO. 25-G—SAFE LATERAL STRENGTH AND REQUIRED PENETRATION OF BOX AND COMMON WIRE NAILS DRIVEN PERPENDICULAR TO GRAIN OF WOOD

SIZE OF NAIL	STANDARD LENGTH (Inches)	WIRE GAUGE	PENETRATION REQUIRED (Inches)	LOADS (Pounds)[1,2] Douglas Fir Larch or Southern Pine	LOADS (Pounds)[1,2] Other Species
BOX NAILS					
6d	2	12½	1⅛	51	See U.B.C. Standard No. 25-17
8d	2½	11½	1¼	63	
10d	3	10½	1½	76	
12d	3¼	10½	1½	76	
16d	3½	10	1½	82	
20d	4	9	1⅝	94	
30d	4½	9	1⅝	94	
40d	5	8	1¾	108	
COMMON NAILS					
6d	2	11½	1¼	63	See U.B.C. Standard No. 25-17
8d	2½	10¼	1½	78	
10d	3	9	1⅝	94	
12d	3¼	9	1⅝	94	
16d	3½	8	1¾	108	
20d	4	6	2⅛	139	
30d	4½	5	2¼	155	
40d	5	4	2½	176	
50d	5½	3	2¾	199	
60d	6	2	2⅞	223	

[1]The safe lateral strength values may be increased 25 percent where metal side plates are used.

[2]For wood diaphragm calculations these values may be increased 30 percent. (See U.B.C. Standard No. 25-17.)

TABLE NO. 25-I—MAXIMUM DIAPHRAGM DIMENSION RATIOS

MATERIAL	HORIZONTAL DIAPHRAGMS Maximum Span-Width Ratios	VERTICAL DIAPHRAGMS Maximum Height-Width Ratios
1. Diagonal sheathing, conventional	3:1	2:1
2. Diagonal sheathing, special	4:1	3½:1
3. Plywood, nailed all edges	4:1	3½:1
4. Plywood, blocking omitted at intermediate joints	4:1	2:1

Note: Tables 25–J, 25–K, and 25–P are presented in Chapter 28.

TABLE NO. 25-Q—NAILING SCHEDULE

CONNECTION	NAILING[1]
1. Joist to sill or girder, toenail	3-8d
2. Bridging to joist, toenail each end	2-8d
3. 1" x 6" subfloor or less to each joist, face nail	2-8d
4. Wider than 1" x 6" subfloor to each joist, face nail	3-8d
5. 2" subfloor to joist or girder, blind and face nail	2-16d
6. Sole plate to joist or blocking, face nail	16d at 16" o.c.
7. Top plate to stud, end nail	2-16d
8. Stud to sole plate	4-8, toenail or 2-16d, end nail
9. Double studs, face nail	16d at 24" o.c.
10. Doubled top plates, face nail	16d at 16" o.c.
11. Top plates, laps and intersections, face nail	2-16d
12. Continuous header, two pieces	16d at 16" o.c. along each edge
13. Ceiling joists to plate, toenail	3-8d
14. Continuous header to stud, toenail	4-8d
15. Ceiling joists, laps over partitions, face nail	3-16d
16. Ceiling joists to parallel rafters, face nail	3-16d
17. Rafter to plate, toenail	3-8d
18. 1" brace to each stud and plate, face nail	2-8d
19. 1" x 8" sheathing or less to each bearing, face nail	2-8d
20. Wider than 1" x 8" sheathing to each bearing, face nail	3-8d
21. Built-up corner studs	16d at 24" o.c.
22. Built-up girder and beams	20d at 32" o.c. at top and bottom and staggered 2-20d at ends and at each splice
23. 2" planks	2-16d at each bearing
24. **Particleboard:**[5] **Wall Sheathing (to framing):** 3/8"-1/2"	6d[3]
5/8"-3/4"	8d[3]
25. **Plywood:**[5] **Subfloor, roof and wall sheathing (to framing):** 1/2" and less	6d[2]
5/8"-3/4"	8d[3] or 6d[4]
7/8"-1"	8d[2]
1 1/8"-1 1/4"	10d[3] or 8d[4]
Combination Subfloor-underlayment (to framing): 3/4" and less	6d[4]
7/8"-1"	8d[4]
1 1/8"-1 1/4"	10d[3] or 8d[4]
26. **Panel Siding (to framing):** 1/2" or less	6d[6]
5/8"	8d[6]
27. **Fiberboard Sheathing:**[7] 1/2"	No. 11 ga.[8] 6d[3] No. 16 ga.[9]
25/32"	No. 11 ga.[8] 8d[3] No. 16 ga.[9]

[1]Common or box nails may be used except where otherwise stated.

[2]Common or deformed shank.

[3]Common.

[4]Deformed shank.

[5]Nails spaced at 6 inches on center at edges, 12 inches at intermediate supports (10 inches at intermediate supports for floors), except 6 inches at all supports where spans are 48 inches or more. For nailing of plywood diaphragms and shear walls, refer to Section 2513 (c). Nails for wall sheathing may be common, box or casing.

[6]Corrosion-resistant siding or casing nails conforming to the requirements of Section 2516 (j) 1.

[7]Fasteners spaced 3 inches on center at exterior edges and 6 inches on center at intermediate supports.

[8]Corrosion-resistant roofing nails with 7/16-inch-diameter head and 1½-inch length for ½-inch sheathing and 1¾-inch length for 25/32-inch sheathing conforming to the requirements of Section 2516 (j) 1.

[9]Corrosion-resistant staples with nominal 7/16-inch crown and 1⅛-inch length for ½-inch sheathing and 1½-inch length for 25/32-inch sheathing conforming to the requirements of Section 2516 (j) 1.

Ductile Moment-resisting Space Frames

Sec. 2625. (a) **General.** Design and construction of cast-in-place, monolithic reinforced concrete framing members and their connections in ductile moment-resisting space frames shall conform to the requirements of this chapter and all the requirements of this section.

> **EXCEPTIONS:** 1. Precast concrete frame members may be used, if the resulting construction complies with all the provisions of this section.
>
> 2. Ductile moment-resisting space frames in Seismic Zones Nos. 1 and 2 may be designed in accordance with the alternate provisions of Section 2626.

All lateral load-resisting frame members shall be designed by the strength method of design except that the alternate design method of Section 2608 (j) may be used, provided that it is shown that the factor of safety is equivalent to that achieved with the strength design method.

Formulas (9-2) and (9-3) of Section 2609 (d) for earthquake loading shall be modified to:

$$U = 1.40\,(D + L + E) \quad \ldots\ldots\ldots\ldots \quad (25\text{-}1)$$

$$U = .90\,D + 1.40E \quad \ldots\ldots\ldots\ldots \quad (25\text{-}2)$$

Members of space frames which are designed to resist seismic forces shall be designed, in accordance with the provisions of this section, so that shear failures will not occur if the frame is subjected to lateral displacements in excess of yield displacements.

(b) **Definitions.**

CONFINED CONCRETE is concrete which is confined by closely spaced special transverse reinforcement which is provided to restrain the concrete in directions perpendicular to the applied stresses.

SPECIAL TRANSVERSE REINFORCEMENT is spirals, stirrup ties or hoops and supplementary cross ties provided to restrain the concrete to make it qualify as confined concrete.

STIRRUP TIES OR HOOPS are continuous reinforcing steel of not less than a No. 3 bar bent to form a closed hoop which encloses the longitudinal reinforcing and the ends of which have a standard 135-degree bend with a 10-bar-diameter extension or equivalent.

(c) **Symbols and Notations.** The following symbols and notations apply only to the provisions of this section:

A_{ch} = area of rectangular or circular core of column measured out-to-out of hoop or spiral, square inches.

A_g = gross area of column, square inches.

A_s = effective cross-sectional area of nonprestressed reinforcement, square inches.

A'_s = effective cross-sectional area of nonprestressed compression reinforcement, square inches.

A_{sh} = total cross-sectional area of hoop reinforcement, including supplementary cross ties having a spacing of s_h and crossing a section with a core dimension of h_c, square inches.

b_w = web width, or diameter of circular section, inches.

d = distance from extreme compression fiber to centroid of tension reinforcement, inches.

d_b = nominal diameter of bar, inches.

f'_c = specified compressive strength of concrete, psi.

f_y = specified yield strength of nonprestressed reinforcement, psi.

f_{yh} = specified yield strength of hoop reinforcement, psi.

h = largest core dimension of rectangular tied column, inches.

P_e = maximum design axial load on a column during an earthquake.

ρ = A_s/bd = ratio of nonprestressed tension reinforcement.

ρ' = A'_s/bd = ratio of nonprestressed compression reinforcement.

s_h = center-to-center spacing of hoops, inches.

V_d = applied shear force at section due to dead load.

V_l = applied shear force at section due to live load.

V_u = total applied design shear force at section.

(d) **Physical Requirements for Concrete and Reinforcing Steel.** 1. **Concrete.** The minimum specified 28-day strength of the concrete f'_c, shall be 3000 pounds per square inch. The maximum specified strength for lightweight concrete shall be limited to 4000 psi.

2. **Reinforcement.** All longitudinal reinforcing steel in columns and beams shall comply with U.B.C. Standard No. 26-4. Reinforcing not conforming to the low-alloy steel requirements of U.B.C. Standard No. 26-4 shall comply with the following additional requirements:

A. Reinforcing shall be limited to billet steel, Grade 40 or 60 bars.

B. The actual yield stress, based on mill tests, shall not exceed the minimum specified yield stress, f_y, by more than 18,000 psi. Retests shall not exceed this value by more than 3000 psi.

C. The ultimate tensile stress shall be not less than 1.33 times the actual yield stress, based on mill tests.

Grades other than those specified for design shall not be used.

(e) **Flexural Members.** 1. **General.** Flexural members shall not have a width-depth ratio of less than 0.3, nor shall the width be less than 10 inches nor more than the supporting column width plus a distance on each side of the column of three fourths the depth of the flexural member. Flexural members framing into columns shall be subject to a rational joint analysis.

2. **Reinforcement.** All flexural members shall have a minimum reinforcement ratio, for top and for bottom reinforcement, of

$$\frac{200}{f_y}$$

throughout their length. The reinforcement ratio, ρ, shall not exceed 0.025. The positive moment capacity at the face of columns shall be not less than 50 percent of the negative moment capacity provided. A minimum of one fourth of the larger amount of the negative reinforcement required at either end shall continue throughout the length of the beam. At least two bars shall be provided both top and bottom.

3. **Splices.** Tensile steel shall not be spliced by lapping in a region of tension or reversing stress unless the region is confined by stirrup-ties. Splices shall not be located within the column or within a distance of twice the member depth from the face of the column. At least two stirrup-ties shall be provided at all splices.

4. **Anchorage.** Flexural members terminating at a column, in any vertical plane, shall have top and bottom reinforcement extending, without horizontal offsets, to the far face of a confined concrete region terminating in a standard 90-degree hook. Length of required anchorage shall be computed beginning at the near face of the column. Length of anchorage in confined regions, including hook and vertical extension, shall be not less than 56 percent of the development length computed by Section 2612 (f) 1, A, B or C, but not less than 24 inches.

> **EXCEPTION:** Where the column resists less than 25 percent of the story-bent shear, at least 50 percent of such top and bottom reinforcement shall be anchored within such column cores and the remainder shall be anchored in regions outside the column core confined as specified herein for columns.

5. **Web reinforcement.** Vertical web reinforcement of not less than No. 3 bars shall be provided in accordance with the requirements of Section 2611, except that:

A. Web reinforcement shall be provided to develop the shears resulting from shear forces at the end of the member computed as

$$V_u \geq \frac{M_a + M_b}{L_{ab}} + 1.4\,(V_d + V_l) \quad \ldots\ldots\ldots\ldots\ldots (25\text{-}3)$$

where M_a and M_b are the ultimate moment capacities of opposite sense at each hinge location of the member and $V_d + V_l$ is the simple span shear. L_{ab} is the distance between M_a and M_b. The ultimate moment capacities shall be computed without the ϕ factor reduction and assuming the maximum reinforcing yield strength based on 25 percent over specified yield.

Ultimate shear capacities shall be computed with the ϕ factor reduction.

B. Stirrups shall be spaced at no more than $d/2$ throughout the length of the member.

C. Stirrup-ties, at a maximum spacing of not over $d/4$, eight bar diameters, 24 stirrup-tie diameters or 12 inches, whichever is least, shall be provided in the following locations:

(i) At each end of all flexural members. The first stirrup-tie shall be located not more than 2 inches from the face of the column and the last, a distance of at least twice the member depth from the face of the columns.

(ii) Wherever ultimate moment capacities or plastic hinges may be developed in the flexural members under inelastic lateral displacement of the frame.

(iii) Wherever required compression reinforcement occurs in the flexural members.

D. In regions where stirrup-ties are required, longitudinal bars shall have lateral support conforming to the provisions of ties for tied columns. Single or overlapping stirrup-ties and supplementary cross-ties may be used.

(f) **Columns Subject to Direct Stress and Bending.** 1. **Dimensional limitations.** The ratio of minimum to maximum column thickness shall be not less than 0.4 nor shall any dimension be less than 12 inches.

2. **Vertical reinforcement.** The reinforcement ratio, ρ, in tied columns shall be not less than 0.01 nor greater than 0.06.

3. **Splices.** Lap splices shall be made within the center half of column height, and the splice length shall be not less than 30 bar diameters. Continuity may also be effected by welding or by approved mechanical devices, provided not more than alternate bars are welded or mechanically spliced at any level and the vertical distance between these welds or splices of adjacent bars is not less than 24 inches.

4. **Special transverse reinforcement.** The cores of columns shall be confined by special transverse reinforcement as specified herein or as required to meet shear requirements.

A. The volumetric ratio of spiral reinforcement shall be not less than that required by Formula (10-3) of Section 2610 (j) nor

$$0.12 \frac{f'_c}{f_{yh}} \quad \text{(25-4)}$$

whichever is greater.

B. The total cross-section area (A_{sh}) of rectangular hoop reinforcement shall be not less than the greater of

$$A_{sh} = 0.30 s_h h_c \frac{f'_c}{f_{yh}} \left(\frac{A_g}{A_{ch}} - 1 \right) \quad \text{(25-5)}$$

or

$$0.12 s_h h_c \frac{f'_c}{f_{yh}} \quad \text{(25-6)}$$

The center-to-center spacing of hoops, s_h, shall not exceed 4 inches.

Single or overlapping hoops may be provided to meet this requirement.

Supplementary cross ties of the same size and spacing as hoops using 135-degree minimum hooks engaging the periphery hoop and secured to a longitudinal bar may be used. Supplementary cross ties or legs of overlapping hoops shall be spaced not more than 14 inches on center transversely.

EXCEPTION: Formula (25-5) need not be complied with if the column design is based on the column core only.

C. Special transverse reinforcement shall be provided in that portion of the column over a length equal to the maximum column dimension or one sixth of the clear height of the column, but not less than 18 inches from either face of the joint.

D. At any section where the ultimate capacity of the column is less than the sum of the shears (ΣV_u) computed by Formula (25-3) for all the beams framing into the column above the level under consideration, special transverse reinforcement shall be provided. For beams framing into opposite sides of the column, the moment components of Formula (25-3) may be assumed to be of opposite sign. For the purpose of this determination the factor 1.4 in Formula (25-3) may be changed to 1.1. For determination of the ultimate capacity of the column, the moments resulting from Formula (25-3) may be assumed to result from deformation of the frame in any one principal axis.

E. Columns which support discontinuous members, such as shear walls, braced frames or other rigid elements, shall have special transverse reinforcement for the full height of the supporting columns.

5. **Column shear.** The transverse reinforcement in columns subjected to bending and axial compression shall satisfy the following requirement:

$$A_v f_y \frac{d_c}{s} = \frac{V_u}{\phi} - V_c \quad \text{(25-7)}$$

V_u shall be computed by using the ultimate moment capacity in the ends of either the beams or columns framing into the connection.

Ultimate moment capacities shall be computed without ρ or other reduction factors and under all possible vertical loading conditions and assuming the maximum reinforcing yield strength based on 25 percent over specified yield. Shear strength shall be computed based on the column core area.

WHERE:

V_c = $v_c A_{ch}$, where v_c shall be in accordance with Section 2611 (e) except that v_c shall be considered zero when

$$\frac{P_e}{A_g} < 0.12 f'_c$$

s = spacing, ≤ ½ minimum column dimension.

d_c = dimension of the column core in the direction of load, in inches.

A_v = total cross-sectional area of special transverse reinforcement in tension within a distance s, except that two thirds of such area shall be used in the case of circular spirals.

(g) **Beam-column Connection.** Special transverse reinforcement shall be provided through the beam-column connection.

1. **Analysis.** The transverse reinforcement through the connection shall be proportioned according to the requirements of Section 2626 (f) 4. The transverse reinforcement thus selected shall be checked according to the provisions specified in Section 2626 (f) 5, with the exception that the V_u acting on the connection shall be equal to the maximum shears in the connection computed by a rational analysis taking into account the column shear and the concentrated shears developed from the forces in the beam reinforcement at a stress assumed at f_y.

Special transverse column reinforcement of one half the amount otherwise required by Section 2626 (g) 1 shall be required within the connection, determined by the depth of the shallowest framing member, where such members frame into all four sides of a column and whose width is at least three fourths the column width. When a corner of a tied column, unconfined by flexural members, exceeds 4 inches, the full special transverse reinforcement shall be provided through the connection and around bars outside of the connection.

Special transverse beam reinforcing shall be provided through the beam column connection to provide confinement for longitudinal reinforcement outside the column core where such confinement is not provided by another beam framing into the connection.

2. **Design limitations.** At any beam-column connection where $P_e/A_g \geq 0.12 f'_c$ the sum of the moment strengths of the columns, at the design earthquake axial load, shall be greater than the sum of the moment strengths of the beam, along their principal planes at that connection.

> **EXCEPTION:** Where certain beam-column connections at any level do not comply with the above limitations, the remaining columns and connected flexural members shall comply and, further, shall be capable of resisting the entire shear at that level accounting for the altered relative rigidities and torsion resulting from the omission of elastic action of the nonconforming beam-column connections.

(h) **Inspection.** For buildings designed under this section, a specially qualified inspector under the supervision of the person responsible for the structural design shall provide continuous inspection of the placement of the reinforcement and concrete, and shall submit a certificate indicating compliance with the plans and specifications.

Ductile Moment-resisting Space Frames—Alternate Provisions for Seismic Zones No. 1 and No. 2

Sec. 2626. (a) **General.** Design and construction of earthquake-resisting reinforced concrete framing members and their connections conforming to the general requirements of this chapter and the special requirements of this section shall be deemed to meet the requirements for a ductile moment-resisting space frame of Section 2312 (j) for buildings in Seismic Zones No. 1 and No. 2, provided the horizontal force factor, K, is not taken less than unity when computing seismic forces under Section 2312 (d).

(b) **Flexural Members.** Web reinforcement shall be required throughout the length of the member. It shall be designed according to Section 2611 except that such web reinforcement shall be not less than 0.15 percent of the area computed as the product of the width of the web and the spacing of web reinforcement along the longitudinal axis of the member. The first stirrup shall be located 2 inches from the column face. The next six stirrups shall be spaced not over $d/4$.

Positive moment reinforcement at the supports of flexural members subject to reversal of moments shall be anchored by bond, hooks or mechanical anchors in or through the supporting member to develop the yield strength of the bar.

Lapped splices located in a region of tension or reversing stress shall be confined by at least two stirrups at each splice.

(c) **Columns.** Additional lateral reinforcement shall be provided for columns as prescribed in this subsection. The spacing of ties at the ends of tied columns shall not exceed 4 inches for a distance equal to the maximum column dimension but not less than one sixth of the clear height of the column from the face of the joint. The first such tie shall be located 2 inches from the face of the joint. Joints of exterior and corner columns shall be confined by lateral reinforcement through the joint. Such lateral reinforcement shall consist of spirals or ties as required at the ends of columns.

Earthquake-resisting Concrete Shear Walls and Braced Frames

Sec. 2627. (a) **General.** Design and construction of earthquake-resisting reinforced concrete shear walls and reinforced concrete braced frames subjected primarily to axial stresses for all buildings shall conform to the requirements of this chapter.

Shear walls and vertical bracing frames shall be designed by the strength design method except that the alternate design method of Section 2608 (j) may be used, provided that the factor of safety in shear is equivalent to that achieved with the strength design method.

Formulas (9-2) and (9-3) of Section 2609 (d) for earthquake loading shall be modified to:

$$U = 1.4\,(D + L) + 1.4E \qquad (27\text{-}1)$$

$$U = 0.9\,D + 1.4E \qquad (27\text{-}2)$$

provided, further, that $2.0E$ shall be used in Formulas (27-1) and (27-2) in calculating shear and diagonal tension stresses in shear walls of buildings other than those complying with requirements for buildings with $K = 0.67$.

(b) **Braced Frames.** Reinforced concrete members of braced frames subjected primarily to axial stresses shall have special transverse reinforcing as specified in Section 2626 (f) 4 throughout the full length of the member. Tension members shall additionally meet the requirements for compression members.

(c) **Vertical Boundary Members for Shear Walls.** Special vertical boundary elements shall be provided at the edges of concrete shear walls in buildings whose lateral force resisting system is as described in Table No. 23-I for a K of .80. These elements shall be composed of concrete-encased structural steel elements of A36,

A441, A500 Grades B and C, A501, A572 (except Grades 60 and 65) or A588, or shall be concrete reinforced as required for columns in Section 2626 (f) with special transverse reinforcement as described in Section 2626 (f) 4 for the full length of the element. The longitudinal reinforcing in these concrete boundary elements shall conform to the requirements of Section 2626 (d) 2.

EXCEPTION: The special transverse reinforcement may be omitted in Seismic Zones No. 1 and No. 2 when the combined dead load, live load and seismic stresses are not over one half of those otherwise allowed.

The boundary vertical elements and such other similar vertical elements as may be required shall be designed to carry all the vertical stresses resulting from the wall loads in addition to tributary dead and live loads and from the horizontal forces as prescribed in Section 2312. Horizontal reinforcing in the walls shall be fully anchored to the vertical elements.

EXCEPTION: In Seismic Zones No. 1 and No. 2 the vertical boundary elements may be designed to carry all vertical stresses resulting from tributary dead and live loads not supported by the shear walls acting as bearing walls.

Similar confinement of horizontal and vertical boundaries at wall openings also shall be provided unless it can be demonstrated that the unit compressive stresses at the opening have a load factor two times that required by Formulas (27-1) and (27-2) above.

(d) **Wall Reinforcement.** Wall reinforcement required to resist wall shear shall be terminated with not less than a 90-degree bend plus a six-bar-diameter extension beyond the boundary reinforcing at vertical and horizontal end faces of wall sections. Wall reinforcement terminating in boundary columns or beams shall be fully anchored into the boundary elements.

Steel Ductile Moment-resisting Space Frames—Seismic Zones Nos. 3 and 4

Sec. 2722. (a) **General.** Design and construction of steel framing in ductile moment-resisting space frames in Seismic Zones No. 3 and No. 4 shall conform to the requirements of the code and to all the requirements of this section. Welding shall comply with U.B.C. Standard No. 27-6.

(b) **Definitions.** 1. **Joints.** The joint is the entire assemblage at the intersections of the members.

2. **Connections.** The connection consists of only those elements that connect the member to the joint.

(c) **Materials.** Structural steel shall conform to A36, A441, A500 (Grades B and C), A501, A572 (Grades 42, 45, 50 and 55), or A588.

EXCEPTION: Structural steel A283 Grade D may be used for base plates and anchor bolts.

(d) **Connections.** Each beam or girder moment connection to a column shall be capable of developing in the beam the full plastic capacity of the beam or girder.

EXCEPTION: The connection need not develop the full plastic capacity of the beam or girder if it can be shown that adequately ductile joint displacement capacity is provided with a lesser connection.

For steel whose specified ultimate strength is less than 1.5 of the specified yield strength, plastic hinges in beams formed during inelastic deformations of the frame shall not occur at locations in which the beam flange area has been reduced such as by holes for bolts.

Note: Table 26–G, Allowable Shear and Tension on Bolts, is presented in Chapter 30.

(e) **Local Buckling.** Members in which hinges will form during inelastic displacement of the frames shall comply with the requirement for "plastic design sections."

(f) **Nondestructive Testing.** Welded connections between the primary members of ductile moment-resisting space frames shall be tested by nondestructive methods for compliance with U.B.C. Standard No. 27-6 and job specifications. This testing shall be a part of the special inspection requirements of Section 306. A program for this testing shall be established by the person responsible for structural design and as shown on plans and specifications.

As a minimum, this program shall include the following: 1. All complete penetration groove welds contained in joints and splices shall be tested 100 percent either by ultrasonic testing or by radiography.

> **EXCEPTION:** When approved, the nondestructive testing rate for an individual welder or welding operator may be reduced to 25 percent, provided the reject rate is demonstrated to be 5 percent or less of the welds tested for the welder or welding operator. A sampling of at least 40 completed welds for a job shall be made for such reduction evaluation. Reject rate is defined as the number of welds containing rejectable defects divided by the number of welds completed. For evaluating the reject rate of continuous welds over 3 feet in length where the effective throat thickness is 1 inch or less, each 12-inch increment or fraction thereof shall be considered as one weld. For evaluating the reject rate on continuous welds over 3 feet in length where the effective throat thickness is greater than 1 inch, each 6 inches of length or fraction thereof shall be considered one weld.
>
> When approved by the building official and outlined in the project plans and specifications, this nondestructive ultrasonic testing may be performed in the shop of an approved fabricator utilizing qualified test techniques in the employment of the fabricator.

2. Partial penetration groove welds when used in column splices shall be tested either by ultrasonic testing or radiography when required by the plans and specifications.

3. Base metal thicker than 1½ inches, when subjected to through-thickness weld shrinkage strains, shall be ultrasonically inspected for discontinuities directly behind such welds after joint completion.

Any material discontinuities shall be accepted or rejected on the basis of the defect rating in accordance with the (larger reflector) criteria of Table No. 27-6-E of U.B.C. Standard No. 27-6.

Steel Ductile Moment-resisting Space Frames—Seismic Zones Nos. 1 and 2

Sec. 2723. (a) **General.** Compliance with this section shall be deemed to meet the requirements for a ductile moment-resisting space frame of Section 2312 (j) and Table No. 23-I for buildings in Seismic Zones No. 1 and No. 2.

(b) **Design and Construction.** The design and construction for steel ductile moment-resisting space frames for buildings located in Seismic Zones No. 1 and No. 2 shall conform to all applicable requirements of this code except Section 2722.

Shear-resisting Construction with Wood Frame

Sec. 4713. (a) **General.** Portland cement plaster, gypsum lath and plaster, gypsum veneer base, gypsum sheathing board and gypsum wallboard may be used on wood studs for vertical diaphragms if applied in accordance with this section. Shear-resisting values shall not exceed those set forth in Table No. 47-I. The effects of overturning on vertical diaphragms shall be investigated in accordance with Section 2303 (b) 3.

The shear values tabulated shall not be cumulative with the shear value of other materials applied to the same wall. The shear values may be doubled when the identical materials applied as specified in this section are applied to both sides of the wall.

(b) **Masonry and Concrete Construction.** Portland cement plaster, gypsum lath and plaster, gypsum veneer base, gypsum sheathing board and gypsum wallboard shall not be used in vertical diaphragms to resist forces imposed by masonry or concrete construction.

(c) **Wall Framing.** Framing for vertical diaphragms shall conform with Section 2517 (g) for bearing walls, and studs shall be spaced not farther apart than 16 inches center to center. Marginal studs and plates shall be anchored to resist all design forces.

(d) **Height-to-length Ratio.** The maximum allowable height-to-length ratio for the construction in this section shall be 2 to 1. Wall sections having height-to-length ratios in excess of 1½ to 1 shall be blocked.

(e) **Application.** End joints of adjacent courses of gypsum lath, gypsum veneer base, gypsum sheathing board or gypsum wallboard sheets shall not occur over the same stud.

Where required in Table No. 47-I, blocking having the same cross-sectional dimensions as the studs shall be provided at all joints that are perpendicular to the studs.

The size and spacing of nails shall be as set forth in Table No. 47-I. Nails shall be spaced not less than ⅜ inch from edges and ends of gypsum lath, gypsum veneer base, gypsum sheathing board, gypsum wallboard or sides of studs, blocking and top and bottom plates.

1. **Gypsum lath.** Gypsum lath shall be applied perpendicular to the studs. Maximum allowable shear values shall be as set forth in Table No. 47-I.

2. **Gypsum sheathing board.** Four-foot-wide pieces may be applied parallel or perpendicular to studs. Two-foot-wide pieces shall be applied perpendicular to the studs. Maximum allowable shear values shall be as set forth in Table No. 47-I.

3. **Gypsum wallboard or veneer base.** Gypsum wallboard or veneer base may be applied parallel or perpendicular to studs. Maximum allowable shear values shall be as set forth in Table No. 47-I.

Note: Table 47–I is presented in Chapter 28.

CHAPTER THIRTY-FOUR

Computer-Aided Design

In case you haven't noticed, we have been invaded, conquered, and enslaved by computers. Typewriters, slide rules, and trig tables are on their way to join the horse collar and the family doctor. This situation constitutes both a boon and a curse for the field of structural design.

Obviously, the ability to crunch large volumes of data, to rapidly manipulate complex mathematical relationships, and to scan inumerable alternatives and find the best answer is a potential benefit. It is hard to imagine a structural designer working today, in whatever role or situation, who is not using a computer in some manner. Toying with the so-called personal computer has trickled down to the nursery school. Programs like STRUDL have now almost a touch of nostalgia.

Seismic design is an area of activity that is certain to expand and improve by quantum leaps as the use of computers becomes routine. Analysis of two- and three-dimensional indeterminate structures, investigation for dynamic behavior, simulation of actual earthquake events, and optimization of design decisions will be within the grasp of the structural designer.

Now for the curse. The trend is in the direction of user ease. You don't have to learn how the computer functions electrically and mechanically. You don't have to learn how to program the computer. You don't have to write your own programs. That is all OK. What is not so good is that you also don't have to understand what the computer is doing with the data and relationships to get from the input to the output.

If you don't understand the problems and the problem-solving processes that the computer is handling for you, you can't exercise your own judgment as to the quality of the computer's output. As stated in the beginning, the purpose of this book is to illuminate the problems, propose approaches to solving them, and explain some present work. Operating the problem-solving processes requires conceptual understanding, hands-on experience, and some facility with the current techniques of practice.

References

1. James Ambrose and Dimitry Vergun, *Simplified Building Design for Wind and Earthquake Forces*, Wiley, New York, 1980.
2. *Uniform Building Code,* 1982 ed., International Conference of Building Officials, 5360 Workman Mill Road, Whittier, CA, 90601.
3. Michael R. Lindeburg, *Seismic Design for the California Civil Professional Engineering Examination*, Professional Engineering Registration Program, San Carlos, CA 94070, 1980.
4. *Seismic Design for Buildings*, Armed Services Manual, United States Government.
5. John A. Blume, Nathan M. Newmark, and Leo H. Corning, *Design of Multistory Reinforced Concrete Buildings for Earthquake Motions*, Portland Cement Association, Skokie, IL, 1961.
6. *Recommended Lateral Force Requirements and Commentary*, Structural Engineers Association of California, San Francisco, CA, 1975.
7. David J. Dowrick, *Earthquake Resistant Design,* Wiley, New York, 1977.
8. Norman B. Green, *Earthquake Resistant Building Design and Construction*, Van Nostrand, New York, 1978.
9. *Reinforced Concrete Structures Subject to Wind and Earthquake Forces,* Publication SP-63, American Concrete Institute, Detroit, MI, 1980.
10. *Proceedings* - World Conference on Earthquake Engineering, (I, II, III, IV, and V). Periodically held conferences; proceedings published by host countries or institutions.
11. Phil M. Ferguson, *Reinforced Concrete Fundamentals*, 4th ed., Wiley, New York, 1979.
12. Paul F. Rice and Edward S. Hoffman, *Structural Design Guide to the ACI Building Code*, 2nd ed., Van Nostrand, New York, 1979.
13. *CRSI Handbook*, 4th ed., Concrete Reinforcing Steel Institute, Schaumburg, IL, 1982.
14. *Building Code Requirements for Reinforced Concrete*, ACI 318-83, American Concrete Institute, Detroit, MI, 1983. (Known as the ACI Code.)
15. Charles G. Salmon and John E. Johnson, *Steel Structures*, 2nd ed., Harper & Row, New York, 1980.
16. *Manual of Steel Construction,* 8th ed., American Institute of Steel Construction, Chicago, 1980. (Known as the AISC Manual.)
17. *Steel Deck Institute Diaphragm Design Manual*, 2nd ed., Steel Deck Institute, 1984.
18. *National Design Specification for Wood Construction,* National Forest Products Association, Washington, D.C., 1982.
19. *Western Woods Use Book*, 2nd ed., Western Wood Products Association, Portland, OR, 1979.
20. *Concrete Masonry Design Manual*, Concrete Masonry Association of California and Nevada, 83 Scripps Drive, Suite 303, Sacramento, CA 95825.
21. Robert R. Schneider and Walter L. Dickey, *Reinforced Masonry Design*, Prentice-Hall, Englewood Cliffs, NJ, 1980.
22. Charles G. Ramsey and Harold R. Sleeper, *Architectural Graphic Standards*, 7th ed., Wiley, New York, 1980.

23. John H. Callender, *Time-Saver Standards for Architectural Design Data,* 5th ed., McGraw-Hill, New York, 1974.

24. Edwin H. Gaylord and Charles N. Gaylord, *Structural Engineering Handbook*, 2nd ed., McGraw-Hill, 1979.

25. Frederick S. Merritt, *Building Construction Handbook*, 3rd ed., McGraw-Hill, New York, 1975.

26. Morton Newman, *Standard Structural Details for Building Construction*, McGraw-Hill, New York, 1968.

27. Joseph De Chiara, *Handbook of Architectural Details for Commercial Buildings*, McGraw-Hill, New York, 1980.

28. Wolfgang Schueller, *High-Rise Building Structures*, Wiley, New York, 1977.

29. Christopher Arnold and Robert Reitherman, *Building Configuration and Seismic Design*, Wiley, New York, 1982.

Glossary

The material presented here constitutes a brief dictionary of words and terms frequently encountered in discussions of the design of structures to withstand wind and earthquakes. Many of the words and terms have reasonably well-established meanings in the scientific and engineering literature. In those cases we have tried to be consistent with the accepted usage. In some cases, however, words and terms are given somewhat different meanings by different authors, by different professional groups, in different fields of study, in different countries, and so on. In these situations we have given the definitions as used by the authors of this work so that the reader may be clear as to our meaning.

In some cases words or terms are commonly misused with regard to their precise meaning. We have generally used such words and terms as they are broadly understood, but we have given both the correct and popular definitions in some cases.

To be as clear as possible in its requirements, the *Uniform Building Code* occasionally gives its own definitions. Reference should be made to these definitions in interpreting *UBC* requirements.

For a fuller explanation of some of these words and terms, the reader should use the Index to find the related discussion in the text.

Acceleration. The rate of change of the velocity, expressed as the first derivative of the velocity (dv/dt) or as the second derivative of the displacement (d^2s/dt^2). Acceleration of the ground surface is more significant than its displacement during an earthquake because it relates more directly to the force effect. $F = ma$, as a dynamic force.

Anchorage. Refers to attachment for resistance to movement; usually a result of uplift, overturn, sliding, or horizontal separation. Tie down, or hold down, refers to anchorage against uplift or overturn. Positive anchorage generally refers to direct fastening that does not easily loosen.

Aseismic. The correct word for description of resistance to seismic effects. Building design actually consists of *aseismic* design, although the term seismic design is more commonly used.

Base shear. See the *UBC* definition of *base* in Section 2312(b).

Battering. Describes the effect that occurs when two elements in separate motion bump into each other repeatedly, such as two adjacent parts of a structure during an earthquake. Also called *pounding* or *hammering*.

Box system. A structural system in which lateral loads are not resisted by a vertical load-bearing space frame but, rather, by shear walls or a braced frame.

Braced frame. Literally, any framework braced against lateral forces. Codes use the term for a frame braced by trussing (triangulation).

Bracing. In structural design usually refers to the resistance to movements caused by lateral forces or by the effects of buckling, torsional rotation, sliding, and so on.

Brittle fracture. Sudden, ultimate failure in tension or shear. The basic structural behavior of so-called brittle materials.

Collector. A force transfer element that functions to collect loads from a horizontal diaphragm and distribute them to the vertical elements of the lateral resistive system.

Confined concrete. Concrete mass that is wrapped by constraining reinforcing (ties, spirals, etc.) so as to develop three-dimensional stress conditions.

Continuity. Most often used to describe structures or parts of structures that have behavior characteristics influenced by the monolithic, continuous nature of adjacent elements, such as continuous, vertical, multistory columns, continuous, multispan beams, and rigid frames.

Critical damping. The amount of damping that will result in a return from initial deformation to the neutral position without reversal.

Damping. The effect that causes a decrease in the amplitude of succeeding cycles of harmonic motion (vibration). May be deliberately increased by devices such as shock absorbers. Occurs naturally in structures as energy is lost through internal strains and friction in joints.

Deflection. Generally refers to the lateral movement of a structure caused by loads, such as the vertical sag of a beam, the bowing of a surface under wind pressure, or the lateral sway of a rigid frame. The total horizontal deflection at the top of a structure caused by lateral loads is also called *drift*.

Degree of freedom. See *Freedom*.

Density. See *Mass*.

Diaphragm. A surface element (deck, wall, etc.) used to resist forces in its own plane by spanning or cantilevering. See *Horizontal diaphragm* and *Shear wall*.

Displacement. Movement away from some fixed reference point. Motion is described mathematically as a displacement–time function. See *Velocity, Acceleration*.

Drag strut. A structural member used to transfer lateral load across the building and into some part of the vertical system. See also *Collector*.

Drift. See *Deflection*.

Ductile. Describes the load–strain behavior that results from the plastic yielding of materials or connections. To be significant, the plastic strain prior to failure should be considerably more than the elastic strain up to the point of plastic yield.

Ductile moment-resisting space frame. Frame that complies with the requirements of Section 2312(j) of the UBC.

Dynamic. Usually used to characterize load effects or structural behaviors that are nonstatic in nature. That is, they involve time-related considerations, such as vibrations, energy effects versus simple force, and so on.

Earthquake. The common term used to describe subterranean ground shocks that result in movement of the ground surface. The shock is usually the result of a ground fault whose location is called the *epicenter* of the quake. The magnitude of the energy release at the epicenter is the basis for rating the size, or relative seriousness, of the quake. This energy release is measured on a scale such as the Richter scale.

Elastic. Used to describe two aspects of stress–strain behavior. The first is a constant stress–strain proportionality, or constant modulus of elasticity, as represented by a straight line form of the stress–strain graph. The second is the limit within which all of the strain is recoverable; that is, there is no permanent deformation. The latter phenomenon may occur even though the stress–strain relationship is nonlinear.

Energy. Capacity for doing work; what is used up when work is done. Occurs in various forms: mechanical, heat, chemical, electrical, and so on.

Epicenter. See *Earthquake*.

Equilibrium. A balanced state or condition, usually used to describe a situation in which opposed effects neutralize each other to produce a net effect of zero.

Equivalent static force analysis. The technique by which a dynamic effect is translated into a hypothetical (equivalent) static effect that produces a similar result.

Essential facilities. Code term for a facility that should remain functional after a disaster such as a major earthquake; affects establishment of the *I* factor for base shear.

Fault. The subterranean effect that produces an earthquake. Consists of a slippage, cracking, sudden strain release, and so on. See *Earthquake*.

Flexible. See *Stiffness*.

Force. An effort that tends to change the shape or the state of motion of an object.

Freedom. In structures usually refers to the lack of some type of resistance or constraint. In static analysis the connections between members and the supports of the structure are qualified as to type, or degree, of freedom. Thus the terms *fixed support, pinned support*, and *sliding support* are used to qualify the types of movement resisted. In dynamic analysis the degree of freedom is an important factor in determining the dynamic response of a structure.

Frequency. In harmonic motion (bouncing springs, vibrating strings, swinging pendulums, etc.), the number of complete cycles of motion per unit of time. See *Vibration*.

Fundamental period. See *Period*.

Geophysical. Refers to the physical behavior characteristics of the ground surface and of subterranean masses.

Hammering. See *Battering*.

Header. Usually used to describe a horizontal element over an opening in a wall or at the edge of an opening in a roof or floor.

Hertz. Same as cycles per second.

Hold down. See *Anchorage*.

Horizontal diaphragm. See *Diaphragm*. Usually a roof or floor deck used as part of the lateral resistive structural system for a building.

Inelastic. See *Stress–strain behavior*.

Interrupted shear wall. Shear wall that is not continuous in its solid construction from its top to the ground.

Lateral. Literally means to the side or from the side. Often used in reference to something that is perpendicular to a major axis or direction. With reference to the vertical direction of the gravity forces, wind, earthquakes, and horizontally directed soil pressures are called lateral effects.

Let-in bracing. Diagonal boards nailed to studs to provide trussed bracing in the wall plane. In order not to interfere with the surfacing materials of the wall, they are usually notched in, or let in, to the stud faces.

Mass. The dynamic property of an object that causes it to resist changes in its state of motion. This resistance is called *inertia*. The magnitude of the mass per unit volume of the object is called its *density*. Dynamic force is defined by $F = ma$, or force equals mass times acceleration. Weight is defined as the force produced by the acceleration of gravity; thus $W = mg$.

Mercalli scale. System used to determine the location of the epicenter of an earthquake on the basis of defining of zones of relative intensity of observed damage and experiences of persons during the quake. This is essentially the basis for the determination of the zones of intensity in the *UBC*.

Modified Mercalli scale. See *Mercalli scale*. Present version of the system.

Moment-resisting space frame. A vertical load carrying space frame in which the members and joints are capable of resisting forces primarily by flexure (*UBC* definition).

Natural period. See *Period*.

Normal. 1. The ordinary, usual, unmodified state of something. 2. Perpendicular; such as pressure normal to a surface, stress normal to a cross section, and so on.

Occupancy. In building code language refers to the use of a building as a residence, school, office, and so on.

Occupancy importance factor (I). *UBC* term used in the basic equation for seismic force: $V = ZIKCSW$. Accounts for possible increased concern for certain occupancies.

Overturn. The toppling, or tipping over, effect of lateral loads.

Parapet. The extension of a wall plane or the roof edge facing above the roof level.

Period (of vibration.) The total elapsed time for one full cycle of vibration. For an elastic structure in simple, single-mode vibration, the period is a constant (called the *natural* or *fundamental period*) and is independent of the magnitude of the amplitude, of the number of cycles, and of most damping or resonance effects. See *Vibration*.

Pier. A short, stocky column with height not greater than three times its least lateral dimension. Section 2417(d) of the *UBC* defines a masonry wall as a pier if its plan length is less than three times the wall thickness.

Plastic hinge. Region where ultimate moment strength of a member may be developed and maintained with corresponding significant inelastic rotation as a result of ductile yielding of steel.

Positive anchorage. See *Anchorage*.

Pounding. See *Battering*.

Pressure. A force distributed over, and normal to, a surface.

Relative stiffness. See *Stiffness*.

Reserve energy. Energy a ductile system is capable of absorbing by plastic strains.

Resonance. See *Vibration*.

Response spectra. See *Spectrum*.

Richter scale. A log-based measuring system for evaluation of the relative energy level of an earthquake at its center of origin (epicenter).

Rigid bent. A two-dimensional (planar) rigid (moment-resisting) frame structure.

Rigid frame. Framed structure in which the joints between members are made to transmit moments between the ends of the connected members. Called a *bent* when the frame is planar and a *moment-resisting space frame* when the frame is three dimensional.

Risk. The degree of probability of loss due to some potential hazard. The risk of an earthquake in a particular geographic area is the basis for the *Z* factor in the *UBC* equation for seismic force: $V = ZIKCSW$.

Rotation. Motion in a circular path. Also used to describe a twisting, or torsional, effect. See *Torsion*.

Seismic. Pertaining to ground shock. See *Aseismic*.

Separation. Often used in structural design to denote situations in which parts of a structure are made to act independently. Partial separation refers to a controlled separation that allows for some interactions but permits independence for other actions, such as a connection that transmits vertical forces but not horizontal ones. The separation may be dimensionally controlled to allow for a specific amount of movement.

Shear. A force effect that is lateral (perpendicular) to the major axis of a structure, or one that involves a slipping effect, as opposed to a push–pull effect. Wind and earthquake forces are sometimes visualized as shear effects on a building because they are perpendicular to the major vertical (gravity) axis of the building.

Shear wall. A vertical diaphragm.

Site response factor *(S)*. A *UBC* term used in the basic equation for seismic force: $V = ZIKCSW$. Accounts for the effect of the period of the ground mass under the building.

Space frame. An ambiguous term, used variously to describe three-dimensional structures. The *UBC* uses a particular definition in Section 2312 in classifying structural systems.

Special ductile frame. A structural frame of reinforced concrete in cast-in-place construction that is designed to accommodate reversible lateral displacements after formation of plastic hinges.

Spectrum. In seismic analysis generally refers to the curve that describes the actual dynamic force effect on a structure as a function of variation in its fundamental period. Response spectra are the family of curves produced by various degrees of damping. This represents the basis for determining the *C* factor in the *UBC* equation for seismic force: $V = ZIKCSW$.

Stability. Refers to the inherent capability of a structure to develop force resistance as a property of its form, orientation, articulation of its parts, type of connections, methods of support, and so on. Is not directly related to quantified strength or stiffness, except when the actions involve the buckling of elements of the structure.

Static. The state that occurs when the velocity is zero; thus no motion is occurring. Is generally used to refer to situations in which no change is occurring.

Stiffness. In structures refers to resistance to deformation, as opposed to strength, which refers to resistance to force. A lack of stiffness indicates a flexible structure. Relative stiffness usually refers to the comparative deformation of two or more structural elements that share a load.

Strain. Deformation resulting from stress. Is usually measured as a percentage of deformation, called *unit strain* or *unit deformation*, and is thus dimensionless.

Stress. The mechanism of force within the material of a structure. Is visualized as a pressure effect (if tension or compression) or a shear effect on the surface of a unit of the material and is quantified in terms of force per unit area.

Stress–strain behavior. The relation of stress to strain in a material or a structure. Is usually visually represented by a stress–strain graph covering the range from no load to failure. Various aspects of the form of the graph define particular behavioral properties. A straight line indicates an elastic relationship; a curve indicates inelastic behavior. A sudden bend in the graph usually indicates a plastic strain, or yield, which results in some permanent deformation. The slope of the graph is defined as the modulus of elasticity of the material.

Tie down. See *Anchorage*.

Torsion. Moment effect involving twisting or rotation that is in a plane perpendicular to the major axis of an element. Lateral loads produce torsion on a building when they tend to twist it about its vertical axis. This occurs when the centroid of the load does not coincide with the center of stiffness of the vertical elements of the lateral load-resisting structural system.

Ultimate strength. Usually used to refer to the maximum static force resistance of a structure at the time of failure. This limit is the basis for the so-called strength design methods, as compared to the stress design methods that use some established stress limit, called the design stress, working stress, permissible stress, and so on.

Uplift. Refers to a net upward force effect; may be due to wind suction, overturning moment, or upward vertical seismic acceleration.

Velocity. The time rate of a motion, also commonly called *speed*.

Vertical diaphragm. See *Diaphragm*. Also called a shear wall.

Vibration. The cyclic, rhythmic motion of a body such as a spring. Occurs when the body is displaced from some neutral position and seeks to restore itself to a state of equilibrium when released. In its pure form it occurs as a harmonic motion with a characteristic behavior described by the cosine form of the displacement–time graph of the motion. The magnitude of linear displacement from the neutral position is called the *amplitude*. The time elapsed for one full cycle of motion is called the *period*. The number of cycles occurring in one second is called the *frequency*. Effects that tend to reduce the amplitude of succeeding cycles are called *damping*. The increase of amplitude in successive cycles is called a *resonant effect*.

Yield. See *Stress–strain behavior*.

Zone. Usually refers to a bounded area on a surface, such as the ground surface or the plan of a level of a building.

Index